BIOMEMBRANES
Volume 6

BIOMEMBRANES

A series edited by
Lionel A. Manson
The Wistar Institute
Philadelphia, Pennsylvania

1971 • Biomembranes • Volume 1
Articles by M. C. Glick, Paul M. Kraemer, Anthony Martonosi,
Milton R. J. Salton, and Leonard Warren

1971 • Biomembranes • Volume 2
Proceedings of the Symposium on Membranes and the Coordination
of Cellular Activities
Edited by Lionel A. Manson

1972 • Biomembranes • Volume 3
Passive Permeability of Cell Membranes
Edited by F. Kreuzer and J. F. G. Slegers

1974 • Biomembranes • Volume 4A
Intestinal Absorption
Edited by D. H. Smyth

1974 • Biomembranes • Volume 4B
Intestinal Absorption
Edited by D. H. Smyth

1974 • Biomembranes • Volume 5
Articles by Richard W. Hendler, Stuart A. Kauffman,
Dale L. Oxender, Henry C. Pitot, David L. Rosenstreich,
Alan S. Rosenthal, Thomas K. Shires, and Donald F.
Hoelzl Wallach

1975 • Biomembranes • Volume 6
Bacterial Membranes and the Respiratory Chain
By N. S. Gel'man, M. A. Lukoyanova, and D. N. Ostrovskii

1975 • Biomembranes • Volume 7
Aharon Katzir Memorial Volume
Edited by Henryk Eisenberg, Ephraim Katchalski-Katzir,
and Lionel A. Manson

BIOMEMBRANES, Volume 6

BACTERIAL MEMBRANES AND THE RESPIRATORY CHAIN

N. S. Gel'man, M. A. Lukoyanova, and D. N. Ostrovskii
A. N. Bakh Institute for Biochemistry
Academy of Sciences of the USSR
Moscow, USSR

Translated from Russian by
Basil Haigh

Springer Science+Business Media, LLC

Library of Congress Cataloging in Publication Data

Gel'man, Nina Samoĭlovna.
  Bacterial membranes and the respiratory chain.

  (Biomembranes; v. 6)
  Translation of Membrany bakteriĭ i dykhatel'naĭa tśep'.
  Bibliography: p.
  Includes index.
  1. Bacterial cell walls. 2. Bacteria—Physiology. 3. Respiration. I. Lukoĭanova,
Marina Arturovna, joint author. II. Ostrovskiĭ, Dmitriĭ Nikolaevich, joint author. III.
Title. IV. Series. [DNLM: 1. Bacteria—Physiology. 2. Cell membrane—Physiology.
3. Respiration. W1 BI858L v. 6/QW52 G319m]
  QH601.B53 vol. 6 [QR77]   574.8'75'08s   [589.9'08'75]
  ISBN 978-1-4684-7711-5   ISBN 978-1-4684-7709-2 (eBook)   75-4531
  DOI 10.1007/978-1-4684-7709-2

The original Russian text, published by Nauka Press in Moscow in 1972, has been corrected by the authors for the present edition. This translation is published under an agreement with the Copyright Agency of the USSR (VAAP).

МЕМБРАНЫ БАКТЕРИЙ И ДЫХАТЕЛЬНАЯ ЦЕПЬ
Н. С. Гельман, М. А. Лукоянова, и Д. Н. Островский.
MEMBRANY BAKTERII I DYKHATEL'NAYA TSEP'
N. S. Gel'man, M. A. Lukoyanova, i D. N. Ostrovskii

© 1975 Springer Science+Business Media New York
Originally published by Plenum Press, New York in 1975
Softcover reprint of the hardcover 1st edition 1975

United Kingdom edition published by Plenum Press, London
A Division of Plenum Publishing Company, Ltd.
Davis House (4th Floor), 8 Scrubs Lane, Harlesden, London, NW10 6SE, England

Foreword

The most valuable service Dr. Gel'man and her colleagues have performed for the many investigators of bacterial membrane systems in producing their first excellent monograph on "The Respiratory Apparatus of Bacteria" in 1966 has been continued and expanded in the preparation of this volume. The authors have brought together in a single volume much of the detail of investigations of bacterial membranes at the ultrastructural level and the chemical and biochemical organizational levels. The approach in bringing together this rapidly increasing volume of discovery has been both comprehensive and systematic, with a constant awareness of the importance of the molecular and functional properties and relationships existing in various bacterial membranes.

The monograph naturally reflects the authors' interest and their own intimate involvement in the elucidation at the molecular level of the respiratory chains organized in the prokaryotic bacterial membrane system. It is entirely appropriate that the chapter devoted to this topic should occupy a substantial proportion of this monograph. Indeed, had this volume been prepared at this very moment, that proportion would have been even greater, as the work in this area has literally exploded in the past two or three years, with the isolation of ATPase and respiratory mutants by Butlin and Gibson and their colleagues, and now also in many other laboratories. This more recent aspect of the respiratory apparatus localized in the bacterial plasma membranes unfortunately emerged too late for this volume. However, many of the questions unanswered in this monograph will continue to be unanswered until the biochemical analysis of the many "respiratory" mutants is complete.

As the authors have realized from their introductory remarks, the work in this field has progressed and will progress even more rapidly in the future. The present volume gives us a balanced perspective of the progress of the work on bacterial membranes and the excellent and abundant bibliography will be a great resource for all of us, not only for the prokaryotic but also for the eukaryotic "membraneologists."

It is a pleasure to be able to see the continuation of the monograph into its present form. It will be greatly appreciated by students for its lucidity and by research workers for its detailed presentations of the various facets of the multifunctional bacterial membrane.

Milton R. J. Salton

Department of Microbiology
New York University School of Medicine
New York, New York

April 1975

Contents

Introduction

The structure and function of biological membranes are fundamental problems in modern biology. The reason for the exceptional interest shown in this problem is that processes of the greatest importance to life take place in the membranes of all cells starting from bacteria and ending with the cells of the human brain. Among the most universal processes connected with biological membranes are the exchange of ions and metabolites between cell and environment and the transformation of energy in electron transport chains during oxidative and photosynthetic phosphorylation. The membrane apparatus of the cell divides it into distinct "compartments," thereby giving the biochemical processes a spatial distribution and creating partition surfaces which play an important role in a number of enzyme reactions. The membranes exercise control over metabolic systems in the cell by regulating the permeability of substrates and of reaction products, as well as of ions as activators or inhibitors of the enzymes. Other specific functions determined by the physiology of the corresponding cell types are also connected with biological membranes. One very important feature is that membranes and, in particular, the enzymes linked together in them are the site of action of cell drugs, poisons, hormones, and some antibiotics.

The fact that the basic physiological functions of membranes are the same in cells at different levels of evolution naturally suggests that membranes are the most primitive biological formations and that the principle of their molecular organization must be the same in different cells.

Oparin (1957, 1960, 1966) expressed the view that in the remote ancestors of modern organisms, which can be imagined as condensations of protoplasm surrounded by a lipoprotein membrane, it performed a number of functions which subsequently were distributed among the different organelles of the cell. In particular, enzymes of the respiratory chain and enzymes of oxidative phosphorylation were evidently located in what at that time was the single external membrane of the most primitive aerobic cells.

1

It seems to the writers of this book that the prime purpose of membrane biochemistry is to investigate molecular organization. An understanding of the action of energy transformation systems and the discovery of the mechanism of transport of materials and other membrane processes are dependent on the solution of this problem.

The term "molecular organization of membranes" must be taken to mean the arrangement and interaction of the proteins and lipids which, together with molecules of water and cations, form the supramolecular structure of the membrane. When the molecular organization of the membranes is studied the choice of test object is most important. Despite the basic similarity between the functions of biological membranes, the morphology of the membrane system differs in cells standing at different levels of the evolutionary ladder. Whereas in bacteria we find a cytoplasmic membrane and an undifferentiated membrane system, in the cells of plants and animals we find differentiated membrane systems consisting of organelles and a network of inner membranes. Naturally because bacterial membranes are the least differentiated, but are functionally closely related to membranes of the cells of higher organisms, they have attracted special attention. Bacterial membranes also are the richest source of material for the study of several processes taking place simultaneously in the same structure, and also for the study of the biogenesis of membranes. Finally, the study of bacterial membranes is interesting from the point of view of evolutionary chemistry, for it can help with the development of ideas regarding the evolution of structure of the cell as a whole.

There are thus weighty arguments in support of the study of bacterial membranes. The number of investigations in this field in fact grows steadily. Modern approaches to the study of the molecular organization of biological, including bacterial, membranes are extremely varied. They include the development of methods of isolating membranes, of determining the composition of their proteins and lipids, of studying the properties of the isolated proteins, and elucidating the nature of the interactions which stabilize the membrane. Considerable attention is being paid to fragmentation of membranes and the study of the properties of the fragments in the hope of reconstructing the membrane in its original native state and of determining the arrangement of its individual zones. Despite much research in these directions, precise results from which a theory of membrane structure and organization could be formulated have not yet been obtained. The main difficulty is that in order to break up the membrane (a system of lipoprotein) into fragments, and in order to obtain individual proteins, powerful solubilizing agents have to be used. Treatment of the membrane in this way gives rise to artefacts which complicate evaluation of the results. For this reason, physical and physicochemical methods have been applied to the study of membranes in order to assess the state of their components *in situ* during their natural interaction. In the last few years, some degree of specialization of these methods has begun to take place. For instance, nuclear magnetic resonance and differential thermal analysis are used to study lipids, while optical rotatory dispersion, circular dichroism, and infrared spectrophotometry are used chiefly to study the state of proteins. Meanwhile, introduction of methods using paramagnetic and fluorescent probes and labels has proved very useful both in the study of the state of lipids and proteins separately and in the elucidation of their interaction.

Electron-microscopic investigations of membranes, especially with the use of new techniques of specimen preparation, are also very important because ultimately they will enable the biochemical data to be correlated with membrane structure.

The purpose of this book is to examine the molecular organization of bacterial membranes. Particular attention will be paid to the properties and organizations of enzymes of the electron transport chain (the respiratory chain) of heterotrophic and chemoautotrophic bacteria. The process of electron transport has been studied comparatively thoroughly, and it can therefore act as "marker" for the study of the topography, if not of the membrane as a whole, at least of those parts of it which contain the components of the respiratory chain.

The Membrane Structures of Bacteria

Morphology of the Bacterial Membrane System

The membrane structures in the cells of about 70 species of bacteria have now been studied.* The results obtained before 1966 have been collected in several surveys (Glauert, 1962; Murray, 1963; Salton, 1964; Van Iterson, 1965; Lascelles, 1965; Gel'man et al., 1966; Kushnarev, 1966a). In recent years the membrane structures of many bacteria, which are listed below, have been studied: *Bacillus anthracis* (Avakyan et al., 1967), *Bacillus licheniformis* (Highton, 1969), *Bacillus fastidiosus* (Leadbetter and Holt, 1968), *Bacillus stearothermophilus* (Walker and Baillie, 1968), *Sarcina ventriculi* and *Sarcina maxima* (Holt and Canale-Parola, 1967), *Sarcina lutea* (Cherni, 1967), *Micrococcus denitrificans* and *Micrococcus halodenitrificans* (Kocur et al., 1968a,b), *Chondrococcus columnaris* (Pate and Ordal, 1967), *Pseudomonas* sp. and *Achromobacter* sp. (Wiebe and Chapman, 1968a,b), *Leptospira* sp. (Kats and Konstantinova, 1966), *Franciscella tularensis, Pasteurella pestis,* and *Bordetella abortus* (Kats, 1966; Avakyan et al., 1967), *Bacteroides insolitus* (Ushijima, 1967), *Haemophilus vaginalis* (Reyn et al., 1966), *Halobacterium halobium* (Cho et al., 1967), *Thermus aquaticus* (Brock and Edwards, 1970), thermophilic sulfur bacteria (Brock et al., 1971), *Caulobacter crescentus* (Cohen-Bazire et al., 1966), *Chloropseudomonas ethylicum* (Holt et al., 1966), *Rhodoteca pundens* and *Rhodopseudomonas* sp. (Cherni et al., 1969), *Ectothiorhodospira mobilis* Pelsh (Remsen et al., 1968), *Ferrobacillus ferrooxidans* (Remsen and Lundgren, 1966a), *Thiobacillus novellus* (Kocur et al., 1968a,b), *Thiococcus* sp. *nov. gen.* (Eimjellen et al., 1967), methane bacteria (Langenberg et al., 1968; Proctor et al., 1969; Davies and Whittenbury, 1970), *Methylococcus capsulatus* and *Methanomonas methanooxidans*

5

(Ribbons et al., 1970), hydrogen bacteria (Repaske, 1966), *Clostridium tetani* and *Clostridium botulinum* (Takagi et al., 1965; Pavlova and Sergeeva, 1969), *Clostridium pectinovorum* (Hoeniger and Hedley, 1968), *Clostridium perfringens* (Hoeniger et al., 1968; Pavlova and Larina, 1969), *Lactobacillus corinoides* (Schotz et al., 1965), *Lactobacillus plantarum* (Kakefuda et al., 1967), *Lactobacillus casei* (J. Brown et al., 1968), and *Lactobacillus acidophilus* and *Lactobacillus bifidus* (Lickfeld, 1967).

Analysis of the electron-microscopic data lies outside the scope of this book; they have been examined in the surveys of Ryter (1969), Van Iterson (1969a,b,c), and Kats (1971). What is important in the present context is that membrane structures, although differing in their complexity, have been found in all bacteria investigated, and membranes are evidently an essential component of the bacterial cell.

The membrane system of bacteria consists of the cytoplasmic membrane and the internal membranous structures. Judging from the electron-microscopic data, the cytoplasmic membrane is similar in shape in all bacteria. Morphological differences between the membrane systems are exhibited at the level of the internal cell structures which are usually formed by invagination of the cytoplasmic membrane.

As Salton (1967b) points out, the membrane system of bacteria exists in two principal forms, lamellar and vesicular, which differ from one another in the arrangement of the Robertson's membrane revealed by fixation with osmic acid. Discrete membranous structures of vesicular or lamellar type in heterotrophs are usually called mesosomes and in photoautotrophs they are called chromatophores. Mesosomes of vesicular type are clearly defined in members of the Bacillaceae family. Mesosomes of lamellar type are observed more frequently in cocci and in certain other bacteria. In many photosynthesizing and chemosynthesizing bacteria there are no discrete membranous structures whatever, but numerous membranes arranged parallel to each other penetrate throughout the thickness of the cell such as, for example, in *Rhodospirillum molishianum* (Giesbrecht and Drews, 1962) and *Nitrosocystis oceanus* (Murray and Watson, 1965). The possibility cannot be ruled out that the arrangement of the membrane in the mesosomes of bacteria is determined to some extent by the conditions of fixation of the cells (pH, fixative, ionic strength, bivalent cations). Burdett and Rogers (1970) describe modification of the structure of the mesosomes in this way under the influence of fixation in the case of *Bacillus licheniformis*. In their opinion the native mesosome consists of a system of vesicles and tubules. The lamellar form of mesosomes observed in the same bacteria and under the same conditions of fixation (Highton, 1970) can possibly be explained by differences in the age of the cells and in the conditions of cultivation. Replacement of vesicular mesosomes by lamellar mesosomes has also been observed during the development of a culture of *Micrococcus lysodeikticus* (Skopinskaya et al., 1972).

Membranous structures are poorly developed in the cells of most Gram-negative bacteria. The exceptions are the cells of the photosynthetic and of some chemosynthetic bacteria. The Gram-positive bacteria, on the other hand, have a well developed system of internal membrane.

The morphological features of the membrane system can be most conveniently described by combining the bacteria into physiological groups: aerobic and anaerobic heterotrophs, photosynthesizing and chemosynthesizing bacteria, as well as special groups such as halophiles, thermophiles, and pathogenic bacteria.

It was naturally expected that the membrane system would be poorly developed in anaerobic bacteria, which belong to the most primitive forms of life on earth (Oparin, 1957; Hall, 1971), particularly because anaerobic fermenters are unable to carry out oxidative phosphorylation, for the organization of which membranes are essential. However, membrane structures in bacterial anaerobes are very well developed indeed, and in some cases they are actually superior to those found in aerobes. Sections through cells of *Lactobacterium pentoaceticum* and *Clostridium oedematiens* are illustrated as examples (Fig. 1).

Membrane structures of strict anaerobes, unlike those in the aerobic bacteria, do not contain oxidoreductases (Tordzhyan and Kats, 1970; Avakyan et al., 1971).

In aerobic bacteria and in facultative anaerobes possessing enzymes of the respiratory chain and generating energy by oxidative phosphorylation, the membrane apparatus is highly variable. In some aerobic bacteria the membrane systems are very well developed, although of different forms. The mesosomes of a *Bacillus subtilis* cell (Granboulan and Leduc, 1967; Ryter, 1968, 1969) illustrated in Fig. 2 or the mesosomes of other bacilli and, in particular, of *Bacillus megaterium* (Ellar et al., 1968) are good examples. Meanwhile the membrane system of *Azotobacter vinelandii* is much less well developed and consists only of diffusely arranged membranes (Tchan and Webber, 1966), although the cells of *Azotobacter* are among those with the most intensive respiration and contain a complete respiratory chain of enzymes. In bacteria of the enteric group (*Escherichia coli* and *Aerobacter aerogenes*), which also contain a respiratory chain of enzymes, the internal membrane systems also are poorly developed (Steed and Murray, 1966; Cota-Robles, 1966; Kennell and Kotoulas, 1967; Nanninga, 1970a,b). Small mesosomes have been found in only a few strains of *E. coli* (Schnaitman and Greenawalt, 1966; Pontefract et al., 1969). The membrane systems of *Proteus vulgaris* are poorly developed (Leene and Van Iterson, 1965).

A striking example of the variety of forms which can be assumed by the membrane system is given by the photosynthesizing bacteria. Whereas the membrane systems of *Rhodospirillum rubrum* and *Chloropseudomonas ethylicum* consist of vesicular structures joined to the cytoplasmic mem-

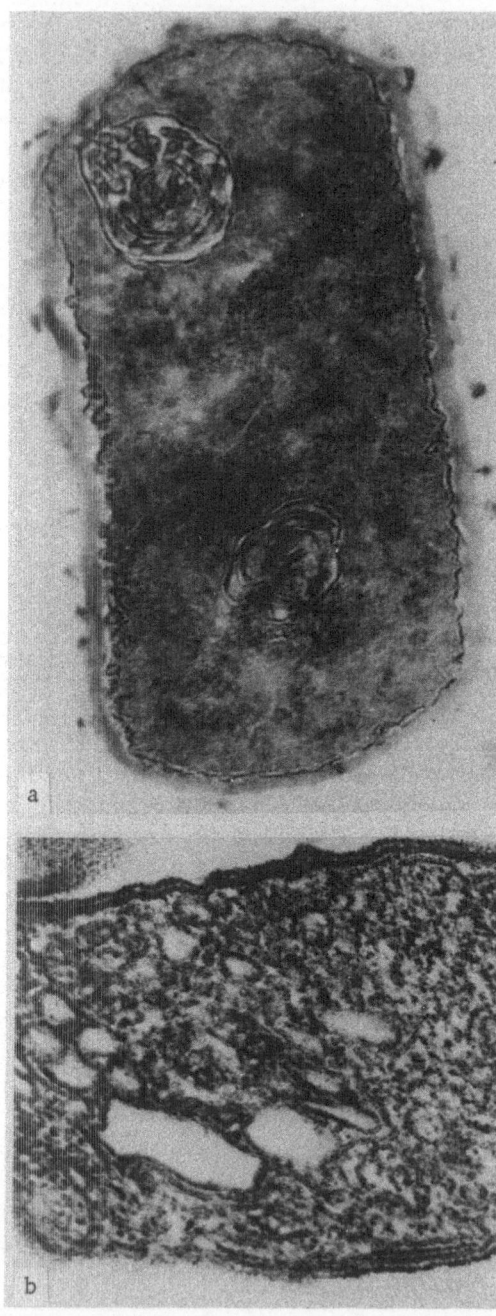

Fig. 1. The membrane system of anaerobic fermentation bacteria: a) *Lactobacterium pentoaceticum*, 140,000 × (Kharat'yan et al., 1967); b) *Clostridium oedematiens*, 90,000 × (Avakyan et al., 1970).

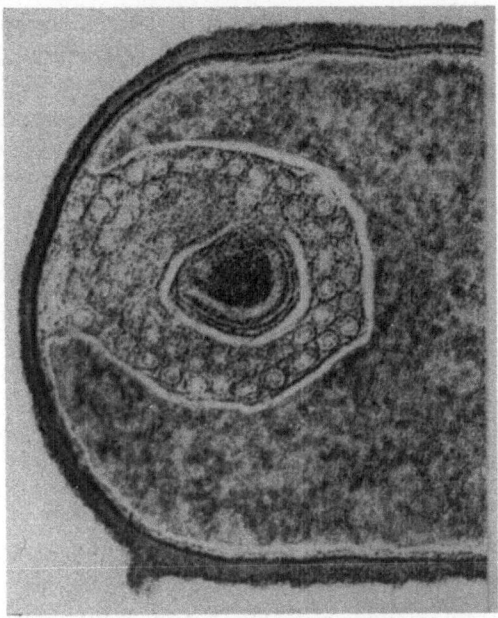

Fig. 2. Mesosomes of *Bacillus subtilis,* 95,000 X
(Ryter, 1969).

brane (Cohen-Bazire and Kunizawa, 1963; Boatman, 1964; Holt, et al., 1966), in *Rhodopseudomonas viridis* and *Rhodopseudomonas palustris* complex lamellar systems of membranes exist, resembling the thylacoids of chloroplasts (Drews and Giesbrecht, 1965; Giesbrecht and Drews, 1966; Tauschel and Drews, 1966; Tauschel and Drews, 1967). An exceptionally complex membrane system is found in the photosynthesizing bacterium *Ectothiorhodospira mobilis* Pelsh (Remsen et al., 1968).

The lamellar membrane system in some chemosynthesizing bacteria (*Nitrobacter agilis, Nitrosomonas europea, Nitrosocystis oceanus*) is extremely well developed (Murray, 1963; Murray and Watson, 1965; Remsen et al., 1967). Complex membrane systems have also been found in *Nitrobacter winogradskii* (Tsien et al., 1968). Another group of chemosynthesizing bacteria — the ferro-oxidizing and thio-oxidizing bacteria — usually do not contain complex membrane systems (Remsen and Lundgren, 1966a,b; Mahoney and Edwards, 1966; Kocur et al., 1968a, b), and small mesosomes have been found only in some strains of the thiobacilli (Shively et al., 1970; Avakyan and Karavaiko, 1970). A distinctive system of internal membranes has been found in the halophilic bacterium *Halobacterium halobium* (Stoeckenius and Rowen, 1967; Cho et al., 1967) and in a marine pseudomonad (Backmuire and MacLeod, 1965). The thermophilic bacterium *Bacillus stearothermophilus* possesses well-developed mesosomes (Walker and Baillie, 1968).

The membrane system of the pathogenic bacteria is poorly developed, as has been shown for *Shigella flexneri* (Pavlova and Pershina, 1966) and for certain other organisms (Avakyan et al., 1967).

On the whole, the size, shape, and structure of the membrane formations in bacteria are extremely varied and their morphology could be used as a taxonomic feature were it not for the fact that the internal membrane structures of the bacterial cell vary with the age of the cells, the composition of the nutrient medium, and certain other factors (Ryter, 1969; Kats, 1971).

Although the internal membrane structures are connected with the cytoplasmic membrane there is no proof that the two are identical. On the contrary, if these two types of membranes are separated it is possible to demonstrate differences in their protein, and, in particular, in their enzymic apparatus (see page 16).

Some extremely interesting electron-microscopic pictures showing differences in the structure of the mesosomal and cytoplasmic membranes have been given by Ryter (1969) (Fig. 3). Evidence of differences in the structure of the cytoplasmic and mesosomal membranes is given by electron-microscopic studies of *Bacillus subtilis* cells, using freeze-etching. The cytoplasmic membrane is covered with numerous small granules, whereas the mesosomal membranes are smooth (Nanninga, 1968, 1970a). Indirect evidence of differences between the two types of membranes is presented by Bertsch et al. (1969). Diphosphatidylglycerol is extracted from *B. megaterium* cells only after their destruction, i.e., evidently from their internal membranes, whereas phosphatidylglycerol and phosphatidylethanolamine are readily extracted from whole cells, i.e., from the cytoplasmic membrane.

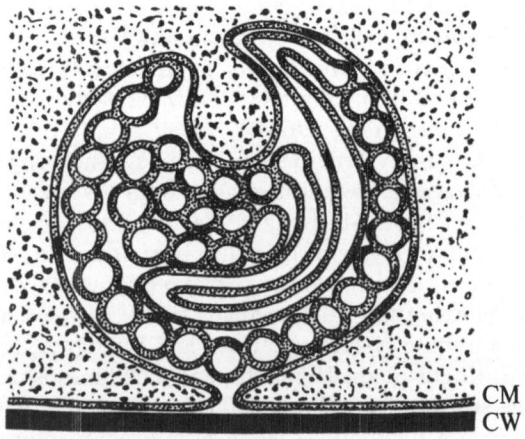

Fig. 3. Diagram of the structure of mesosomes (Ryter, 1969). CM) Cytoplasmic membrane; CW) cell wall.

The variety of forms of internal membrane structures in bacteria is evidence of their changeable, evolutionary form in the bacterial cell. This is one of the chief differences distinguishing membrane structures of bacteria from mitochondria and chloroplasts. Bacterial membranes are not like the endoplasmic reticulum: they occupy a relatively small part of the cytoplasm and they form something resembling a membranous net only in the cells of *Azotobacter* and certain chemosynthesizing and photosynthesizing bacteria.

Comparison of the morphology of the membrane system in the different physiological groups of bacteria shows that the development and structure of the membrane systems in the bacterial cell are not determined by the method of obtaining energy, for membrane systems are present both in anaerobes and in aerobes, both in heterotrophs and in photo- and chemoautotrophs. Nor is the development of the membrane system determined by the method of cell reproduction; membrane formations are found in both spore-forming and non-spore-forming bacteria. While the extreme diversity of the membrane system in bacteria is an undeniable fact, it is impossible to construct an evolutionary series to reflect the correlation between antiquity of origin, development of metabolism, and form of the membrane system in bacteria.

Methods Used to Isolate Membranes

To study the molecular organization of bacterial membranes, they must first be isolated from the cell in a sufficiently pure form. Methods of destruction of the bacterial cell and isolation of the membranes or their fragments are examined in several surveys (Hughes and Cunningham, 1963; De Ley, 1964; Salton, 1964; Stolp and Starr, 1965; Strominger and Ghuysen, 1967; Work, 1967; Murray, 1968).

One of the milder methods most commonly used to destroy bacterial cells is hydrolysis of the cell wall with lysozyme (Weibull, 1953, 1956). The walls of some Gram-positive bacteria are completely removed by the action of lysozyme; protoplasts are preserved in an isotonic medium, and by subsequent osmotic shock the membranes can be isolated from them. In hypotonic medium, hydrolysis of the cell wall is accompanied by lysis of the cell and liberation of the membranes. However, the cells of some Gram-positive bacteria are resistant to lysozyme (Murray, 1968). In these cases it is useful to treat the cell first by heat or by pancreatic lipase, as has been done for two strains of *Listeria monocytogenes* (Ghosh and Murray, 1967). Preliminary treatment with trypsin is recommended to produce lysis of certain lactic acid bacteria (Neujahr, 1970; Barker and Thorne, 1970). Work on *Bacillus stearothermophilus* and *Bacillus coagulans* has shown that autolysis of the cells facilitates removal of the cell wall and the formation of protoplasts (Golovacheva et al., 1968).

The walls of Gram-negative bacteria differ in their structure from those of Gram-positive bacteria and are accordingly resistant to lysozyme (Salton, 1964). However, in some Gram-negative bacteria such as *E. coli* the action of lysozyme, in the presence of EDTA and in an isotonic medium, leads to the formation of spheroplasts which preserve two membranes: a membrane belonging to the cell wall and the cytoplasmic membrane (Repaske, 1958; Nagata et al., 1966; Voss, 1967; Birdsell and Cota-Robles, 1967; Gumpert and Nermut, 1967). The cell wall membrane is not necessarily preserved over the entire surface of the spheroplast. Like protoplasts, spheroplasts are disintegrated by osmotic shock, liberating the membranes. The work of Mizushima et al. (Miura and Mizushima, 1968, 1969; Mizushima and Ito, 1968) showed that the cell wall membrane of *E. coli* can be separated from the cytoplasmic membranes if lysed spheroplasts are centrifuged in a sucrose gradient. Fragments of the cell wall and membranes of *E. coli* can be separated by electrophoresis (White et al., 1972).

Cells of the halophilic bacteria are a special case, for their walls are fragmented in medium containing $1.4 M$ NaCl while their cytoplasmic membrane disintegrates if the NaCl concentrations is reduced to $1 M$ or below (Onishi and Kushner, 1966; Stoeckenius and Rowen, 1967). Cells of the psychrophilic bacteria also undergo lysis in water or in weak salt solutions, when their membranes are fragmented (Korngold and Kushner, 1968). A method of obtaining membranes from helophilic bacteria in $0.02 M$ $MgCl_2$, which prevents the membrane from breaking up into very small fragments, has been suggested by Brown et al. (1965).

Membranes of several mycoplasmas are highly sensitive to a decrease in the ionic strength of the medium, but lysis of the cells is not accompanied by fragmentation of the membranes, and indeed this method can be used to isolate the membranes from the cells (Razin, 1964).

The quantity of membranes obtained by lysis of bacteria varies from 10 to 30% of the dry weight of the cell; these differences in yield are probably attributable to the degree of development of the membrane systems (Salton and Freer, 1965).

To purify the membranes from cytoplasmic proteins and ribosomes, repeated washing with buffer solution (Salton, 1967a) or treatment with deionized water at pH 8.5 (Mizushima et al., 1967) is recommended. However, these procedures cause partial destruction of the membranes. The degree of purity is usually estimated from the content of components of the cell wall (hexosamine, diaminopimelic acid, or rhamnose) or components of the cytoplasm (RNA, soluble enzymes, e.g., polynucleotide phosphorylase) in the membranes.

Lysis of the cell naturally is accompanied by visible disturbances in the arrangement of the membrane system of the cell. This phenomenon is clearly illustrated in the case of cells of *Thiococcus* sp. *nov. gen.* during osmotic shock, when the closely packed membranes are unrolled and leave the cell in the form of a long tube (Eimjellen et al., 1967). The membranes of *Micro-*

coccus lysodeikticus lose their lamellar arrangement during lysis of the cells, they become straightened out, and are liberated from the cell in the form of closed vesicles of different sizes (Lukoyanova et al., 1961; Salton and Chapman, 1962). The changes produced by the removal of the cell wall in the meso-somes of protoplasts of various bacilli can easily be seen (Fig. 4). During lysis of the cell the mesosomes are discharged into the space outside the cyto-plasmic membrane and undergo partial degradation (Ryter and Jacob, 1963; Ryter and Landman, 1964; Fitz-James, 1964; Weibull, 1965; Ryter et al., 1967; Ghosh and Murray, 1967; Silva, 1967; Golovacheva et al., 1969).

In some bacteria the membrane system does not suffer such severe changes during lysis of the cell. Isolated membrane systems from *Azotobacter agilis,* for instance, are similar to those of the intact cell (Pangborn et al.,

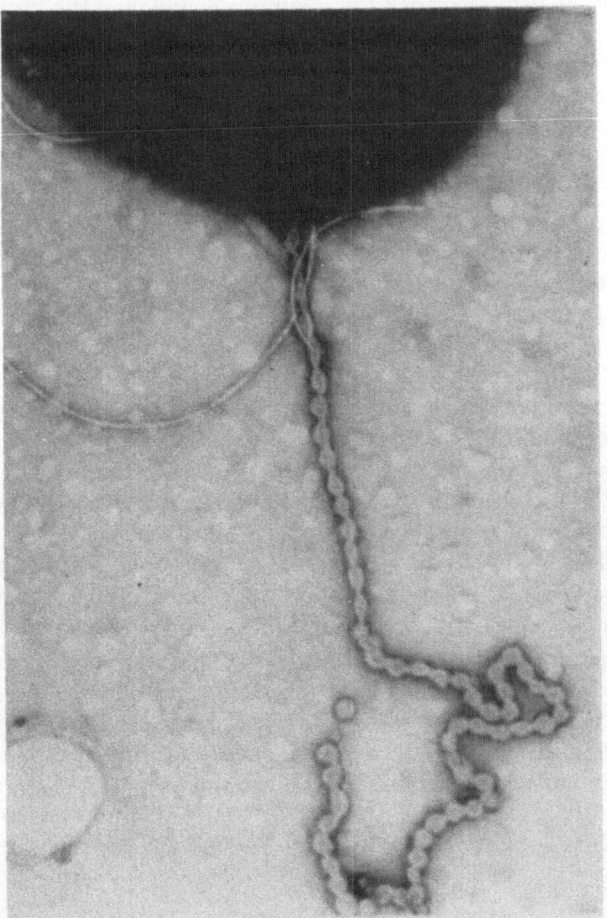

Fig. 4. Protoplasts of *Bacillus subtilis* with mesosomal membranes attached to them, 33,000 X (Ryter, 1969).

1962; Robrish and Marr, 1962). Isolated chromatophores of *Chlorobium thiosulfatophilum*, shaped like rectangular vesicles, resemble these structures in the intact cell (Sykes et al., 1965; Cohen-Bazire et al., 1964).

The most probable causes of disturbance of the structure of bacterial membrane systems during lysis of the cell are the absence of formed organelles and the connection between the mesosomes and the cytoplasmic membrane.

Investigators are usually satisfied by the fact that membranes can be isolated in a comparatively pure form by lysis of the cell and they pay no heed to the processes taking place when this is done. In this connection some interesting experiments have been carried out with the protoplasts of *B. megaterium* (Corner and Marquis, 1969).

After comparing the effects of molecules of different sizes on the rate of lysis of the protoplasts, Corner and Marquis draw the following picture of protoplasts during osmotic shock: in a hypotonic medium the protoplasts swell and the membrane pores are stretched. When the pores are large enough to allow the passage of the stabilizing solution, a cascade process of swelling begins. Mechanical tensions develop in the membrane which give rise to small ruptures, to liberation of substances from the cell, and to a decrease in pressure. Proteins evidently play a leading role in determining the mechanical properties of membranes: urea and glycerol aggravate the osmotic instability of the membrane, while formaldehyde, conversely, strengthens the membrane.

Other evidence of deformation of the membrane during lysis of the cell is given by changes in the activity of the respiratory chain. Respiration of protoplasts is usually indistinguishable from cell respiration. Many substrates were oxidized at the same rate in the protoplasts of *Acetobacter peroxidans* and *Gluconobacter liquefaciens* as in intact cells (De Ley and Dochy, 1960a,b). Some degree of inhibition of oxidation in protoplasts usually occurs during osmotic shock. For instance, when protoplats of *A. vinelandii* are converted into membranes, oxidation of succinate and malate is considerably depressed (Jose and Wilson, 1959). Inhibition of oxidation of amino acids also was observed by Nermut (1965) during lysis of the spheroplasts of *P. vulgaris*. Meanwhile, oxidation of many substrates took place at a high rate in membranes of *G. liquefaciens* obtained from protoplasts (Stouthamer, 1961).

Interesting information on the effect of swelling and the attendant deformation of the membrane on the respiratory chain is given by L. Smith (1962). Spheroplasts of *B. subtilis* obtained in $0.5\,M$ sucrose were added to sucrose solutions in concentrations of between 0.2 and $0.8\,M$. It was found that oxidation of amino acids was inhibited while oxidation of $NADH_2$ was increased in sucrose solutions ranging from 0.2 to $0.4\,M$. Smith assumed that deformation does not extend to parts of the membrane carrying the respiratory chain, and she explains the inactivation of oxidation of amino acids by detachment of the loosely bound dehydrogenases.

By contrast with $NADH_2$ dehydrogenase, whose activity in cell membranes is usually unchanged by osmotic shock, in some bacteria another enzyme of the respiratory chain — succinate dehydrogenase — is inactivated (Repaske, 1954; Owen and Freer, 1970; R. Tucker, 1960).

Inhibition of succinate oxidase during lysis of the cells of protoplasts of *M. lysodeikticus* is evidently explained by inactivation of succinate dehydrogenase (Gel'man et al., 1959; Gel'man et al., 1963a).

It is not always possible to isolate membranes by lysis of the cell, and sometimes more vigorous treatment with ultrasound or mechanical disintegration is necessary. Under these conditions the membrane breaks up into fragments, frequently so small that they can be harvested only be centrifugation at between 100,000 and 150,000 *g*. By such methods of cell destruction a mixture of fragments of the wall and fragments of the membranes is obtained; it has been shown that these particles can be separated (Gray and Thurman, 1967; Cusanovich and Kamen, 1968). Separation of the cell wall and cytoplasmic membrane of *E. coli* can be done by centrifugation of fragments obtained by destroying the cells in a press in a sucrose gradient (Schnaitman, 1970a,b). Purification of the cytoplasmic membrane of *E. coli* and isolation of a fraction with a high concentration of cytochromes, succinate dehydrogenase, and transport systems has been described by Fox et al. (1970).

As might be expected, during disintegration of the cell the enzyme systems of the membranes are injured. Oxidation of glucose took place just as rapidly in particles obtained at 100,000 *g* from a homogenate of *Pseudomonas* sp. cells as in the cells, but the oxidation of other sugars was sharply inhibited (Bentley and Schlechter, 1960). Correlation between the physical state of the cytoplasmic membrane and respiratory activity has been demonstrated in *E. coli* (Henneman and Umbreit, 1964). It has been reported that destruction of *P. vulgaris* cells in a Hughes press considerably reduced the oxidation of lactate, succinate, and certain other substrates (Jones and King, 1964). During sonication of *Haemophilus parainfluenzae* cells, 29% of the succinate dehydrogenase, 36% of the $NADH_2$ dehydrogenase, and 92% of the lactate dehydrogenase went into solution (White, 1965b). Oxidation of lactate in *Corynebacterium diphtheriae* cells disintegrated by ultrasound took place at only 10% of the rate in intact cells; meanwhile no injury to succinate oxidase and to $NADH_2$ oxidase took place (Scholes and King, 1965a).

After sonication of respiratory chain-containing particles from *Ferrobacillus ferrooxidans* the link between cytochromes c and a was injured (Din et al., 1967). Correlation exists between the size of the soniated fragments of *Nitrobacter* membranes and their nitrite-oxidase activity. The least active enzyme-containing particle has an area of 400 nm^2 (Tsien and Laudelout,

1968). Freezing and thawing cells of *Nitrosomonas* leads to separation of a group of cytochromes from the membranes (Hooper et al., 1972).

During destruction of the cell and isolation of the membranes or their fragments, one further factor of considerable importance to maintenance of the native state of the membrane must be taken into account: bivalent cations, especially magnesium.

Membranes obtained by osmotic shock from the protoplasts of *Pseudomonas* differed whether magnesium ions were present or not. Membranes obtained in the absence of magnesium did not oxidize succinate and malate; those which were obtained in a medium containing Mg^{++} ions (5×10^{-3} M) oxidized these substrates readily (Mizuno et al., 1961). Magnesium ions and spermin stabilize oxidation in the membranes of *M. lysodeikticus* (Ishikawa and Lehninger, 1962). It has been shown in the case of membranes of *Alcaligenes faecalis* that stability of the membrane depends not so much on the nature of the cation as on the ionic strength of the solution; K^+ ions also stabilize the membrane and prevent detachment of the weakly bound factors of phosphorylation (Adolfson and Moudrianakis, 1971a,b). To prevent destruction of the membranes in the course of their isolation from the cell and their purification, treatment with 0.1-0.5% glutaraldehyde is recommended; under these conditions mild fixation of the membranes takes place without substantial inactivation of the enzyme proteins (Cruden and Stanier, 1970; Ellar et al., 1971). Often preparations of membranes are frozen or lyophilized for preservation. However, the stability of the bond between the protein components and the membrane may be altered by treatment with various agents. For instance, lyophilic drying of a preparation of *E. coli* membranes enables 50% of the $NADH_2$ dehydrogenase to be removed from it simply by water, whereas the enzyme was not extracted from the original preparation even by digitonin after prolonged sonication or treatment with phospholipase A (Gutman et al., 1968). It is interesting to note that, as these workers observed, the destructive action of factors such as freezing and drying on the isolated bacterial membrane had almost no effect on the properties of the membranes in the intact cell. The membranes evidently lose some of their essential components during isolation, and this argument can be carried a step further by saying that very pure membranes are in fact no longer membranes.

Separation of the Cytoplasmic Membrane and Mesosomes

One of the objects of the biochemistry of bacterial membranes is to separate the cytoplasmic membrane from the internal membranes so that the chemical, structural, and functional properties distinguishing these structures can be determined.

In the few investigations which have so far been carried out in this direction, the phenomenon of discharge of the mesosomes from the cytoplasm into the medium in the course of protoplast formation has been used. Mesosomal membranes from *B licheniformis* differ from the cytoplasmic membrane in their protein composition. These differences do not apply to the cytochromes, the content of which is identical in the cytoplasmic membrane and mesosomes. Meanwhile $NADH_2$ oxidase and succinate oxidase have been found only in the cytoplasmic membrane (Reaveley, 1968; Reaveley and Rogers, 1969). However, according to Ferrandes and Chaix (1972), the mesosomes and cytoplasmic membrane of *B. subtilis* differ in the composition of their cytochromes. Succinate oxidase is found only in the cytoplasmic membrane while $NADH_2$ oxidase is found in both types of membranes (Ferrandes et al., 1970).

Some progress has been made in the separation of mesosomal and cytoplasmic membranes from the protoplasts of *B. megaterium* by centrifugation in a sucrose gradient. The fraction of mesosomal membranes incorporated P^{32}, Fe^{59}, and C^{14} more intensively, although some incorporation also took place into cytoplasmic membranes (Fitz-James, 1968). However, according to other observations the mesosomes and cytoplasmic membrane of *B. megaterium* and *B. subtilis* do not differ in the rate of metabolism of their components (Daniels, 1971; Patch and Landman, 1971).

By centrifugation of lysed protoplasts of *Listeria monocytogenes* in a Ficoll gradient, a fraction in which succinate dehydrogenase, lactate dehydrogenase, and $NADH_2$ dehydrogenase were concentrated was obtained, together with a fraction with negligible activity of these enzymes but containing active $NADH_2$ oxidase and ATPase. Morphologically the first fraction corresponds to mesosomes and the second to the cytoplasmic membrane. The mesosomal fraction incorporates P^{32} more actively than the fraction of cytoplasmic membranes (Ghosh and Murray, 1969). Fragments of the cytoplasmic membrane and of the mesosomes can be separated if cells of *B. subtilis* are broken down in a disintegrator in the presence of DNase and lysozyme (Durner and Mach, 1970). The procedure used to separate mesosomes from the cytoplasmic membrane of *Staphylococcus aureus* has been described in detail by Popkin et al. (1971).

The cytoplasmic membrane has been successfully separated from the thylacoid membranes. Unlike the thylacoids, the cytoplasmic membrane of *Rhodospirillum rubrum* has no bacteriochlorophyll (Ketchum and Holt, 1970). Differences in the composition of the proteins of the cytoplasmic membrane and thylacoids are clearly defined in 16-h cells of *R. rubrum* grown under anaerobic conditions and in light. The cytoplasmic membrane has a high content of respiratory chain enzymes and it has a higher $NADH_2$ oxidase/bacteriochlorophyll ratio than the thylacoids (Throm et al., 1970; Oelze and Drews, 1970b). The cytoplasmic membrane differs in its protein, lipid, and enzyme composition from membranes containing the photosynthetic apparatus in *Chlorobium* (Cruden and Stanier, 1970).

It is very difficult to judge the structural biochemical features distinguishing the cytoplasmic membrane from the internal membranes with the comparatively few facts available. Nevertheless, certain conclusions can already be drawn. It is interesting to note that the cytoplasmic membrane of heterotrophic bacteria contains a complete respiratory chain, while the mesosomes contain only the components of the chain: cytochromes or dehydrogenases. In this context it is essential to note that in both structural elements of the cells of *S. aureus* succinate dehydrogenase has been identified by cytochemical methods (Tordzhyan and Kats, 1969). The cytoplasmic membrane also exhibits some degree of autonomy in the photoautotrophs, in which, like the cytoplasmic membrane of the heterotrophs, it contains $NADH_2$ oxidase.

Electron Microscopy of the Membranes

Membranes when in the cell and isolated membranes, after fixation with osmic acid or permanganate and subsequent shadow casting of the sections, have a thickness of 75-80 Å and thus correspond to the ordinary unit membrane.

The use of other methods of fixation of bacterial membranes has revealed structures of two different types: knob-like subunits and globular structures with parameters that differ from those of the knob-like subunits. In membranes of *M. lysodeikticus,* for instance, knob-like subunits with a head measuring 90 Å in diameter, and attached to the membrane by a pedicle 20 Å in thickness, have been found by negative staining (Fig. 5a) (Biryuzova et al., 1964; Biryuzova and Meisel', 1964). Similar structures have been found on the internal membranes of *Eubacterium* sp., the cytoplasmic and internal membranes of *Bacillus stearothermophilus* (Bladen et al., 1964), on the membranes of *Bacillus pumilis, Bacillus licheniformis, Bacillus brevis, Bacillus circulans, E. coli* C, *Proteus vulgaris,* and *Shigella dysenteriae* Y6R (Abram, 1965), and also on the membranes of *Rhodospirillum rubrum* (Löw and Afzelius, 1964), of *N. oceanus* (Remsen et al., 1967), *Nitrobacter winogradskii,* (Tsien et al., 1968), *B. megaterium* (Ellar et al., 1968), *A. vinelandii* (Jones and Redfearn, 1967a), and *Lactobacillus casei* (Barker and Thorne, 1970).

The presence of knob-like subunits on bacterial membranes indicates that these membranes are similar in structure of the mitochrondria. Nevertheless, there are undoubted differences between the morphology of the mitochondria and these membranous structures. In bacteria these structures are invaginations of the cytoplasmic membrane and there is no line of

Fig. 5. Membranes of *Micrococcus lysodeikticus* (negatively stained) (Biryuzova et al., 1964; Lukoyanova et al., 1967, 1971): a) original membranes, 120,000 ×; b) membranes after proteolysis, 80,000 ×; c) membranes after treatment with deoxycholate, 60,000 ×.

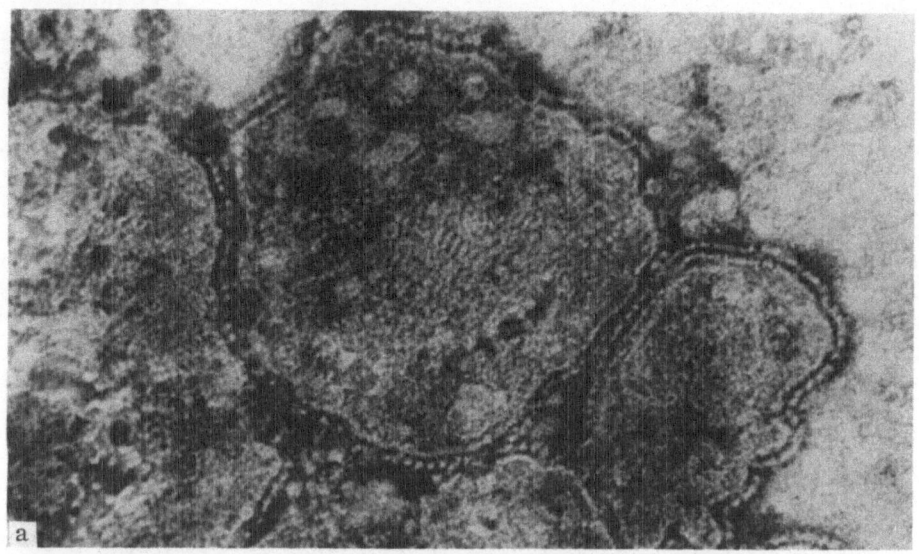

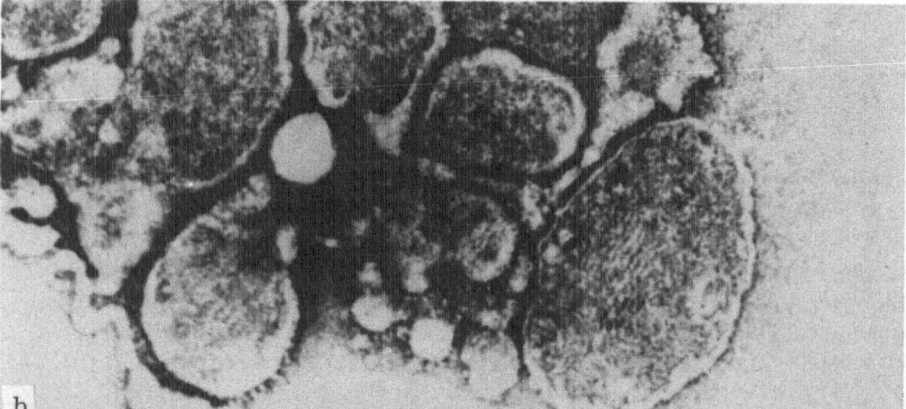

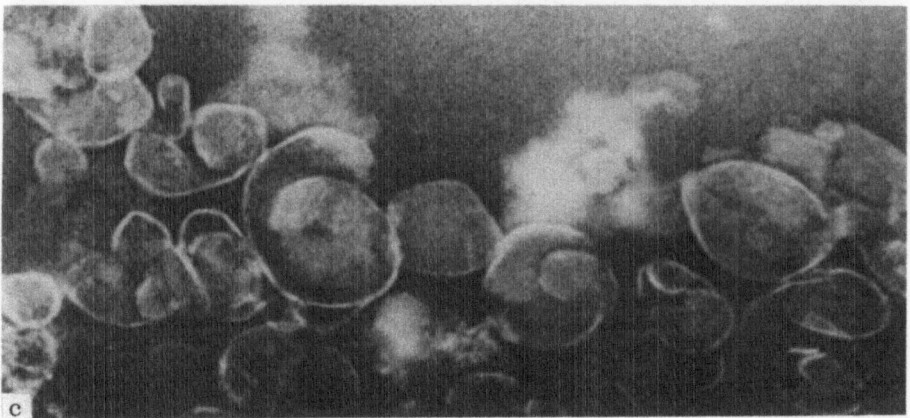

demarcation between them and the cytoplasm. Knob-like subunits located on the cytoplasmic membrane and in the internal membranes are in direct contact with the cytoplasm. Mitochondrial cristae are invaginations of the internal mitochondrial membrane and are buried in the matrix of the mitochondrion. The function of the knob-like subunits is examined on page 83. Negative staining of the membranes of *Bacillus stearothermophilus* and *Bacillus coagulans* reveals globular structures with are smaller in size than the fungoid outgrowths (Golovacheva et al., 1968).

Globular structures have also been identified in bacterial membranes examined under the electron microscope in specimens prepared by other methods. The method of freeze-etching has been found particularly useful.

According to Nanninga (1968, 1970a) the mesosomal membranes of *B. subtilis* (4- to 5- h cultures) have a smooth surface, while the cytoplasmic membrane is covered with globules, the number and arrangement of which differ on its outer and inner surfaces. Remsen (1968), who also worked with *B. Subtilis,* reached similar conclusions. However, Remsen notes that at a certain period of development of the cell the smooth mesosomes are replaced by mesosomes bearing globules. A globular structure has also been found on the spore membranes of *B. subtilis* (Remsen, 1966). Globules have been found by freeze-etching on the cytyoplasmic membrane of *E. coli.* The distribution of the globules differs in the membranes of cells during the logarithmic and stationary phases (Bayer and Remsen, 1970). Working with the same method, Nanninga found asymmetry of the membranes of *E. coli* as shown by the fact that the globules were more numerous on the outer than on the inner surface (Nanninga, 1970a,b). Kushnarev (1966b) showed earlier that globular structures are present in the cytoplasmic membrane of *E. coli.* Globules measuring 80 Å in diameter were found on the chromatophore membranes of *Rhodospirillum rubrum* after fixation with permanganate and negative staining with uranyl acetate (Fuchs, 1966).

It is difficult to compare the electron-microscopic pictures of objects fixed and stained by different methods, each of which involves the risk of artefacts. Remsen et al. (1967) attempted to compare the results obtained by negative staining and by freeze-etching. The fungoid outgrowths revealed on membranes of *N. oceanus* by negative staining are evidently identical with the structures found on these membranes by the freeze-etching method. Together with fungoid outgrowths, another system of globules with a periodicity of 40 Å was revealed by freeze-etching in the membranes of *N. oceanus.*

Although information on the existence of globular structures in bacterial membranes is still scanty and frequently contradictory, the simple fact that this globularity of the membranes has been discovered is itself of considerable importance, for it usggests that they are heterogeneous in structure and that different types of membranes possess different structural details. Evidence of these structural differences is given by the observations of Nanninga (1968,

1970a), who demonstrated differences between the structure of the mesosomal and cytoplasmic membranes of *B. subtilis*. Differences between the
structure of these two types of membranes were mentioned earlier by Biryuzova
et al. (1964).

The possibility cannot be ruled out that the inner and outer surfaces of
the cytoplasmic membrane have different structures, i.e., that the membrane
is asymmetrical. Support for this view is given by observations on the distribution of globules in the cytoplasmic membrane of *E. coli* (Fiil and Branton,
1969; Nanninga, 1970a,b).

The Biogenesis of Membranes

The biogenesis of membranes is a complex process and it is therefore being
studied in various directions, including by the synthesis of the components
and mechanism of assembly of the membrane, and variation in the composition and structure of the membrane depending on the conditions. The chief
difficulty in the way of interpretation of the comparatively few experimental
data is that usually the synthesis of only a few components of the membrane
has been studied, and these components — the cytochromes, dehydrogenases,
and individual lipids — are only of minor importance, and no attempt has
been made to link this process with changes in the composition, morphology,
and functions of the membrane system as a whole.

Some information on the biogenesis of membranes can be obtained from
investigations into the effect of the conditions of cultivation on the morphology of the membrane system as a whole. In response to a decrease in
the intensity of aeration the number of mesosomes in the cells of *Caulobacter
crescentus* rises and this is accompanied by an increase in the heme concentration in the cells (Cohen-Bazire et al., 1966). An increase in the number
of mesosomes during reduced aeration of *A. vinelandii* cells is described by
Tchan and Webber (1966). According to Voelz (1965) the development of
mesosomes in the cells of *Myxococcus xanthus* is determined by the age
of the culture and the conditions of growth. Mesosome formation is stimulated
under suboptimal conditions such as, for example, in anaerobiosis. The cause
of the increase in number of mesosomes and, hence, of the total surface area
of the membrane during anaerobiosis is unknown. Admittedly, in some
bacteria a decrease in the intensity of aeration stimulates the synthesis of
one or, less frequently, of several cytochromes (see p. 179). However, it
must not be forgotten that the total content of cytochromes in the bacterial
membrane as a rule does not exceed 5%, and for this reason even a certain
increase in the content of one of them cannot explain a large increase in the
content of membrane material in the cell. However, the possibility cannot
be ruled out that cytochrome synthesis is accompanied by the synthesis of
other membrane proteins and lipids, with the result that there is an increase
in the total content of membrane material. This last hypothesis emerges

directly from the experiments of White and Tucker (1969a,b), who found that
a decrease in the partial pressure of oxygen during cultivation of *Haemophilus
parainfluenzae* is accompanied by an increase in the synthesis not only of
cytochromes, but also of the phospholipids of the membrane.

The development of the membrane system is not stimulated by a decrease
in aeration in all bacteria with a respiratory chain. On the contrary, if
Spirillum serpens is grown anaerobically, it contains fewer membrane structures than the corresponding aerobic cultures; the cultivation of *M. lysodeikticus*
under anaerobic conditions in general causes no change in the membrane
structures (Murray, 1963). Anaerobiosis does not change the morphology of
the membrane structures in *Listeria monocytogenes* also (Edwards and
Stevens, 1963). The mesosomes of *B. subtilis* are actually destroyed if the
bacteria are kept for 2 h at 4°C without aeration (Ryter, 1969). Unfortunately, changes in the morphology of the membrane structures have not
been compared with their content of cytochromes and other components
in all the bacteria listed above. Such data are available only for *Bacillus
macerans* and *Staphylococcus epidermidis,* although no significant differences
were found in the morphology of the mesosomes of cells grown under
aerobic or anaerobic conditions and differing in their cytochrome content
(Jacobs and Conti, 1965; Conti et al., 1968; Jacobs et al., 1967a,b). With
a change from anaerobic to aerobic conditions there is a sharp increase in
the intensity of biosynthesis not only of cytochromes, but also of phospholipids, glycolipids, and menaquinone in *S. aureus* (Frerman and White, 1967;
White and Frerman, 1967). Increased synthesis of membrane proteins and
lipids and the probable formation of new portions of the respiratory chain
ought to be accompanied by a general increase in size of the membrane apparatus, but there is no evidence of this in the papers by Conti and co-workers
describing their experiments. Incidently, according to the observations of
Kats and Tordzhyan (1968), changes appear in the structure of the membrane
apparatus of *S. aureus* depending on the conditions of oxygenation.

The mesosomes in general are relatively labile structures whose development depends on a number of physiological conditions. Their dependence
on the composition of the nutrient medium has been demonstrated for *E. coli*
(Fishman and Weinbaum, 1967). If the nutrition of the cells was disturbed
in certain ways, three proteins were found to be deficient in the membranes
(Weinbaum et al., 1970). A decrease in the concentration of valine or
threonine in the nutrient medium is accompanied by destruction of the
mesosomes of *Streptococcus faecalis* (Higgins and Shockman, 1970). Lowering the temperature of cultivation of *B. subtilis* from 37 to 15°C leads to the
discharge of mesosomes into the space outside the cytoplasmic membrane
(Neale and Chapman, 1970). One factor which regulates membrane formation
is evidently the intracellular pH (Kahane and Razin, 1970; Gould and Lern-
narz, 1970), the value of which depends on the pH of the medium (Kashket

and Wong, 1969). The secretion of penicillinase by staphylococcal cells is accompanied by destruction of the mesosomes (Beaton, 1968). Cultivation of the obligate anaerobe *Lactobacillus pentoaceticum* under aerobic conditions also leads to destruction of the membranes (Kats and Kharat'yan, 1969).

Investigations of the development of membranes during the various phases of the growth of the culture are very promising, but few such invesgations have been undertaken. In young *B. subtilis* cells no mesosomes are present, but only a system of vesicles connected with the cytoplasmic membrane. These formations are replaced 36 h later, before sporulation, by typical mesosomes (Remsen, 1968). Only invaginations of the cytoplasmic membrane have likewise been found in *Vibrio marinus* cells in the early logarithmic phase, and complex systems of membranes do not appear until later (Felter et al., 1970). The existence of mesosomes in young *B. subtilis* cells seems doubtful in light of the discovery by Nanninga (1971) of mesosomes in these cells by the freeze-etching method only if the cells are prefixed by the Ryter-Kellenberger method.

No final conclusions regarding the mechanism of membrane biogenesis can be drawn from the facts described above. The most likely solution to the problem is to postulate that membrane synthesis takes place irregularly, perhaps in loci whose rate of formation is determined by the synthesis of individual components and, perhaps, by other physiological conditions. Experimental results showing that the incorporation of C^{14}-glycine into the weakly bound membrane proteins of *M. lysodeikticus* and, in particular, into ATPase is not inhibited by chloramphenicol, whereas its incorporation into the firmly bound proteins is inhibited (Vambutas and Salton, 1970a,b), are interesting in this connection.

The fact that the content of individual proteins and pigments is sharply increased suggests that hypertrophy and invagination of the membrane take place unevenly as a result of the insertion of proteins, phospholipids, and pigments into the corresponding areas of the pre-existing membrane (Oelze et al., 1970). The synthesis of membrane lipids evidently does not have to be synchronized with the synthesis of membrane proteins: inhibition of protein synthesis by chloramphenicol leads to an increase in the lipid content of the membranes in *Mycoplasma*. Despite the change in the protein-lipid composition of the membranes, the cells still remain capable of division and the membranes still contain active ATPase (Kahane and Razin, 1969a). The lack of close coordination between the synthesis of membrane proteins and lipids is also shown by the experiments of Mindich (1970), who observed that the content of membrane proteins was doubled in the absence of lipid synthesis. Such membranes contained only 12% of lipids instead of the usual 20-21%. Inhibition of protein synthesis by chloramphenicol causes the conversion of the mesosomes into a structure of myelin type (Giesbrecht and Ruska, 1968). Differences in the rate of C^{14} and P^{32} metabolism in the phospholipids of

S. aureus also indicate the nonuniformity of synthesis of the membrane components (Short and White, 1970). A similar conclusion can be drawn from results showing that a fragment of the membrane of *Haemophilus parainfluenzae* extracted with EDTA differs in the composition of its phospholipids and in the intensity of their metabolism from the rest of the membrane (Tucker and White, 1970b,c). The existence of zones with different rates of formation in the membrane is also illustrated by results showing an increase in the rate of formation of enzyme proteins, especially succinate dehydrogenase, unaccompanied by any significant changes in the content of other membrane proteins. The enzyme proteins are evidently built into certain sites on the membrane (Schnaitman, 1970a,b).

Some interesting results shedding light on the connection between the development of the membrane system and the synthesis of its components has been obtained by the study of photosynthesizing bacteria in which the change from heterotrophic to autotrophic growth is accompanied by the development of a membrane system. The formation of the membrane system in the cells of *Rhodopseudomonas spheroides,* when transferred from aerobic, dark conditions to growth in light and under anaerobic conditions, is accompanied by the accumulation of membrane proteins and bacteriochlorophyll (Gray, 1967). The synthesis of bacteriochlorophyll and phospholipids is coordinated during the formation of the chromatophores of *R. spheroides* (Lascelles, 1968; Lascelles and Szilagyi, 1965). The membranes of the photosynthesizing bacterium contain 157 nmoles of lipid phosphorus/mg protein, while those of the heterotrophic organism contain only 90 nmoles. Thylacoid formation in *R. rubrum* and in *Rhodopseudomonas capsulata* during the change to autotrophic growth correlates with an increase in the fatty acid and lipid phosphorus content in the chromatophores (Schröder and Drews, 1968). At the same time, intensive synthesis of proteins and bacteriochlorophyll takes place in cultures of *Rhodopseudomonas palustris* and *R. rubrum* if the partial pressure of oxygen is lowered, even during growth in darkness (Tauschel and Drews, 1967; Biedermann and Drews, 1968). It is suggested that the chromatophores arise by the incorporation of new pigments and proteins into the cytoplasmic membrane, its outgrowths, and its invaginations (Gorchein et al., 1968a,b; Gorchein, 1968a,b; Oelze and Drews, 1969, 1970a). Since the molar ratio between the components varies within wide limits during biosynthesis of the membranes, bacteriologists consider that the hypothesis of stoichiometrically constant complexes of respiratory carriers, as postulated for the mitochondria (Green and Perdue, 1966), does not apply to bacterial membranes. (See the survey by Wimpenny, (1969.) It is considered more likely that the bacterial membranes are a mosaic of zones of different composition; membrane formation takes place by the insertion of new proteins, new lipids, or their aggregates.

In order to understand the biogenesis of membranes it is very important to know whether the membrane grows from many points simultaneously or in only one region, such as the region of formation of the cell septum. An interesting study from this point of view was carried out by Koch and Boniface (1971), the results of which show that newly formed permeases of carbo-hydrates in growing *E. coli* cells were incorporated into those parts of the membrane which had previously performed transport functions, or in other words, they were incorporated uniformly throughout the membrane. To judge from the distribution of newly formed labeled proteins and lipids in the *E. coli* membrane, growth of the membrane presumably takes place by the insertion of proteins and lipids at many points of the membrane (Tsukagoshi et al., 1971; Wilson and Fox, 1971a).

The Chemical Composition
of Bacterial Membranes

Lipids

Since the membrane lipids of comparatively few bacteria have been studied, it will be advisable to give a brief outline of the lipids of intact cells. It is not within the scope of this book to give a full account of this subject, and the reader is referred to the relevant surveys (Asselineau, 1962; O'Leary, 1962; Macfarlane, 1964a,b; Kates, 1964, 1966; Bloch, 1965; Lennarz, 1966; Ikawa, 1967; Bergel'son, 1967; Goldfine, 1968; Kates and Wassef, 1970).

With some reservations the data on the lipid composition of Gram-positive bacteria can be applied to their membranes, for up to 80% of the total lipids of the cell in these bacteria are located in the cytoplasmic membrane and its derivatives, and they are present in the cell wall in only very small quantities or not at all. In Gram-negative bacteria, the situation is not quite so simple because of the high content of lipids in the cell wall.

Phospholipids and neutral lipids of bacteria include a wide variety of compounds, but in individual taxonomic groups the range of lipid components is limited (Kates, 1964). The most constant phospholipids in bacteria are phosphatidylglycerol and phosphatidylethanolamine. Phosphatidylcholine, phosphatidylinositol, and diphosphatidylglycerol are less common. Lipo-amino acids are found in certain Gram-positive bacteria; very recently glucosaminylphosphatidylglycerol, 1,2-dihydrophytylglycerophosphoryl glycerophosphate, phosphatidylglucose, and sphingolipids have been found in individual species of bacteria.

A good example to illustrate the specificity of the composition of bacterial phospholipids is the ratio between the two most widely occurring phospholipids: phosphatidylglycerol and phosphatidylethanolamine. In the phospholipids of *Thiobacillus thiooxidans* phosphatidylethanolamine accounts for 20%, phosphatidyl-*n*-monomethylethanolamine for 36%, and phosphatidyl-

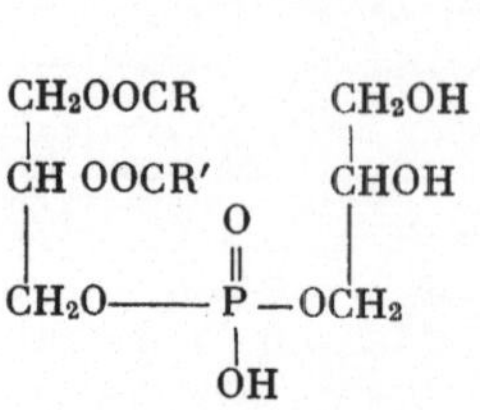

$$CH_2OOCR \quad CH_2OH$$
$$CH\ OOCR' \quad CHOH$$
$$\qquad\qquad\qquad O$$
$$\qquad\qquad\qquad \|$$
$$CH_2O\!\!-\!\!-\!\!-\!\!P\!-\!OCH_2$$
$$\qquad\qquad\quad |$$
$$\qquad\qquad\quad OH$$

Phosphatidylglycerol

$$\qquad\qquad\qquad\qquad\qquad O$$
$$\qquad\qquad\qquad\qquad\qquad \|$$
$$CH_2OOCR \quad CH_2O\!\!-\!\!P\!\!-\!\!OCH_2$$
$$CHOOCR' \quad CHOH \quad OH \quad CHOOCR''$$
$$\quad O$$
$$\quad \|$$
$$CH_2O\!\!-\!\!-\!\!P\!-\!OCH_2 \qquad\qquad CH_2OOCR'''$$
$$\qquad\quad |$$
$$\qquad\quad OH$$

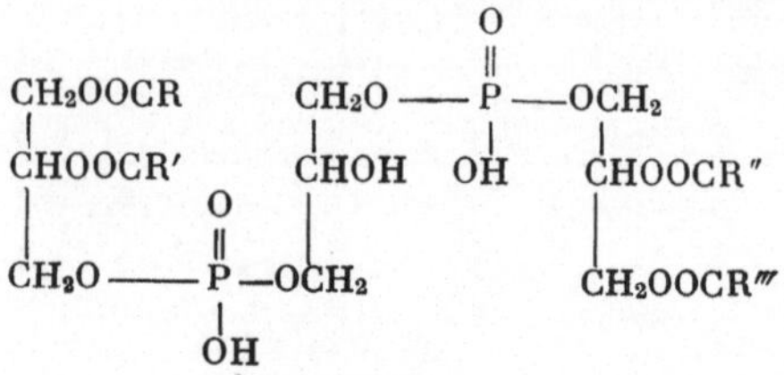

Diphosphatidylglycerol

$$CH_2OOCR$$
$$CH\ OOCR'$$
$$\qquad\qquad\quad O$$
$$\qquad\qquad\quad \|$$
$$CH_2O\!\!-\!\!-\!\!-\!\!P\!\!-\!\!-\!OCH_2CH_2\overset{+}{N}H_3$$
$$\qquad\qquad\quad |$$
$$\qquad\qquad\quad O^-$$

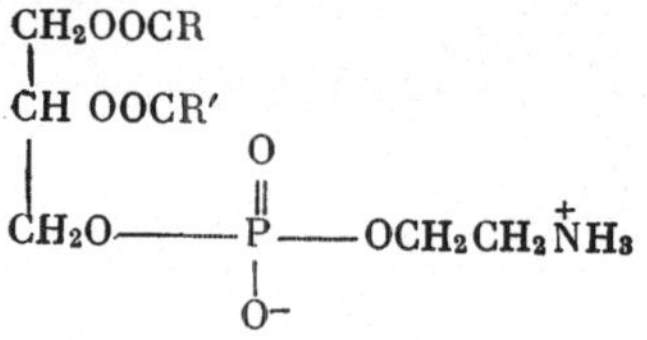

Phosphatidylethanolamine

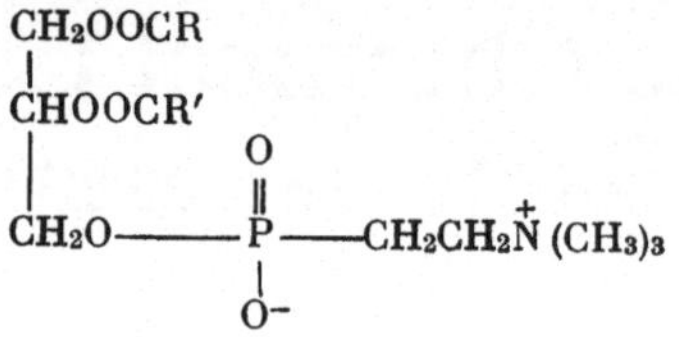

$$CH_2OOCR$$
$$CHOOCR'$$
$$\qquad\qquad\quad O$$
$$\qquad\qquad\quad \|$$
$$CH_2O\!\!-\!\!-\!\!-\!\!P\!\!-\!\!-\!CH_2CH_2\overset{+}{N}(CH_3)_3$$
$$\qquad\qquad\quad |$$
$$\qquad\qquad\quad O^-$$

Phosphatidylcholine

$$CH_2OOCR \qquad\qquad CH_2O$$
$$CH\ OOCR' \qquad\qquad CHO\Big\} C_6H_{11}O_5NH_2$$
$$\qquad\qquad O$$
$$\qquad\qquad \|$$
$$CH_2\!\!-\!\!O\!\!-\!\!P\!\!-\!\!O\!\!-\!\!CH_2$$
$$\qquad\qquad |$$
$$\qquad\qquad OH$$

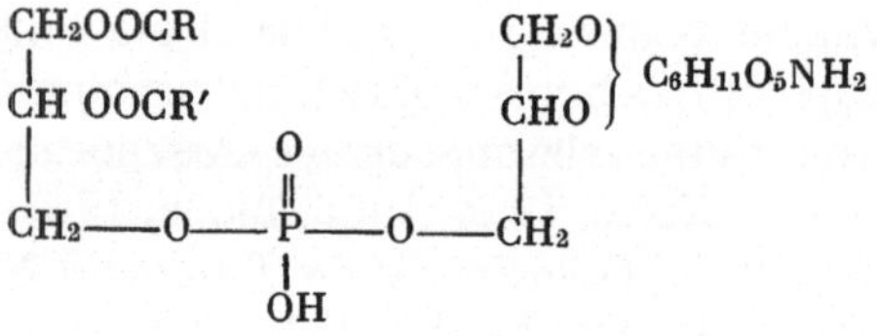

Glucosaminylphosphatidylglycerol

$$CH_2OOCR \qquad\qquad CH_2O$$
$$CH\ OOCR' \qquad\qquad CHO\Big\} \!\!-\!COCH\!-\!R''$$
$$\qquad\qquad O \qquad\qquad\qquad\qquad |$$
$$\qquad\qquad \| \qquad\qquad\qquad\qquad\ NH_2$$
$$CH_2\!\!-\!\!O\!\!-\!\!P\!\!-\!\!O\!\!-\!\!CH_2$$
$$\qquad\qquad |$$
$$\qquad\qquad OH$$

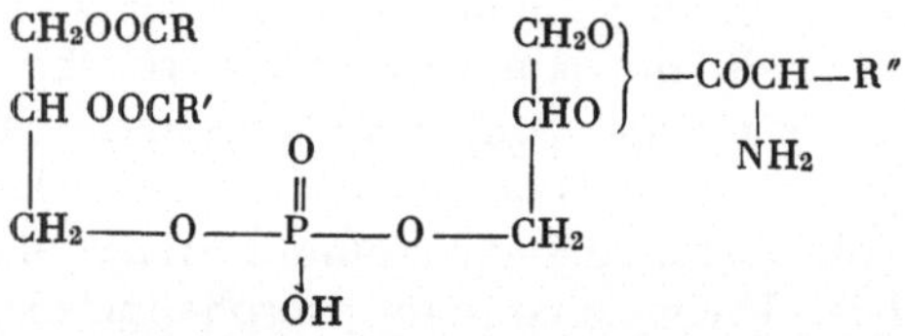

Lipoamino acid: R'' denotes amino acid residue

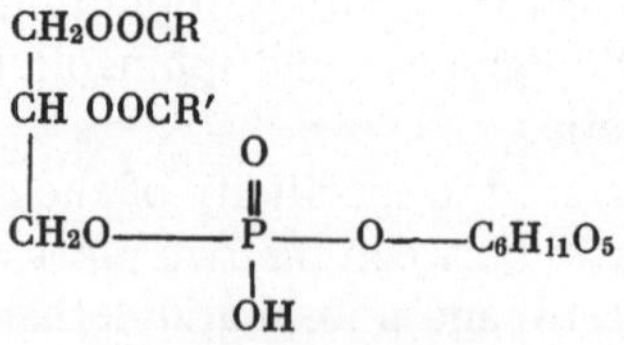

$$CH_2OOCR$$
$$CH\ OOCR'$$
$$\qquad\qquad\quad O$$
$$\qquad\qquad\quad \|$$
$$CH_2O\!\!-\!\!-\!\!-\!\!P\!\!-\!\!O\!\!-\!\!C_6H_{11}O_5$$
$$\qquad\qquad\quad |$$
$$\qquad\qquad\quad OH$$

Phosphatidylinositol

$$\begin{array}{ll}
CH_2OR & CH_2O-\overset{\overset{O}{\|}}{P}-(OH)_2 \\
CHOR- & CHOH \\
\quad\quad \overset{O}{\underset{\|}{}} & \\
CH_2O-\underset{\underset{OH}{|}}{P}-OCH_2 &
\end{array}$$

$$CH_3-\left[-CH-(CH_2)_3-\underset{CH_3}{}\right]_3-CH-CH_2-CH_2-\\ \quad\quad\quad\underset{CH_3}{|} \quad\quad\quad\quad R \quad\quad\quad \underset{CH_3}{|}$$

1,2-Dihydrophytylglycerophosphoryl glycerophosphate

glycerol for 37% of the total (Shively and Benson, 1967). Approximately the same ratio is found in *Ferrobacillus ferrooxidans* (Short et al., 1969). The phospholipids of *Hyphomicrobium vulgare* contain 60% phosphatidylethanolamine and phosphatidyl-*n,n*-dimethylethanolamine, 10% phosphatidylglycerol, and about 30% phosphatidylcholine (Goldfine and Hagen, 1968). In the cells of *Bacillus cereus* 50% of the phospholipids consist of phosphatidylethanolamine, compared with about 90% for the cells of *Ps. aeruginosa* and *E. coli,* and 84% for those of *Salmonella typhimurium* (Houtsmuller and Van Deenen, 1963; Singha and Gaby, 1964; Kaneshiro and Marr, 1962; Macfarlane, 1962a,b. Some bacteria (*M. lysodeikticus, Staphylococcus aureus, Streptococcus faecalis, Arthrobacter simplex*), on the other hand, contain only phosphatidylglycerol and diphosphatidylglycerol (Macfarlane, 1961a,b; Vorbeck and Marinetti, 1965; White and Frerman, 1967, 1968; Ward and Perkins, 1968; Yano et al., 1970). Kates (1964) points out that there is more phosphatidylethanolamine than phosphatidylglycerol in the Gram-negative bacteria, while the opposite is true of the Gram-positive bacteria. However, there are exceptions. The phospholipids of Gram-positive anaerobes (*Clostridium butyricum*) contain a higher proportion of phosphatidylethanolamine and *n*-methylphosphatidylethanolamine, and a smaller proportion of phosphatidylglycerol (Baumann et al., 1965). Kates (1964) suggests that the high content of phosphatidylethanolamine in the Gram-negative bacteria is associated with the cell wall membrane. Evidence for the participation of phosphatidylethanolamine in the synthesis of lipopolysaccharides of the cell wall of *Salmonella typhimurium* is interesting in this connection (Rothfield and Horecker, 1964; Rothfield et al., 1969).

Another example of differences in the phospholipid composition of bacteria is given by the distribution of lipoamino acids (*O*-esters of amino acids and phosphatidylglycerol). These compounds are found only in certain Gram-positive bacteria: *Clostridium welchii, Staphylococcus aureus, Bacillus cereus, Lactobacillus* sp., and *S. faecalis* (Macfarlane, 1962b, 1964a,b; Houtsmuller and Van Deenen, 1964; Ikawa, 1963; Kocur, 1970; Dos Santos Mota et al., 1970). No lipoamino acids are found in the phospholipids of *M. lysodeikticus.*

Some bacteria have a very characteristic phospholipid composition. They include the obligate halophiles, in which up to 95% of the total phospholipids consists of 1,2-dihydrophytylglycerophosphoryl glycerophosphate; only a small proportion consists of phosphatidylglycerylglucosamine (Kates et al., 1966, 1967). The genus *Mycoplasma* has a distinctive phospholipid composition, represented in most species by phosphatidylglycerol and its glucosyl derivatives (P. Smith, and Henrikson, 1965; P. Smith, et al., 1965; Shaw et al., 1970). In strain *Mycoplasma* S 743, which is not dependent on steroids, a sphingolipid has been found together with phosphatidylglycerols (Plackett et al., 1970).

The phospholipids of *Bacteroides melaninogenicus* consist mainly of sphingolipids: ceramide phosphorylethanolamine, ceramide phosphorylglycerol, and ceramide phosphorylglycerophosphate (Rizza et al., 1970). In the mycobacteria and corynebacteria, glycosylphosphatidylinositol has been taken to be the dominant phospholipid. In some mycobacteria, however, up to 50% of the phospholipids consist of diphosphatidylglycerol, 10% of phosphatidylethanolamine, and only 40% of mannosylphosphatidylinositol (Akamatsu and Nojima, 1965).

Until recently the view was held that phosphatidylcholine does not exist in bacteria, and that bacteria do not therefore possess enzymes of choline synthesis; the only exceptions were *Agrobacterium radiobacter* and *Agrobacterium rhizogenes* (Goldfine and Ellis, 1964). However, many bacteria have subsequently been shown to contain phosphatidylcholine. Phosphatidylcholine, in fact, accounts for up to 20% of the total phospholipids of *Rhodopseudomonas spheroides* and *Rhodomicrobium vannielli,* 37% of the phospholipids of *Thiobacillus novellus,* and 30% of the phospholipids of *Hyphomicrobium* NQ 521 (Lascelles and Szilagyi, 1965; Hagen et al., 1966; Park and Berger, 1967; Barridge and Shively, 1968). The phospholipids of *Brucella abortus* also have a high phosphatidylcholine content (Thiele and Busse, 1968).

The composition of the phospholipids differs not only in the major taxonomic groups of bacteria but also in closely related species and strains; instability of composition has been found even in the same bacteria, depending on the age of the culture and the conditions of cultivation.

Species differences in phospholipid composition have been found for various members of the thiobacilli (Barridge and Shively, 1968). The phospholipids of 6-hour cells of *Bacillus natto* contain 27% (of the lipid phosphorus) phosphatidylglycerol and 22% phosphatidylethanolamine, while 44-hour cells contain 51% and 8%, respectively (Urakami and Umetani, 1968). The phosphatidylglycerol and diphosphatidylglycerol content in the cells of *S. aureus* is considerably altered with the change from anaerobic to aerobic conditions (Frerman and White, 1967). The ratio between individual phospholipids also changes with age in the cells of certain Gram-negative bacteria (Randle et al., 1969) and also in *S. aureus* (Short and White, 1970).

No general principle governing the distribution of phospholipids in individual groups of bacteria can be detected from a comparison of their phospholipid composition. Nor has it been possible to prove that the phospholipid composition depends on the morphology and function of the membrane system. Even in the membrane structures of *Hyphomicrobium* NW 521, *Nitrosocystis oceanus,* and *Nitrosomonas europea,* which are very similar in their morphology, the lipid composition is different. In *Hyphomicrobium,* for instance, 22% of the phospholipid phosphorus is accounted for by phosphatidylethanolamine, 36% by phosphatidyl-*n*-methylethanolamine, 30% by phosphatidylcholine, and 10% by polyglycerophosphatides. In *N. oceanus* 67% is accounted for by phosphatidylethanolamine and 28% by polyglycerophosphatides, while in *N. europea* 78% is accounted for by phosphatidylethanolamine and 17% by polyglycerophosphatides (Hagan et al., 1966). Even in physiologically closely related bacteria, such as the aerobes, which possess respiratory chain enzymes and carry out oxidative phosphorylation, the phospholipid composition may differ considerably. This is the case, for example, with *M. lysodeikticus, Thiobacillus thiooxidans, Pseudomonas aeruginosa,* and *E. coli.*

An interesting feature distinguishing the bacterial lipids, including phospholipids, is that they contain predominantly saturated fatty acids (chiefly C_{14}-C_{18}) with a straight or branched chain. Unsaturated acids constitute a very small proportion, which consists chiefly of monoene or, less frequently, diene acids. No polyene acids are found in bacteria. In some families, cyclopropane acids are found, mainly *cis*-11,12-methylene octadecane (lactobacillic) acid (Hoffman, 1962; O'Leary, 1962; Abel et al., 1963; Knoche and Shively, 1969).

$$CH_3(CH_2)_5 - \overset{\displaystyle H}{\underset{\displaystyle |}{C}} - \overset{\displaystyle H}{\underset{\displaystyle |}{C}} - (CH_2)_9COOH$$
$$\underset{\displaystyle CH_2}{\diagdown \diagup}$$

cis-11,12-Methylene octadecane acid (lactobacillic acid)

Molecular species of phospholipids have been shown to exist, for example, for the phosphatidylethanolamine from *E. coli* (Van Golde and Van Deenen, 1967):

Fatty acid		Total composition	First position	Second position
14 : 0		2.5	2.4	—
16 : 0		43.5	75.2	6.9
16 : 1		22.0	2.9	44.1
17 : 0	(cyclopropane)	5.5	—	10 6
18 : 0		—	3.2	—
18 : 1		26.6	16.1	38.4

Molecular species have also been found for the phosphatidylglycerol and diphosphatidylglycerol from *E. coli* (Kanemasa et al., 1967), the diphosphatidylglycerol and phosphatidylethanolamine from *Mycobacterium tuberculosis* and *Mycobacterium phlei* (Okuyama et al., 1967), the phosphatidylglycerol from *Mycoplasma hominis* (P. Smith and Koostra, 1967), and the phosphatidylglycerol from *Mycoplasma laidlawii* B. (McElhaney and Tourtellotte, 1970).

The content of saturated and unsaturated.acids varies in the lipids of different bacteria (Erwin and Bloch, 1964). In *B. megaterium,* for instance, monoene acids account for 4% of the total, compared with 20% in *M. lysodeikticus* and 45% in *Corynebacterium diphtheriae* (Fulco et al., 1964).

The composition of fatty acids may also vary to some extent, depending on the temperature of cultivation of the cells (see page 70). Changes in the content of individual fatty acids with age of the culture have been found in *E. coli* (Cronan, 1968; Cronan and Vagelos, 1972), and such changes have also been found depending on aeration in *S. aureus* (White and Frerman, 1968). During development of a culture of *Pseudomonas aeruginosa,* changes were found in the fatty acid composition in both the first and the second generations, affecting different phospholipids. The widest range of variations was observed for the 18:1 and 19 (cyclopropane) acids, namely from 10% to 40% of the total fatty acids, respectively (Hancock and Meadow, 1969).

Changes in the fatty acid composition have been observed during the appearance of new forms. In the "rugose" variant of *Vibrio cholerae* the 16:0 and 16:1 fatty acids account for 43% of the total acids, but in the original cells they accounted for 73%. The content of the 18:1 acids is more than doubled in the rugose variant (Brian and Gardner, 1969). In the transition from the bacillary form to the L-form, the composition of the fatty acids also changes, as has been shown in the case of *Proteus vulgaris* (Nesbitt and Lennarz, 1965).

In general, the fatty acid composition is highly specific not only for the families and species of bacteria, but even for individual strains. Differences in the content of particular fatty acids have been found in bacilli of the species *Bacillus thuringensis, Bacillus anthracis,* and *Bacillus cereus* (Kaneda, 1968) and int two strains of *Neisseria* (Lewis et al., 1968). Strains may also differ in their content of iso- and anti-iso forms (Kaneda, 1967).

One of the features distinguishing bacterial lipids from lipids of all other organisms is that bacteria contain no steroids. The presence of cholesterol has been reliably established only in some strains of Mycoplasma (Smith and Rothblat, 1960; Morowitz et al., 1962).

In recent years, the glycolipids of mycobacteria and certain other Gram-positive bacteria have been closely studied (Shaw, 1970). In the mycobacteria these compounds are represented by glycosyl glycerides, carotenyl glucoside, and triacylglucose (Lederer, 1967; Smith and Mayberry, 1968). In some Gram-positive bacteria diglycosyl diglycerides have been found, in

TABLE 1. Composition of Bacterial Membranes of Heterotrophs (in per cent of dry weight)

Bacteria	Proteins	Lipids	Nucleic acids	Carbo-hydrates	Literature citations
Microcossu lysodeikticus	50	28	0.2	15-20 (Mannan)	Gilby et al., 1958 Macfarlane, 1964a,b
Sarcina lutea	40	29	3.6	10	Brown, J., 1961
Bacillus megaterium	63-69	16-20	1.5	1-10	Weibull and Bergström, 1958
Bacillus megaterium	50	26	24	2	Godson et al., 1961
Bacillus megaterium KM	70	25	–	1	Yudkin, 1962, 1966
Bacillus megaterium	65	20	5.40	8.45	Yamaguchi et al., 1967
Ditto, after alkaline treatment	58	27	2.30	1.20	
Bacillus subtilis	62	16	21	–	Bishop et al., 1967a
Staphyloccus aureus	41	22.5	–	2	Mitchell and Moyle, 1956, 1959
Streptococcus faecalis	49-55	28	4	0.1	Schockman et al., 1963
Mycoplasma sp.	–	36	–	–	Razin et al., 1963
Mycoplasma laidlawii	–	25-30	–	7-8	Smith, P., 1964
Micrococcus denitrificans	63	32	–	0.62	Scholes and Smith 1968a
Listeria monocytogenes	50	32	1.6	0.4	Ghosh and Carroll, 1968
Pseudomonas fluorescens	50	16-18	5.7	–	Hunt et al., 1959
Bifidobacterium bifidum	70	7.8	8.3	12	Exterkate et al., 1970
Pseudomonas ATCC 19855	62.8	30.5	–	2	Martin and Macleod, 1970
Pseudomonas BAL-31	50	25	10	10	Franklin et al., 1971

which the carbohydrate components may be glucose, galactose, or mannose (Shaw and Baddiley, 1968). Examples of these include: galactosyl-α-glucosyl (1 → 1)-diglyceride in the case cf *Lactobacillus casei*, β-glucosyl-β-glucosyl (1 → 1)-diglyceride in *B. subtilis* and *S. aureus*, dimannosyl diglyceride in *M. lysodeikticus*, and diglucosyl diglyceride in *B. cereus* (Brundish et al., 1966; Lennarz, 1964; Lennarz and Talamo, 1966; Saito and Mukoyama, 1971). The content of glycolipids varies within wide limits, from a few per cent of the total cell lipids in *M. lysodeikticus* to 50% in *Microbacterium lacticum* and *M. laidlawii*. Although these investigations were carried out with lipids of whole cells, they can also be applied to the membranes, because it has been shown that the lipids of Gram-positive bacteria are localized in the membranes and not in the cell wall. The hypothesis that hydrophilic groups of the glycolipid molecules form pores in the membrane through which small ions and water-soluble metabolites can pass therefore deserves close attention (Brundish et al., 1967; Shaw and Baddiley, 1968). In Gram-negative bacteria, only the lipopolysaccharides of the cell wall are known; these substances are therefore of no interest in connection with the organization of the cytoplasmic membrane of bacteria. Recently glucosyl diglycerides have been found in the cells of *Pseudomonas diminuta*, but their localization is unknown (Wilkinson, 1969).

The content of lipids in the membranes of most bacteria is 30% (Table 1). Meanwhile, the chromatophore membranes of *Chromatium* contain 70% of lipids (Cusanovich and Kamen, 1968), while the membranes of the gas vacuoles of the halobacteria are virtually lipid-free (Stoeckenius and Engelman, 1969). The lipid:protein ratio may vary with the phase of growth of the culture (Schockman et al., 1963; A. Brown and Pearce, 1969). This ratio also depends on the pH of the medium: with a change of pH from 7.0 to 8.5, the density of the membranes of *M. laidlawii* increases because of an increase in their protein content (Kahane and Razin, 1970). The lipid: protein ratio for membranes of growing *S. aureus* cells has also been shown to depend on the pH of the medium (Gould and Lennarz, 1970).

The content of protein, lipids, carbohydrates, and nucleic acids in the membranes is determined to some extent by the method of obtaining them and, in particular, by the Mg^{++} concentration. Mizushima and co-workers have observed that lysis of *B. megaterium* in 5×10^{-3} M $MgCl_2$ leads to an increase in the RNA content in the membrane, on account of the ribosomes which it contains. Results obtained by the same workers showed that the relative proportions of proteins, lipids, and other components of the membranes may vary slightly in different strains of bacteria (Mizushima et al., 1966a,b). In membranes from *E. coli* the nucleic acid content has also been found to vary with the Mg^{++} concentration (Nagata et al., 1967).

The phospholipids of bacterial membranes have attracted particular attention, and this becomes perfectly clear considering the role of these compounds in the structure of biological membranes. The phospholipids of the membranes and (for Gram-negative bacteria) of the membranes with fragments of the wall consist of the following components: phosphatidylglycerol, diphosphatidylglycerol, phosphatidylethanolamine, phosphatidylcholine, phosphatidylinositol, and esters of phosphatidylglycerol with amino acids, phytol, or glucosamine (Table 2). Generally, phospholipids of the cells and membranes isolated from them are similar. Differences are found in the composition of the phospholipids from membranes of bacteria belonging to different taxonomic groups and also to individual strains, as has previously been noted for phospholipids of whole cells.

Just as with intact cells, examination of the phospholipids of isolated membranes reveals different assortments of components in different bacteria, but the importance of this phenomenon to the molecular organization to the membrane has not yet been adequately explained.

Studies of the phospholipids of isolated membranes have demonstrated their variation with the conditions of cultivation and, in particular, with the pH of the medium and the age of the culture. The ratio between the content of aminoacylphosphatidylglycerol and phosphatidylglycerol rises with a fall in the pH of the medium in membranes from *S. aureus, S. faecalis, B. subtilis,* and *M. laidlawii* B (Houtsmuller and Van Deenen, 1965; Op den Kamp et al., 1969; Koostra and Smith, 1969). However, this is not a general principle. For instance, phosphatidylglucosamine acculumates in the mem-

TABLE 2. Phospholipids of Membranes and Membranes plus Wall of Bacteria*

Phospholipid	Micrococcus lysodeikticus (Macfarlane, 1961a,b)	Staphylococcus aureus (White and Frerman, 1967)†	Streptococcus faecalis (Abbott and Abrams, 1964a,b)‡	Mycoplasma laidlawii (Smith et al., 1965)	Mycobacterium phlei (Akamatsu et al., 1966)	Bacillus stearothermophilus (Cardet et al., 1969)	Bacillus subtilis, pH 7.0 (Op den Kamp et al., 1969)	Bacillus subtilis, pH 5.0 (Op den Kamp et al., 1969)	Bacillus megaterium M (Weibull, 1957)	Bacillus megaterium KM (Yudkin, 1962, 1966)	Bacillus megaterium MK 10D pH 7.2 (Op den Kamp et al., 1963, 1961)	Bacillus megaterium MK 10D pH 5.0 (Op den Kamp et al., 1963, 1961)	Bacillus megaterium NRLL (Mizushima et al., 1966a)	Pseudomonas ovalis Chester (Francis and Phizackerley, 1965; Phizackerley et al., 1966)	Pseudomonas putida (Baginsky and Rodwell, 1966)	Escherichia coli K_{12} (Nagata et al., 1967)	Haemophilus parainfluenzae (White, 1968)	Halobacterium halobium (Marshall, Brown, 1968)	Rhodopseudomonas spheroides (Gorchein, 1968a,b)	Azotobacter vinelandii (Jurtshuk, Schlech, 1969)
Phosphatidylglycerol	72	37	70	60	—	28.3	36	18	90	—	40	8	19.5	24	—	25	18	—	34	—
Diphosphatidylglycerol	4	24	—	—	52	49.4	12	13	—	—	5	5	40.5	65	14	—	3	—	35	5
Phosphatidylethanolamine	—	—	—	—	10	19.4	30	27	—	95	40	40	28	—	86	70	78	—	23	64
Phosphatidylcholine	—	—	—	—	—	—	—	—	—	—	—	—	—	—	—	—	0.4	—	—	—
Phosphatidylinositol	12.9	—	—	—	38	—	—	—	—	—	—	—	—	—	—	—	—	—	—	—
o-Amino acid esters of phosphatidylglycerol	—	40	—	—	—	—	22	42	—	—	15	15	—	—	—	—	—	—	—	—
Phosphatidylglycerol-glucosamine	—	—	—	—	—	—	—	—	—	—	—	32	—	6.0	—	—	—	—	—	—
Phosphatidylglucose	—	—	—	30	—	—	—	—	—	—	—	—	—	—	—	—	—	—	—	—
Di-o-dihydrophytylglycero-phosphoryl glycero-phosphate	—	—	—	—	—	—	—	—	—	—	—	—	—	—	—	—	—	90	—	—

*Macfarlane (1961a,b) and White and Frerman (1967a,b) give the phospholipid content in per cent of phospholipid phosphorus; all other authors, in per cent of total phospholipids.

†Gould and Lennarz (1970) found no diphosphatidylglycerol.

‡According to Dos Santos Mota et al. (1970), besides containing phosphatidylglycerol, the cells of S. faecalis also contain diphosphatidylglycerol, alanyl-PG, lysyl-PG, and diglucosyl-PG.

branes of *B. megaterium* MK 10D grown at pH 5.0, but the content of amino-acylphosphatidylglycerol remains unchanged (Op den Kamp et al., 1967). Synthesis of the membrane phospholipids in *S. aureus* is stimulated by aeration; under these circumstances the content of phosphatidylglycerol and diphosphatidylglycerol is doubled, while the content of lysylphosphatidyl-glycerol remains unchanged (Frerman and White, 1967).

Bacterial membranes contain no cholesterol. The only exceptions are certain species of *Mycoplasma,* whose cytoplasmic membrane has been found to contain up to 70% of the total cell cholesterol (Tourtellotte et al., 1963; Argaman and Razin, 1965).

Several investigations have demonstrated the presence of carotenoids and hydroxycarotenoids in bacterial membranes (Jackson and Lawton, 1958; Mathews and Sistrom, 1959; Stephens and Starr, 1963; Sasaki, 1964; Salton and Ehtisham-ud-Din, 1964). Smith (1963) found a carotenoid of the neurosporin type (tetrahydrolycopene) and a hydroxycarotenoid, both in the free state and as an ester with a fatty acid in membranes of the genus *Mycoplasma.* Hydroxycarotenoids have also been found in the membranes of other bacteria. Seven carotenoids, all with the same chromophore group but differing in polarity, have been found in the membranes of *M. lysodeikticus.* It has been suggested that these hydroxylated carotenoids are of the neurosporin type (Rothblat et al., 1964).

According to Smith, carotenoids in bacterial membranes, like steroids in other biological membranes, play an important role in maintaining the structure of the membrane (Smith, 1963). However, Razin and Cleverdon (1965) showed that there is no correlation between the content of carotenoids and of cholesterol in the membrane of *M. laidlawii.* The strength of the membrane of *M. laidlawii* is determined not by cholesterol and not by carotenoids, but by the unsaturated fatty acids of the phospholipids (Razin et al., 1966). Meanwhile, the carotenoid-free membranes of *M. Lysodeikticus* are less stable than normal membranes (Salton and Schmitt, 1967).

Bacterial membranes contain quinones: menaquinones and ubiquinones with side chains of different lengths. The content of quinones may reach 0.5-1 nm/g dry weight. The question of the quinones of bacterial membranes will be discussed more fully later in connection with the composition of the respiratory chain enzymes (page 137).

The glycerides of the membranes have received little study; mainly a diglyceride has been found in the neutral lipids of the membranes of *M. lysodeikticus,* and 77% of its acids consists of C_{15} acid (Macfarlane, 1961a). Diglycerides account for about 50% of the neutral lipids in the membranes of *B. megaterium* NRLL, while the rest consists of monoglycerides and free fatty acids, the content of which is 7% of the total fatty acids of the membranes (Mizushima et al., 1966a).

TABLE 3. Fatty Acid Content of Membrane Lipids of Gram-Positive Bacteria and of Membrane plus Wall of Gram-Negative Bacteria (in per cent of total fatty acids)

Acid	Lactobacillus casei	Micrococcus lysodeikticus	Bacillus megaterium	Bacillus licheniformis	Bacillus stearo-thermophilus	Sarcina lutea	Bacillus subtilis	Normal	L-form	Mycoplasma laidlawii	Aerobacter aerogenes	Escherichia coli B	Pseudomonas sp.	Salmonella gallinarum	Chlorobium thio-sulfatophilum	Haemophilus parainfluenzae
Below 12	0.1	2.9	2.8	–	–	–	–	–	–	3.3	–	–	–	–	–	–
12:0 Unbranched	0.3	0.5	0.7	–	–	–	–	0.79	1·59	–	–	–	2.6	2.9	–	0.26
12:1	0.2	0.8	1.1	–	–	–	–	–	–	–	–	–	–	–	–	0.03
13:0 Unbranched	0.2	0.3	0.5	–	–	–	–	–	–	–	–	–	–	5.9	–	0.02
14:0 Branched	0.5	1.1	3.0	4.2	2.9	6.0	0.9	1.27	3.71	20.3	–	6.8	3.5	9.1	14.6	9.85
14:0 Unbranched	2.2	0.8	2.6	–	4.2	1.8	3.6	1.06	0.83	20.3	5.7	2.1	2.9	3.8	7.1	0.56
14:1	0.8	–	–	–	–	–	–	–	–	–	–	–	–	–	–	–
15:0 Branched (iso)	0.4	71.7	26.0	–	–	–	13	–	–	–	–	–	–	–	–	0.46
15:0 Branched (anti-iso)	0.5	71.7	29.0 ·	50.4	30.7	80.2	36	–	–	–	–	–	–	–	–	0.46
15:0 Unbranched	0.9	1.5	2.9	–	–	–	–	0.23	–	–	–	–	–	–	–	0.01
16:0 Branched (iso)	0.5	0.6	1.3	–	–	–	2.6	0.20	0.38	–	–	–	–	–	–	–
16:0 Branched (anti-iso)	–	1.7	2.1	–	–	–	2.6	0.201	–	–	–	0.8	4.2	3.8	2.3	–
16:0 Unbranched	21.5	5.1	7.3	3.9	3.1	1.5	16.6	31.18	26.43	55.3	56.4	42	49.5	49	8.4	39.93
16:1	7.2	2.0	3.4	12.0	23.5	1.3	24.9	19.39	8.74	–	5.7	2.3	18.3	4.7	60	41.32
17:0 Branched	1.1	3.3	2.6	28.2	33.2	1.3	24.9	0.73	0.71	–	–	–	–	–	–	0.04
17:0 Unbranched	1.6	0.2	0.4	–	–	–	–	0.51	0.77	–	3.8	–	–	–	–	–
17:1	4.5	0.6	1.3	–	–	–	–	0.50	0.90	–	–	–	8.1	–	–	–
18:0 Branched (iso)	1.1	0.6	0.7	0.2	1.2	2.5	–	–	–	–	–	–	–	–	2.5	0.07
18:0 Unbranched	6.8	2.3	2.6	1.1	1.2	3.6	1.6	7.20	8.43	4.1	9.6	1.7	8.6	2.0	1.6	0.05
18:1	21.7	2.4	8.8	–	–	–	–	26.11	31.85	1.9	18.8	25.7	2.3	2.0	1.4	2.37
18:2	–	–	–	–	–	–	–	6.58	12.28	–	–	–	–	–	–	1.56
19:0 Branched (anti-iso)	3.0	–	–	–	–	–	–	–	–	–	–	–	–	–	–	0.57
19:propane	16.6	–	–	–	–	–	–	–	–	–	–	–	–	–	–	–
20:1	–	2.1	–	–	–	–	–	–	–	–	–	–	–	–	–	–
20:0 Unbranched	8.3	0.4	2.0	–	–	–	–	–	–	–	–	–	–	–	–	–

Note. L. casei, M. lysodeikticus, B. megaterium (Thorne and Kodicek, 1962); B. licheniformis, B. stearothermophilus, S. lutea, A. aerogenes, E. coli, Pseudomonas sp, S. gallinarum (Cho and Salton, 1964, 1966);S. faecalis Normal (Freimer, 1963);L-form (Panos et al., 1966); M. laidlawii (Razin et al., 1966, data for phospholipids); C. thiosulphatophilum (Schmitz, 1967); H. parainfluenzae (White and Cox, 1967, data for phosphatidylethanolamine); B. subtilis (Bishop et al., 1967a).

Glycolipids of the membranes have not yet been adequately studied. In the membranes of *S. faecalis,* a glucosyldiglyceride and a glycosylgalactosyldiglyceride have been found (Vorbeck and Marinetti, 1965).

As Table 3 shows, neither the membranes of bacteria nor the intact cells contain polyene fatty acids, and the content of diene and monoene acids is low and varies in different bacteria. Saturated acids, with both straight and branched chains, and cyclopropane acids are characteristic of the lipids of bacterial membranes. A small proportion of the fatty acids (2-10%) may exist in the membrane in the free form. It will be recalled that the area occupied by phospholipid molecules in a monolayer depends on the length and structure of the fatty acid residues (Fig. 6). Some such relationship probably obtains also in biological membranes, in which it determines the intermolecular distances and the degree of orderliness of the lipid molecules in the membrane (Van Deenen, 1965, 1966, 1968; O'Brien, 1967).

Although theoretical calculations and model experiments indicate a role of fatty acids in the organization of the membrane, it is difficult to reconcile them with the variability of the fatty acid composition of the lipids in biological membranes (see O'Brien, 1967). In bacteria, unlike in mitochondria and chloroplasts, there are no polyene fatty acids; the predominant acids are saturated and in a branched chain. The role of polyene fatty acids in bacteria is evidently performed by branched-chain fatty acids and cyclopropane acids, giving the bacterial membrane its flexibility and elasticity while allowing dense packing of the components (Kodicek, 1963; Kates, 1964). To maintain a definite ratio between the fatty acids in the membrane lipids of certain bacteria, such as *E. coli,* enzymic mechanisms exist to regulate the relative proportions of saturated and unsaturated acids and their *cis-* and *trans-*forms (Esfahami et al., 1969). Meanwhile the fatty acid composition of the lipids of the bacterial membrane may vary within wide limits. The ratio between palmitic, stearic, and octadecene acids in the lipid fractions of the membrane plus wall of *Brevibacterium flavum* varied when glucose was replaced by a different source of carbon, namely acetate (Otsuka and Shio, 1968). By adding various fatty acids to the nutrient medium it is possible to obtain membranes of *M. laidlawii* with polar lipids that differ sharply in their fatty acid composition. For example, the content of elaidic acid (18:1) may reach 30 moles % of the total fatty acids; that of stearic acid (18:0), 65 moles %; and that of isopalmitic acid (16:0), 78 moles %. Changes in the fatty acid composition of the membranes are reflected in the cell morphology, although the cells still remain viable (McElhaney and Tourtellotte, 1969).

Changes in the fatty acid composition within such wide limits are difficult to reconcile with the view that a strictly definite physical state of the lipids is essential for the formation of a physiologically perfect membrane.

Attempts to develop generalizations of any kind governing the composition of the lipids of bacterial membranes from both the taxonomic and the

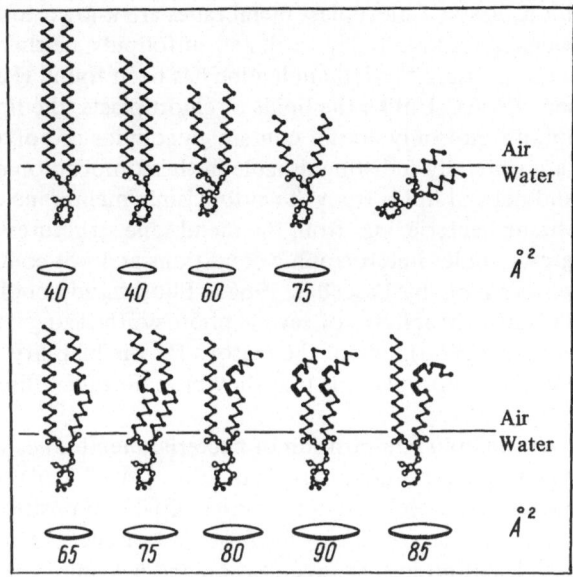

Fig. 6. Area occupied by phosphoglyceride molecules on the air—water interphase, depending on the nature of the fatty acid residues (in $\mathring{A}^2$) (Van Deenen, 1965).

physiological aspect have encountered great difficulties. The number of permutations and combinations of phospholipids, neutral lipids, and fatty acid components which have so far been classified is enormous. How is it possible to reconcile data on the composition of phospholipids in aerobic bacteria whose membranes contain a chain of electron transfer enzymes and which are the functional analogues of mitochondria? In some bacteria the membrane lipids consist mainly of phosphatidylglycerol, in others of phosphatidylethanolamine, and in a third group of 1,2-dihydrophytylglycerophosphoryl glycerophosphate. In contrast with bacterial membranes, the phospholipid composition of mitochondrial membranes is very similar even in distantly related organisms, and it includes several components. In the mitochondria of animal cells, on the average, phosphatidylcholine accounts for 40% of the total phospholipids, phosphatidylethanolamine for 36%, diphosphatidylglycerol for 20%, and phosphatidylinositol for 3% (Ansell and Hawthorne, 1964; Rouser et al., 1968). The same components are found in the mitochondria of yeast, although in rather different quantitative proportions (Wallace et al., 1968).

The special features of the lipid composition of bacterial membranes are revealed not only by comparison of membranes carrying an electron transport chain with mitochondria, but also by comparison of membranes of chromatophores and chloroplasts. Chloroplast membranes are known to contain galactosylglycerides, phosphatidylglycerol, and sulfolipids; neither phosphatidylcholine nor phosphatidylethanolamine has been found (Park, 1966; Weier and Benson, 1966). Unlike the lipids of chloroplasts, the lipids of bacterial chromatophores not only do not contain glycolipids or polyene fatty acids, but they are generally indistinguishable in the composition of their phospholipids and neutral lipids from the cytoplasmic membrane of the cells of photosynthesizing bacteria, and from the membrane structures of the same bacterial grown under heterotrophic conditions and not containing chlorophyll (see the survey by Lascelles, 1968). Phosphatidylcholine has been found in the chromatophores of several photosvnthesizing bacteria. According to Lennarz (1966), one of the reasons for the inability of bacteria to form structures like chloroplasts is that they have no galactolipids or α-linolenic acid.

Comparison of the lipid composition of bacterial membranes with that of their functional analogues (mitochondria and chloroplasts) leads to the unexpected conclusion that membranes of identical lipid composition are not required for the organization of analogous processes of energy transformation. If the principles of molecular organization and action of the electron-transport chains are assumed to be similar in prokaryotes and eukaryotes, the only possible explanation for the lack of standardization of the lipid composition of bacterial membranes is that the lipid components do not determine the assembling of enzyme chains in them, and that this role is played by the membrane proteins. The lipid composition is to some extent determined by the structure and composition of the proteins, which require "complementary" lipids in order to create a membrane.

Results showing the existence of a correlation between changes in the composition of proteins and lipids during the growth of *Halobacterium halobium* cells deserve attention from this point of view (Marshall and Brown, 1968; Brown and Pearce, 1969).

The general morphology of the membranes, the globular structures, and the behavior of protoplasts during osmotic shock likewise are independent of lipid composition. The results obtained by Op den Kamp and co-workers (Op den Kamp et al., 1967, 1968; Van Iterson and Op den Kamp, 1969) deserve particular examination in the respect. These workers showed that the shape of the protoplasts and the character of their lysis during osmotic shock differ in protoplasts of *B. megaterium* and *B. subtilis* grown at acid or neutral pH values and differing in their phospholipid composition. Apparently, it should be concluded from these results that the organization of the membranes nevertheless depends on the lipid composition, for replacement of phosphatidylglycerol by glucosaminylphosphatidylglycerol in *B. megaterium* or by lysylphosphatidylglycerol in *B. subtilis* causes changes in the properties of the membranes. However, in light of observations that the properties of protoplasts are generally similar in bacteria which differ considerably in their lipid composition, the appearance of new properties of the membranes after a change in their lipid composition must be interpreted as a type of molecular disorganization of the membrane uncompensated by synthesis of the corresponding protein components.

Proteins. Structural Protein

Investigations of membrane proteins began only very recently, stimulated to a large extent by progress in methods of protein chemistry. Work on membrane proteins can be conventionally divided into two categories: isolation and determination of the properties of individual proteins, and the study of conformation of proteins in the membrane by various physical methods.

In this section we shall examine only the first group of investigations; details of the conformation of membrane proteins will be described in the next chapter, which deals with the molecular organization of membranes. These investigations have been mainly of the proteins of mitochondrial membranes, but information has more recently been published on bacterial proteins also.

Not long ago it was considered that up to 25% of the composition of the proteins of mitochondrial membranes consisted of enzyme proteins — cytochromes, dehydrogenases, and ATPases, — and that the rest consisted of what was called structural protein. Similar calculations also were applied to proteins of those bacterial membranes which, like mitochondria, contain a chain of electron transport enzymes. However, methods of electrophoresis in polyacrylamide gel in vigorous deaggregating systems of solvents have revealed large numbers of proteins in mitochondrial membranes. It must be remembered that the minor protein components, accounting for less than 2-3% of the total membrane proteins, cannot generally be detected by electrophoresis in polyacrylamide gel. When the results of electrophoresis of membrane proteins in polyacrylamide gel are interpreted, the possibility of dissociation of the proteins into subunits, due to the use of strong solubilizing agents such as phenol, urea, or dodecyl sulfate, must also be borne in mind.

The number of proteins in the membranes of different mycoplasmas has been studied in considerable detail by electrophoresis in polyacrylamide gel by the method of Takayama et al. (1966). As a result, more than ten components were detected, and the position and size of the bands differed for individual strains (Rottem and Razin, 1967). Approximately the same number of proteins has been found by gel electrophoresis in the membranes of *M. laidlawii* solubilized by dodecyl sulfate (Morowitz and Terry, 1969).

Differences in protein composition have been found by comparing the membranes of mycoplasmas with those of certain other bacteria (Razin and Rottem, 1968). Biedermann and Drews (1968) showed differences in the protein composition of chromatophores of bacteria belonging to the

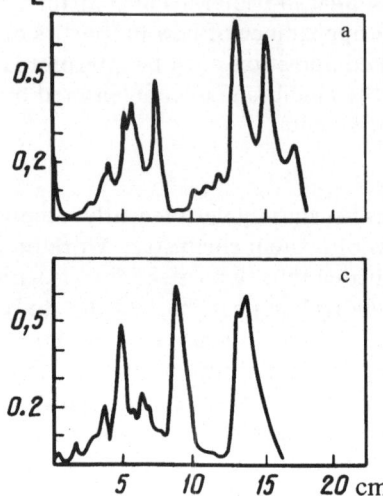
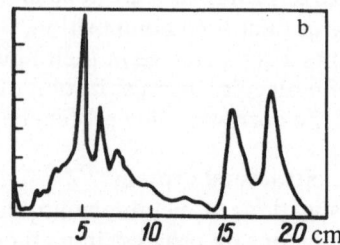

Fig. 7. Distribution of thylacoid
proteins in three bacteria belonging
to the Athiorhodaceae family
(Biedermann and Drews, 1968): a)
Rhodospirillum rubrum; b) *Rhodopseudomonas viridis;* c) *Rhodopseudomonas capsulata.*

Athiorhodaceae family: *Rhodospirillum rubrum, Rhodopseudomonas
viridis,* and *Rhodopseudomonas capsulata,* organisms which differ both in
the morphology of their membrane system and in the composition of their
pigments (Fig. 7). By gel electrophoresis, differences were found in the protein
composition of the thylacoids and cytoplasmic of *R. rubrum* (Biedermann and
Drews, 1968; Oelze and Drews, 1969, 1970b), and also in the protein composition of the cytoplasmic membrane and membranes containing the photosynthetic apparatus in *Chromatium* (Cruden and Stamier, 1970). Fragments
of thylacoids (molecular weight about 5.6×10^5), purified from other cell
membranes, contain only four proteins (Biedermann, 1971).

More than 20 proteins have been found in the membranes of *M. lysodeikticus* (Salton et al., 1967; Grula and Savoy, 1971). By comparing the
mobility of these components in gels of different porosity it has been
shown that the membrane proteins of *M. lysodeikticus* form a series with
molecular weights from 12,000 to 160,000 (Fig. 8), and components with
molecular weights of 23,000, 57,000, and 61,000 are present in relatively
higher contents (Ostrovskii et al., 1969). About 30 proteins have been
found in the cytoplasmic membrane of *E. coli,* but none is more prevalent
than the rest, whereas in the cell wall a protein with a molecular weight of
44,000 is predominant (Schnaitman, 1970a,b).

Numerous protein components have also been found by electrophoresis
in polyacrylamide gel in the membranes of *Halobacterium halobium* after
defatting and solubilization in 8 *M* urea (Brown and Pearce, 1969). Sharply
different results were obtained in the experiments of Patterson and Lennarz

(1970), who showed that one protein in the membranes of *Bacillus* sp. accounts for 90% of the total membrane proteins. Probably, further experiments will show whether this "monoprotein" membrane has unique functional features.

Since fractionation of the membrane proteins takes place in vigorous solubilizing systems, it is difficult to judge their functions. Intriguing results have recently been obtained by two groups of workers (Inouye and Pardee, 1970a,b; Shapiro et al., 1970), who studied the connection between the synthesis of membrane proteins and the attachment of bacterial DNA to the membrane. By the use of a double labeling technique (C^{14} and H^3) and of proteins in the presence of SDS, these workers were able to demonstrate substantial quantitative changes in two membrane proteins when DNA synthesis and division of *E. coli* cells were suppressed. It is tempting to deduce that these proteins are directly concerned with the process of attachment of DNA to the membrane and its replication, although the experiments described are evidently only a beginning.

One method of measuring the size of membrane proteins *in situ* is by radiation inactivation, and this has shown that a fragment of membranes of *M. lysodeikticus* contains malate dehydrogenase and $NADH_2$ dehydrogenase,

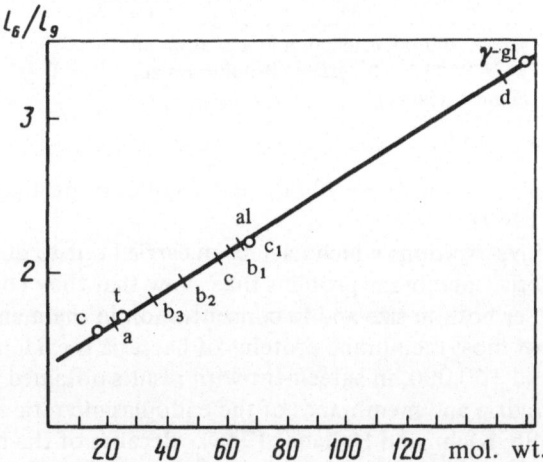

Fig. 8. Determination of molecular weight of proteins of *Micrococcus lysodeikticus* from results of electrophoresis in polyacrylamide gel (Ostrovskii et al., 1969): l_6/l_9) ratio between lengths of migration in 6% and 9% gels; c) cytochrome c; t) trypsin; al) serum albumin; γgl) human γ-globulin; a, b_1, b_2, b_3, c, c_1; d) designation of protein bands on gel strips; molecular weight of protein a 23,000, b_1 61,000, c 57,000.

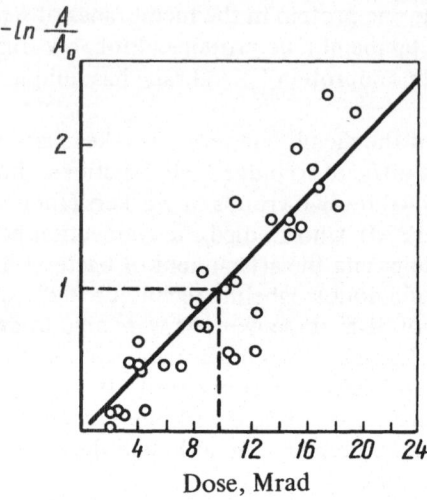

Fig. 9. Graphic calculation of size of D_{37} after irradiation of malate dehydrogenase in fragments of membrane of *Micrococcus lysodeikticus* by fast electrons (Ostrovskii et al., 1969): A and A_0) dehydrogenase activity before and after irradiation. Molecular weight of dehydrogenase 70,000, calculated by equation mol. wt. $= (0.72 \times 10^{12})/D_{27}$ (Hutchinson and Pollard, 1961).

the molecular weights of which are 70,000 and 73,000, respectively (Fig. 9) (Ostrovskii et al., 1969).

Even the few investigations which have been carried out to determine the properties of bacterial membrane proteins thus show that they contain many proteins which differ both in size and in concentration in the membrane. The molecular weight of most membrane proteins of bacteria is evidently low, between 20,000 and 160,000, in agreement with results obtained for membranes of mitochondria and membranes of the endoplasmic reticulum (Schnaitman, 1969b; Kiehn and Holland, 1968). Because of the heterogeneity of membrane proteins, the problem of the existence of the dominant protein component (the structural protein) must evidently be considered.

It was observed in the early 1960's that much (up to 50%) of the membrane protein of mitochondria possesses no enzyme activity and constitutes a homogeneous substance with a marked affinity for phospholipids (Criddle et al., 1962). It was postulated that this protein fraction has the function of a structural agent which, by combining with enzyme proteins and lipids, unites these components into functioning groups. This protein was called structural protein. One way in which the localization of this protein in the mitochondria was assumed to take place was that it formed a layer covering enzyme complexes, such as the complex I which contains $NADH_2$ dehydrogenase (Hatefi and Stempel, 1969).

The concept of a single structural protein opposed to some extent the concept of the structurally determining role of the bimolecular lipid layer (Danielli and Davson, 1935). At first it achieved wide popularity, although it quickly fell into oblivion, having lost much of its experimental basis. By electrophoresis in polyacrylamide gel and by other methods, the considerable heterogeneity of the material which had previously been regarded as homogeneous structural protein was demonstrated. It was also found that enzymically active proteins, such as mitochondrial ATPase, can enter the fraction of structural protein through denaturation (Schatz and Saltzgaber, 1969). Finally, Schnaitman (1969b), after careful fractionation of the protein of the external and internal membrane of the mitochondria and reticulum, found that these membranes contain virtually no identical proteins despite a total number of 12-23 proteins present. The most widely represented proteins account for not more than 15% of the total, and the component present in the largest amount differs in different membranes. A similar investigation has also been carried out with proteins of nuclear membranes and the cytoplasmic membrane (Kiehn and Holland, 1968), in which the absence of a dominant component also was demonstrated.

Attempts have been made to find structural protein in bacterial membranes also. Mirsky (1969) isolated a protein fraction from the membranes of B. megaterium KM which resembled the so-called structural protein of mitochondria in its properties. This bacterial protein had $S_{20,w} = 1.96 \pm 0.04$ when centrifuged in 0.5% sodium dodecyl sulfate (a small, rapidly sedimenting component was present); it moved in one band during electrophoresis in gel (pH 9.2; 7% gel), and in its solubility it corresponded to the description of structural protein. However, during electrophoresis in polyacrylamide gel in 37% acetic acid with 2 M urea, as described by Takayama et al. (1966), the preparation formed six or seven bands.

By using a completely different method, Mirsky isolated a fraction from the same membranes which was homogeneous on electrophoresis in a vigorous solvent system, but the functions of this new protein have not yet been explained. Mirsky suggests that the idea of a single structural protein as applied to bacterial membranes is too great a simplification. At best it is possible to speak of a group of proteins playing some form of structural role. Evidence of the absence of a single structural protein is also given by gel electrophoresis of the membrane proteins of M. lysodeikticus previously labeled with C^{14}. As shown by the content of label in eluates of the individual fractions, their protein content was shown not to exceed 5-10% of the total membrane protein (Sofronova et al., 1971). No structural protein was found in the thylacoids of *R. rubrum* (Oelze and Drews, 1969), and these workers suggest that all the proteins are enzymic in character. In connection with the structural role of individual proteins, mention must be made of the interesting work of Schnaitman (1969a), who showed that mutants of *E. coli* deficient in certain membrane enzymes (nitrate reductase, etc.) do not contain a protein with molecular weight of 20,000 which is typical of the membranes of this bacterium. Schnaitman suggests that this protein, which accounts for about 10% of the total membrane proteins, plays the role of an agent giving the polyenzymic complex its structural conformation. Cessation of the synthesis of this protein probably gives rise to breakdown of the complex and the release of its components into the cytoplasm during synthesis in a state in which their incorporation into the membrane is no longer possible.

The existence of such an agent was predicted originally by Azoulay and co-workers (Azoulay et al., 1967; Azoulay and Puig, 1968), who first obtained *E. coli* mutants which were unable to synthesize individual components of the nitrate reductase system of the membranes and who demonstrated that this system could be reconstructed from components in solution.

A group of workers using homogenates of a culture of *Hydrogenomonas facilis* (Kuehn et al., 1969) consider that this bacteria could secrete a structural protein as a protein–lipid complex. The complex consisted of 70% lipids (phosphatidylethanolamine, phosphatidylserine, etc.) and 30% protein, giving a single band on electrophoresis in gel –pH 9.5; 8 M urea) and four bands on electrophoresis by the method of Takayama et al. (1966). These workers consider that these four bands correspond to oligomers of the structural protein, but the question of the membrane origin of the protein remains unsettled because the lipoprotein complex was isolated, not from the membrane fraction, but from the supernatant after centrifugation of the cell homogenate at 105,000 g for 1 h.

A structural protein in the original sense of the term thus probably does not exist in bacterial membranes, but the idea of its existence continues to be a fruitful source of useful hypotheses and reflections (Wallach, 1968).

The hypothesis regarding the existence of structural protein in all biological and bacterial membranes is supported, in particular, by Lenard and Singer (1966), who directed attention to the extremely close similarity between the spectra of circular dichroism and optical rotatory dispersion for different membranes. In their opinion, this similarity can most easily be explained by the presence of one predominant component, common to all membranes, occupying the inner part of the membrane. Besides enzymically active proteins, it is very likely that regulatory proteins, also exist in bacterial membranes to maintain contact between the various enzyme systems in the membrane and between the membranes and other parts of the cell. In particular, it has been suggested (Maeda and Nomura, 1966; Nagel de Zwaig and Luria, 1967) that chains of these regulatory proteins are responsible for transmitting the lethal impulse from bacteriocins to physiologically important centers of the cell.

Another approach to the study of membrane proteins is determination of their amino-acid composition. These results are shown in Table 4, in which the composition of some membrane proteins of mitochondria and chloroplasts is given for comparison.

According to the calculations of Hatch and Bruce (1968), the ratio between hydrophilic and hydrophobic amino acids is about 1.4 for the membrane proteins of mitochondria and chloroplasts, 1.8 for the lipoproteins of blood serum, and 1.7-2.1 for soluble proteins. The hydrophilic amino acids include aspartic and glutamic acids, serine, threonine, arginine, lysine, and histidine. The hydrophobic amino acids include valine, leucine, isoleucine, phenylalanine, and methionine. The membrane proteins of mitochondria and chloroplasts thus have a higher content of hydrophobic amino acids. This is true also of the membrane proteins of some bacteria (Table 4). Meanwhile the membrane proteins of *Bifidobacterium bifidum* and *Listeria monocytogenes* have approximately the same ratio between their hydrophilic and hydrophobic amino acids as soluble proteins. In addition, as Hatch and Bruce point out, membrane proteins of mitochondria and chloroplasts have a comparatively low content of charged amino acids (aspartic and glutamic acids, arginine and lysine); the membrane proteins of *E. coli* and *M. laidlawii* have a relatively low content of charged amino acids, 28.58,and 30.42 moles %, respectively.

The simplest explanation of the role of proteins in membrane structure could be that they increase the content of hydrophobic amino acids and decrease that of charged and polar amino acids in the membrane proteins.

However, this ratio between the amino acids in membrane proteins, although observed in some cases, is by no means an invariable rule. A final solution to the problem of individual differences between membrane proteins must await determination of the amino acid sequence in individual proteins, and also their secondary and tertiary structure and the effect of the molecular environment of the protein in the membrane on these structural characteristics.

The main difficulty encountered in the study of membrane proteins is their isolation from the membranes and their purification. Without attempting in this section to examine the factors determining the assembly of membrane proteins and their links with each other and with lipids, it will simply be stated that some proteins are weakly bound and can be extracted with aqueous solutions at certain pH values or by chelating agents, while other proteins are firmly bound and can be extracted only by detergents or by chaotropic agents. The first group of agents extracts proteins linked together in the membrane by electrostatic forces, while the second group extracts proteins linked together mainly by hydrophobic interactions.

TABLE 4. Amino Acid Composition of Membrane Proteins (mole %)

Amino acid	Structural protein of chloroplasts (Criddle, 1966)	Membranes of Bacillus megaterium (Yamaguchi et al., 1967)	Membranes of Escherichia Coli (Nagata et al., 1967)	Membranes of Listeria monocytogenes (Ghosh and Carroll, 1968)	Proteins of chloroplast lamellae (ji et al., 1968)	Membrane dehydrogenases of Micrococcus lysodeikticus (Zbukova et al., 1966)	Membranes of Mycoplasma (Morowitz and Terry, 1969)	Membranes of Mycoplasma (Choules and Bjorklund, 1970)	Membranes of Bifidobacterium bifidum (Exterkate et al., 1970)	Membranes of Bacillus stearothermophilus (Hachimori et al., 1970)	Membranes of Bacillus subtilis (Hachimori et al., 1970)	ATPase of membranes of Bacillus stearothermophilus (Hachimori et al., 1970)
Lysine	8.17	5.1	5.20	7.9	5.5	3.7	5.77	7.1	5.87	5.4	3.75	4.55
Histidine	2.2	1.5	–	1.4	1.4	3.1	1.32	1.4	1.76	1.73	1.015	1.82
Arginine	1.04	3.9	6.47	3.5	7.2	7.1	2.76	3.6	5.77	3.0	2.23	4.95
Aspartic Acid	13.95	6.5	9.8	10.8	8.8	7.5	12.30	12.5	11.4	9.09	6.5	9.45
Threonine	3.74	4.4	5.23	6.6	4.7	6.0	7.90	7.5	6.48	5.57	3.88	5.45
Serine	5.47	4.7	4.23	7.7	5.7	4.0	7.44	6.4	6.30	4.09	3.49	4.79
Glutamic acid	13.08	6.3	11.68	10.8	9.2	9.0	9.59	9.4	9.60	13.16	14.4	12.88
Proline	5.14	4.1	2.73	5.2	5.9	5.8	4.29	3.9	4.46	4.63	2.73	4.83
Glycine	9.02	8.6	–	11.3	10.5	10.3	8.31	8.1	9.33	7.64	5.36	7.94
Alanine	10.40	9.5	–	10.1	9.6	13.9	9.62	8.1	12.4	10.55	15.04	9.23
Cystine		8.2	–		0.7		0.12					
Valine	7.43	6.9	8.73	5.6	6.5	8.5	5.18	7.0	6.53	8.02	4.96	8.26
Methionine	0.15	2.9	0.27	–	1.7	1.8	2.15	2.3	1.70	3.67	10.5	2.49
Isoleucine	6.61	6.5	6.53	5.3	5.3	3.6	5.27	7.0	4.50	6.86	5.24	8.08
Leucine	11.38	8.4	10.60	8.7	11.0	9.0	8.96	9.5	7.86	8.65	5.69	8.29
Tyrosine	0.71	4.1	–	1.9	3.8	2.1	4.36	4.1	2.68	3.67	2.09	2.92
Phenylalanine	0.23	4.8	3.53	3.6	6.15	2.9	4.70	4.8	3.48	4.10	3.87	3.93
Tryptophan	–	3.5	–	–	–	–	–	2.3	–	–	–	–
Hydrophilic	–	–	–	–	–	–	–	–	–	–	–	–
Nonpolar	1.60	1.10	1.30	2.19	1.34	1.57	1.8	1.6	1.92	1.35	1.14	1.41

The proportion of weakly bound proteins in the membranes is small and it varies with the type of membranes.

Evidently, some soluble membrane proteins are firmly bound neither with lipids nor with other proteins, and these can be obtained in a highly purified form. They include several cytochromes of the c type from various bacteria (see Chapter V).

The membrane ATPases of bacteria, which are also among the weakly bound membrane proteins, have been studied in considerable detail. An ATPase, consisting of a homogeneous proteins ($S_{20,w}$ = 13.4) with a molecular weight of 385,000, has been extracted from the membranes of *S. faecalis* with water; the enzyme is composed of 12 subunits (mol. wt. 33,000). The ATPase is held in the membranes by salt bridges involving Mg^{++} or Ca^{++} (Abrams, 1965; Abrams and Baron, 1968a,b; Schnebli et al., 1970). The ATPase in the membranes of *M. lysodeikticus* is also bound by bivalent cations and it can be extracted by lowering the ionic strength of the tris buffer (from 0.1 M to 0.0003 M tris) and can readily be purified; this protein is homogeneous and its sedimentation coefficient is 14-15 S (Munoz et al., 1968, 1969). The ATPase of *Bacillus stearothermophilus* is completely extracted from the membranes by EDTA. The homogeneous protein has $S_{20,w}$ = 11.9, mol. wt. 280,000, and it contains 20% of α-helices. Measurement of the activation energy showed that at 55°C the molecule of the purified enzyme undergoes a conformational change (Hachimori et al., 1970). The membrane ATPase of *Lactobacillus casei* is extracted with EDTA (Barker and Thorne, 1970). However, not in all bacteria is the ATPase weakly bound in the membrane. For instance, the enzyme cannot be extracted at all with water from the membranes of *S. aureus* (Gross and Coles, 1968); ATPase can be solubilized from the membranes of *E. coli* by detergent (Evans, 1970).

The penicillinase in the membranes of *Bacillus licheniformis* and the alkaline phosphatase in the membranes of *B. subtilis* are also linked by ionic bonds (Lampen, 1967; Sargent and Lampen, 1970; Wood and Tristram, 1970).

Up to 30% of the membrane $NADH_2$ dehydrogenase of *B. megaterium* can be extracted by treatment of the membranes with weak alkali (pH 8.5), The enzyme, when purified by salting out with ammonium sulfate and centrifugation in a sucrose gradient, contained flavin adenine dinucleotide (FAD) (1 × 10^{-8} mole/mg protein) and 100 μg phospholipids/mg protein. Its sedimentation constant was 28 S (Mizushima, 1968; Ishida and Mizushima, 1969).

The strongly bound proteins of the membrane, which account for most of its mass, are naturally most interesting from the point of view of its molecular organization.

Many methods can be used to separate membrane proteins from lipids. However, separation of the lipid still does not imply purification of the protein. Because of their ability to aggregate, membrane proteins easily form insoluble complexes, and solubilizing agents have to be used to isolate the individual proteins.

To obtain membrane proteins in a form free from lipids and in a solubilized state, several methods are used. The membranes can be treated with various detergents such as dodecylsulfate (Engelman et al., 1967) or with 2-chloroethanol (Zahler and Wallach, 1967), sodium iodide (Kolber and Stein, 1967), and aqueous pyridine (Blumenfeld, 1968). The lipids can be removed by solvents and the proteins then solubilized with dodecylsulfate, formic acid, or guanidine, or they may be succinylated (Rosenberg and Guidotti, 1968). A group of proteins can be extracted from mitochondria by treatment with 1.4% acetic acid (Zahler et al., 1968); a homogeneous protein not containing lipids is extracted with 8 M urea (Fleischer et al., 1968). In some cases preliminary treatment of the membranes with phospholipase A is effective. Curtis (1969) recommends that membrane proteins be solubilized after removal of lipids from the membranes by treatment of the residue with a mixture of acetone, chloroform, and methanol acidified with $HClO_4$. Solubilization of the proteins is usually followed by fractionation with ammonium sulfate, gel filtration, and purification on ion-exchange cellulose derivatives or Sephadex.

Unfortunately, all methods for isolating firmly bound membrane proteins have the same disadvantage: they cause denaturation of the proteins and loss of enzyme activity.

During solubilization of the membrane by detergents, protein-detergent complexes are formed which prevent determination of the molecular weight and other physicochemical parameters of the protein. The results of sedimentation analysis of solubilized membrane proteins must be interpreted with particular care. For example, a protein-binding lactose solubilized from membranes of *S. aureus*, according to the results of analytical centrifugation, has a molecular weight of 100,000, but according to gel filtration its molecular weight is about 1 million (Hengstenberg, 1970). Investigations have shown that a mixture of proteins obtained from membranes behaves in the analytical centrifuge as a homogeneous protein (Rodwell et al., 1967; Engelman et al., 1967; Rosenberg and Guidotti, 1968).

Difficulties in the isolation of membrane proteins are the only reason it has so far proved impossible to obtain homogeneous, highly purified preparations of all the cytochromes, dehydrogenases, and other proteins from bacterial membranes. Moreover, the membrane proteins of bacteria are no exception because the proteins of mitochondrial membranes likewise have not been obtained in the pure form, with the exception of cytochrome oxidase, cytochromes c_1 and b, $NADH_2$ dehydrogenase, and succinate dehydrogenase (Okunuki, 1966; Singer, 1966, 1968; Davis and Hatefi, 1971a,b), although these proteins differ in their properties and lipid content, depending on the method used to isolate them from the membranes.

The isolation and purification of firmly bound proteins from bacterial membranes by the use of strong solubilizing agents to free them from lipids and to rupture the bonds between individual proteins have been carried out in only a few investigations. Information on the purification of firmly bound cytochromes and dehydrogenases is given in Chapter V. Besides components of the respiratory chain, attempts have also been made to purify other membrane proteins. A protein participating in lactose transport and firmly bound in the membrane of *E. coli* has been extracted by dodecylsulfate and separated from other proteins by electrophoresis in polyacrylamide gel. The molecular weight of this protein is 30,000, although the possibility cannot be ruled out that it had been split up into subunits, so that its molecular weight may be a multiple of 30,000 (Kennedy, 1969). Treatment of membranes of *A. vinelandii* with butanol caused solubilization of the proteins, after which a flavoprotein was isolated (Shetna et al., 1966). Membranes of *E. coli* were solubilized with NaI, and after fractionation on Sephadex G-75 columns, a protein possessing β-galactoside permease activity was isolated (Kolber and Stein, 1967). Proteins participating in vectorial phosphorylation and transport of sugars (Kaback, 1970a,b) and amino acids (Gordon et al., 1972) and Mg^{++}, Ca^{++}-dependent ATPase (Evans, 1970) have been isolated from the membranes of *E. coli*. Nitrate reductase, a nonheme ferroprotein with molecular weight of 160,000, has been extracted from fragments of membranes of *Micrococcus denitrificans* by alkaline acetone. The protein has a high content of aspartic and glutamic acids (Forget, 1971). A lactose-binding protein has been isolated from the membranes of *S. aureus* (Korte and Hengstenberg, 1971).

Although the study of membrane proteins is still only in its initial phase, certain conclusions can be drawn from the results obtained. The most interesting discovery from our point of view is the exceptionally great variety of protein components discovered by recent work on the membrane proteins of bacteria. This was most marked of all in the aerobic bacteria, in whose membranes are localized the enzymes of the respiratory chain, ATPase, transport proteins, proteins participating in a wide variety of biosyntheses, and — finally, in some bacteria — flagellin. Many proteins have also been found in the membranes of other physiological groups of bacteria, notably the obligate anaerobes and photosynthesizing bacteria. The molecular organization of the membrane must therefore be such that each protein is sufficiently firmly bound and lies next to those proteins with which it interacts. Further, each enzyme must be so positioned in the membrane that it is accessible for its substrate. Another, no less interesting fact contributing to our understanding of the organization of proteins in the membrane is the existence of strongly and weakly bound proteins, a distinction which probably results both from the properties of the proteins and from the special features distinguishing the protein-lipid region of the membrane where they are attached. A third important fact is the absence of any quantitatively predominant structural protein. Finally, as a fourth noteworthy fact, the hydrophobia of the membrane proteins and the readiness with which they undergo aggregation may be mentioned.

In the field of study of highly purified membrane proteins, as has already been pointed out, the investigator is beset by difficulties due to the crisis in methods of protein chemistry and, in particular, methods not adapted for work with hydrophobic proteins with a marked tendency to aggregation. Be that as it may, the study of membrane proteins is a central problem in the study of the molecular organization of biological membranes, and it is vital to the understanding of the mechanisms by which they function. The efforts of chemists and biochemists in the next decade will therefore, in all probability, be concentrated on the study of membrane protein chemistry.

Other Components of Membranes

Until recently only the major components of bacterial membranes, namely proteins and lipids, received attention. However, evidence has been obtained that bacterial membranes also contain polysaccharides and nucleic acids which are usually detectable by ordinary analysis. These results must be interpreted very cautiously, for nucleic acids and also polysaccharides could be cytoplasmic in origin and could "adhere" to the membrane during isolation. This applies in particular to RNA, for ribosomes are often left connected with the membrane and are visible in the electron microscopy. We have already mentioned the difficulty of separating the cytoplasmic membrane and internal membranes of bacterial cells from the cell wall in Gram-negative and some Gram-positive bacteria. The existence of polysaccharide components can therefore be discussed only in bacterial membranes which can be obtained after complete removal of the cell wall. Membranes of certain Gram-positive bacteria from which the wall can be completely removed have been analyzed for the presence of carbohydrates and nucleic acids (Table 1). However, the results were contradictory and no definite conclusions can be drawn. In fact, the results for the content of nucleic acids and polysaccharides in membranes of *B. megaterium* obtained by different workers are highly inconsistent. There are also surprising differences in the content of nucleic acids and carbohydrates even in the membranes of bacteria such as *B. subtilis* and *B. megaterium*, or *M. lysodeikticus* and *S. aureus*, whose structural organizations are so closely similar.

Cations are another group of components of the membrane which participate in its structure and stabilization. Both the composition of the cations of membranes and their role in the maintenance of membrane structure have been studied from this point of view. In this section we shall discuss only the concentration of cations in the membrane; their role in the stabilization of its structure will be examined later in the section on the molecular organization of membranes as a whole.

It has been shown that Mg^{++} ions are linked with the membrane lipids of certain species of *Pseudomonas* and that all the Mg^{++} is found in the acid phospholipid fraction, where the ratio $P:Mg^{++} = 2.2:1$. Gordon and MacLeod (1966) consider that Mg^{++} ions can form chelates with phospholipids, and thereby increase their packing density in the membrane. Mg^{++} ions are also found in phospholipids from the halophilic bacterium *Halobacterium cutirubrum* in the ratio $P:Mg^{++} = 1.8:1$, i.e., 1 mole magnesium is bound with 1 mole phosphatidylglycerophosphate (Gordon and Macleod, 1967). Although the lipids were obtained from the membrane plus wall fraction, these results are of substantial importance to the understanding of the role of magnesium ions in other membranes also.

In membranes of *Mycoplasma* the concentration of Mg^{++} is 1.30×10^{-3} mg/mg membrane protein (Engelman and Morowitz, 1968). Teichoic acids participate in the supplying of magnesium ions to the membranes of *B. licheniformis* (Hughes et al., 1971). The quantitative content of cations in a membrane is determined to some extent, it will be remembered, by their concentration in the medium. It would probably be advantageous to determine only those cations whose binding by chelating agents leads to the disturbance of a particular function of the membranes, such as electron transport.

An important role in the construction of biological membranes is played by water molecules (Klotz, 1964; Salem, 1964; Ling, 1965; Hechter, 1965; Finean et al., 1968; Tait and Franks, 1961). The state of water in membranes is examined in Chapter III.

The Molecular Organization of Bacterial Membranes

Introduction. Problem and Methods

The "molecular organization of membranes," in the wide sense of this term, means the arrangement of the molecular components of the membrane and the mechanism of their interaction during the performance of its biological functions.

The investigation of the molecular organization of membranes can be subdivided into a series of stages or a series of intermediate problems. The most important of these, in our opinion, are the following:

1. Determination of the state of each component of the membrane in terms describing different levels of structural organization and phase transitions.
2. Assessment of the relative arrangement of the components.
3. Determination of the character of the bond between the components and their influence on each other.

The character of the problem largely determines the methods used by the investigator. These methods can be classified in accordance with the above-listed stages of investigation, as follows:

1. Spectral methods (circular dichroism, optical rotatory dispersion, infrared spectrophotometry, nuclear magnetic and electron paramagnetic resonance), differential thermal analysis, radiation invactivation, etc.;
2. Action of enzymes on the membrane, determination of the accessibility of reaction groups for chemical modification, electron microscopy, and x-ray diffraction;
3. Treatment with agents disturbing a certain type of bond (variation of the pH or ionic strength of the solutions; action of surface-active substance, of organic solvents, etc.).

The last-mentioned method must inevitably cause disruption of the membrane into fragments of different sizes and complexity, of which the most interesting are those which still retain some of the catalytic properties of the original membrane. The resulting fragments are more convenient in many respects for investigation than whole membranes. Accordingly, the study of enzyme complexes from membranes has grown into one of the major sections of membrane biology. Because of its accessibility to biologists who lack expensive spectrometric and other apparatus, this approach has become the most widespread from the quantitative point of view.

We shall discuss each of these problems as they apply to the investigation of bacterial membranes. However, certain difficulties of a fundamental character must be pointed out.

The catchphrase coined by the French microbiologist, Monod, "What is valid for the bacterium is also valid for the elephant," has been widely quoted. It is quite evident, however, that the bacterium is not an elephant and that an extremely important branch of medicine, the chemotherapy of infectious diseases, is based on this fact. Therefore, when we begin to discuss the facts and hypotheses which may shed light on the molecular organization of bacterial membranes we shall not assume the unity of the "world of membranes" as axiomatic; i.e., we shall refrain from the unreserved extrapolation of ideas concerning the membrane of the erythrocyte to the membrane of *E. coli*. To carry our logic to the end, we must not place the membranes of different bacteria or different membranes (e.g., mesosomal, cytoplasmic) of the same organism in the same category, although we simply do not have sufficient factual material available at present to make such a distinction because bacterial membranes have only recently become objects of more than usual interest to investigators (Harold, 1970).

Another inevitable restriction at the present time is that all biological membranes, and bacterial membranes in particular, are multifunctional and consist of many different components (See Chapters II and IV). The different physical and chemical parameters which change during the function of individual components of the membrane are parameters which apply to the membrane in general. We shall thus discuss the molecular organization of bacterial membranes in the narrower meaning of the term: on the level of general description of the most stable states of the membrane as a whole.

Characteristics of the State of Individual Components of the Membrane

General Remarks

Before discussing the state of proteins, lipids, and other substances of the membrane, it will be worthwhile mentioning in what state they may be. The general details of the levels of organization of biopolymers and of cooperative processes in some biological systems are familiar to biologists from publications on molecular biology (see, for example, Bresler, 1966; Konev et al., 1970). We shall discuss levels of organization only in connection with the organization of bacterial membranes.

Proteins. When it evaluates the state of membrane proteins, membrane biochemistry uses the ideas and instruments of polypeptide and protein chemistry, themselves an extensive section of organic chemistry (see *Current Problems in Peptide and Protein Chemistry,* Moscow, Nauka, 1969). It would be most logical to begin the investigation of molecular organization of membranes with the discovery of the primary structure of membrane proteins, for it is the primary structure which determines the higher levels of organization of the protein molecule and, consequently, its enzymic properties. However, purely technical difficulties arising during the attempt to separate hydrophobic membrane proteins have so far not permitted this to be done. Only in few cases has the amino-acid composition of some proteins of bacterial membranes been successfully determined. The problems were examined more fully in Chapter II; here we shall only mention that membrane proteins of some bacteria have a relatively high content of nonpolar amino acids which evidently facilitates their interaction with lipids. At the same time, an increased content of hydrophobic amino acids is not essential for membrane proteins, as is clear from the example of the proteins of *Listeria monocytogenes* and *Bifidobacterium bifidum* (see Table 4).

So far as the secondary protein structure is concerned, the most probable states of the polypeptide chain are the α-helix, the β-form (the antiparallel arrangement of the polypeptide chain), and the state of a random coil, although other forms are also possible in principle.

The nature of the secondary structure of a protein is determined from the character of its optical rotatory dispersion (ORD) spectra, its circular dichroism (CD) curves, and its absorption spectra in the infrared region. In investigations of this type it is usual to compare these parameters with those for synthetic polypeptides whose structure has been established by several independent methods. Semiempirical formulas exist for calculating secondary structure from the results of measurement of optical activity, but the possibilities of their application are limited (Jirgensons, 1970). For this reason, spectral analysis is frequenlty reduced simply to the detection of shifts in the position of the extrema by comparison with the spectra of known polypeptides and proteins.

Lipids. In contrast with proteins, the chemical composition and metabolism of the lipids of bacterial membranes have been studied much more completely (Chapter II), for with the exception of some lipopolysaccharides, these lipids are relatively simple substances. The extremely great variety of the bacterial lipids and their difference from the membrane lipids of higher organisms are striking. When we contemplate the state of lipids in a membrane, we involuntarily enter a region controlled by the physical chemistry of liquid crystals. The following two characteristics of the state of lipids are the most important: first, the mobility of their nonpolar groups ("liquidity" or orderliness); and second, the character of the microassociations of the lipid molecules (micelles, lamellae, hexagonal packing, etc.). The literature on these problems is extensive (Luzzati, 1962; Reiss-Husson and Luzzati, 1967; Lucy, 1969), since artificial lipid films have until now been regarded as the best models of biological membranes (Tien and Diana, 1968; Deborin, 1967), although the true state of lipids in the membrane has been determined only recently as the result of the development of nuclear magnetic resonance, paramagnetic probing, and differential thermal analysis methods as applied to the investigation of membranes (Chapman, 1969).

Other Substances in the Membrane. The data on polysaccharides and also on nucleic acids, polyamines, inorganic ions, and other minor components of bacterial membranes are too few in number to be considered in this section. Only a few comments will be given on the state of the membrane water, again because of the limited amount of information.

Optical Activity and Conformation of Protein

When a beam of plane-polarized light, consisting of the sum of two beams, one circularly polarized to the right (R) and the other circularly polarized to the left (L), is passed through the solution of a protein two phenomena take place. First, because of the difference between the refractive indices for the R-beam and L-beam, they emerge from the solution with a phase shift. The plane of polarization of the merging combined beam will therefore be inclined relative to the plane of the incident beam. The optic rotation of the protein at wavelength λ is usually calculated as the mean rotation in degrees per optically active unit and is designated $[m]\lambda$.

Second, the L and R components of plane-polarized light not only differ in their rate of passage through a medium, but also in their absorption by protein molecules. For this reason, their sum on leaving the solution, because of the difference in amplitudes, is no longer a plane-polarized but an elliptically polarized beam (circular dichroism). This property of a protein is measured by the degree of ellipticity of the emerging light $[\theta]_\lambda$, calculated per optically active unit.

In the study of suspensions of biological membranes and their fragments, a third phenomenon also plays an important role: this is the dispersion of light, which leads to very undesirable distortions of optical rotation and circular dichroism. Every attempt must be made to avoid it.

The optical activity of a protein membrane depends on its conformation, and it can serve to some extent as the basis of a method of assessing the secondary, tertiary, and quaternary structures of a protein. The most important information which it gives is not so much the values of $[m]$ and $[\theta]$ at the same wavelength, as curves reflecting these parameters as functions of λ in the region of absorption of the peptide bond itself or of the other protein chromophores (185-240 nm; 240-600 nm). The link between optical activity and the relative amount of the α-helix, β-form, and random coil structure in the protein molecule is expressed quantitatively by equations of which that most frequently used is the equation of Moffitt and Young (1956):

$$[\alpha] = a_0 \frac{\lambda_0^2}{\lambda^2 - \lambda_0^2} + b_0 \frac{\lambda_0^4}{(\lambda_2 - \lambda_0^2)^2}$$

where $[\alpha]$ is the specific rotation of the plane of polarization, α_0, β_0, and λ_0 are constants, and λ is the wavelength.

Some workers (Maddy and Malcolm, 1965; Wallach and Zahler, 1966) have used this equation in an attempt to calculate the relative number of peptide groups of membrane proteins in a particular structural form. Estimation of the structural state of membrane proteins in this way was extremely important as a means of testing some of the assumptions in the widely used Davson–Danielli model of the membrane, according to which the protein lies on the surface of the membrane in an unfolded state.

The following discoveries were made:

1. In membranes of widely different origin (from bacteria to erythrocytes), the proteins contain many α-helical segments. For instance, Maddy (1967), who analyzed the proteins of erythrocyte membranes extracted by butanol, showed that the value of b_0 for these proteins is 90, corresponding to 17% of the total number of peptide groups being helical in form. (For a 100% α-helical polypeptide chain, Maddy assumed: b_0 = 535; λ = 216 nm). According to calculations by other workers, α-helical segments account for about 40% of the total in erythrocyte membranes and about 25% in fragments of mitochondrial membranes (see the survey by Beychok, 1968), while in the cytoplasmic membrane of the animal cell the figure may reach 60% (Wallach, 1969). For the protein in bacterial membranes (*B. subtilis*), it is assumed from the optical rotatory dispersion data that the degree of helicity is 25-33% and that the remainder of the molecule has no particularly marked structure (Lenard and Singer, 1966).

By the use of a new method of calculation of circular dichroism spectra, up to 40% of the β-form was detected in the membrane proteins of *Mycoplasma* (Choules and Bjorklund, 1970).

2. The extremes on the optical rotatory dispersion and circular dichroism curves are displaced by 2-6 nm toward the long-wave (red) region compared with observation for pure proteins.

3. The absolute values of the optical activity were much below those of the optical activity of soluble proteins and polypeptides.

These observations do not yet rest on a sufficiently precise theoretical basis, but their most probable causes are considered to be as follows: either a substantial part of the protein molecules of the membrane is in a nonpolar environment, suggesting that hydrophobic lipid-protein interactions most probably take place, or the membrane proteins assume special forms of secondary structure which do not exist outside membranes. However, Urry et al. (1970) consider that any peculiarities of the optical rotatory dispersion spectra of membrane preparations are simply due to artifacts caused by dispersion of light by the particles. If an appropriate correction is introduced for turbidity, the spectra of membranes come to resemble more closely in character those of soluble proteins with an α-helix.

Since the state of the chromophore groups in a protein in different conformations and the mechanism of interaction between these groups have not yet been fully explained, the existing equations of optical rotatory dispersion are only approximate in character. On this account, Wallach (1969) considers that at present it would be better to use optical acitivity as a type of indicator, as a sensitive method of detecting conformation changes associated with various physiological changes in the membrane, and to avoid making quantitative calculations of the content of α- and other forms of secondary structure.

Proteins in bacterial membranes and in membranes of mycoplasmas show the same optical activity as those in other membranes with similar characteristic properties (Lenard and Singer, 1966; Steim and Fleischer, 1967; Steim et al., 1969). Statistics from the paper by Lenard and Singer (1966) for the optical activity of bacterial membranes are reproduced in Table 5, and the optical rotatory dispersion curve for erythrocyte membranes which, according to these same workers, is indistinguishable from that for membranes of *B. subtilis,* is given in Fig. 10.

The maxima on the ORD and CD curves of bacterial and erythrocyte membranes are shifted by 2-6 nm toward longer wavelengths than those of the synthetic polypeptides. It is interesting to note that displacement toward the long-wave region does not take place on solubilization of the membranes by detergent (SDS) or after removal of the lipid followed by solution of the protein in 2-chloroethanol.

By the action of phospholipase A on erythrocyte membranes or cytoplasmic membranes of other animal cells, the CD maximum is shifted into the blue region (to 193 nm), the negative ellipticity at 202 nm is increased, and the band at 223 nm is narrowed. All these may mean that on destruction of some of the phospholipids the environment of the peptide chromophores becomes more polar. Narrowing of the band around 223 nm with no change in amplitude also suggests some decrease in the degree of α-helicity (Wallach, 1969). During the action of phospholipase C, on the other hand, nonpolar interactions are strengthened and this is manifested as displacement of the maximum onto the CD curve into the red region. On the basis of these observations it should be concluded that the lipids of membranes are in a state of powerful interaction with the membrane proteins, but Glaser et al. (1970) were unable to find any changes in the CD curves during the action of phospholipase C on eryth-

TABLE 5. Parameters of Optical Rotatory Dispersion
(ORD) and Circular Dichroism (CD) for Some
Biological Membranes (Lenard and Singer, 1966)

Parameter	Membranes of *Bacillus subtilis*			Erythrocyte membranes
	0.15 MNaCl 0.008 *M* phosphate, pH 7.7	0.05 *M* DDS-Na	2-Chloro-ethanol, 100%	0.008 *M* phosphate pH 7.7
ORD				
λ_{min}	235-237.5	233	233	235-235.5
$\lambda_{crossover}$	227	221-222	224	226-227
λ_{max}	201-203	198-199	–	201-202
CD				
λ_{min}	224	222	221-222	224
λ'_{min}	221-212	208	221	210
$\theta_{min} \times 10^{-3}$	-12.9	-20.7	-32.4	-13.2
$\theta'_{min} \times 10^{-3}$	-7.8	-23.8	-28.5	-6.1

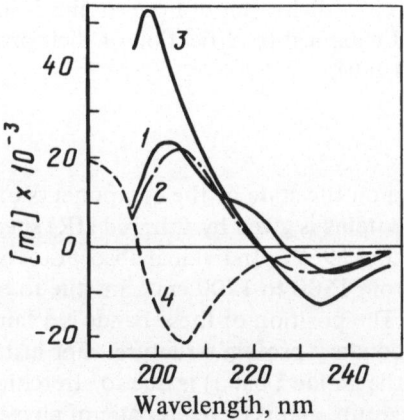

Fig. 10. Dispersion of optical activity of
suspension of erythrocyte membranes and
poly-L-lysine in 0.008 *M* phosphate buffer,
pH 7.7 (Lenard and Singer, 1966): 1) erythrocyte membranes; 2-4) poly-L-lysine:
2) β-form, 3) α-helix, 4) random coil.

rocytes, although this enzyme splits more than 75% of sensitive phospholipids. These workers also point out that on heating from 20 to 80°C there
is a marked decrease in helicity of the membrane proteins (the proportion of
helical areas falls from 43 to 23%); this process, moreover, does not depend
on the integrity of the phospholipids.

The proteins of fragments of mitochondrial membranes undergo appreciable changes with a change in the pH of the medium. The position of
the minimum on the ORD curve is displaced from 234 to 237.5 nm as the
result of neutralization of the buffer solution (pH 10 → pH 6.7). Submito-
chondrial particles swell if the medium is made alkaline, but what happens
to the protein under these circumstances is uncertain. It is considered that
some loosening of the membrane structure takes place through deaggregation
of the protein (Wrigglesworth and Packer, 1969).

Model experiments by Gulik-Krzywicky et al. (1969) showed that circular
dichroism and optical rotatory dispersion of the proteins vary considerably
during interaction between the proteins and lipids. However, it is evident
that by no means all conversions in the lipoproteins of a membrane are accompanied by changes in the secondary structure of the protein and, cor-
respondingly, by changes in their optical activity. It has been shown, for
instance, that the CD curves of cytochrome c_4 from *Azotobacter vinelandii*
are unchanged during the transition of this membrane protein from the oxidized to the reduced state (Van Golde et al., 1968), that optical activity in
blood lipoproteins is unchanged even after removal of the lipid (Granda and
Scanu, 1966), and that circular dichroism of lipoproteins is independent of
the pH of the medium and chemical modification of their proteins (succinylation) (Scanu et al., 1969).

Infrared Spectroscopy

Important information on the state of the components of a membrane
and, in particular, of its proteins is given by infrared (IR) spectroscopy
(Chirgadze, 1965; Kobyakov, 1969). Individual absorption bands in the
region of wave numbers from 1500 to 1700 cm^{-1} are due to oscillations of
atoms in peptide groups. The position of these bands is a fairly sensitive
indicator of changes in secondary protein structure. For instance, absorption at 1652-1655 cm^{-1} (the amide I band) is due to stretching vibration of the
C=O bond in the peptide group, and is characteristic of an α-helical conformation and a random polypeptide chain. On conversion of the protein into
the β-form the absorption maximum is shifted to 1630 cm^{-1}, and a "shoulder"
also appears at 1695 cm^{-1}. Conversion to the β-form is also accompanied
by the appearance of an absorption band at 1520 cm^{-1} instead of 1535-1540
cm^{-1} in the case of the α-helix. This region (amide II) is connected with
bending of the N—H bond in the peptide group, but in view of the super-
position of absorption bands of other groups, it is only an additional characteristic of structural transition.

Usually the spectra are recorded in dry material, for water gives too high and too heterogeneous a background. A thick suspension of membranes is applied as a uniform layer on disks of AgCl, germanium, or other suitable material and dried in air or from a frozen state *in vacuo*. So that wet membranes can be studied, their ordinary water is replaced by heavy water (D_2O), which has no absorption band in the 1500-1700 cm^{-1} region. This way, the natural conformation conversions of the membrane proteins can be studied as they perform certain of their functions, and changes in the secondary and tertiary structure of the proteins can be identified as a result of treatment by various agents (Verzhbinskaya and Pershina, 1971). The maximum of the amide I band for the α- and β-forms in heavy water is displaced to 1648 and 1632 cm^{-1}, respectively.

As a result of analyzing the IR spectra of whole membranes and their components, Maddy (1967), Chapman and Wallach (1968), and other investigators have concluded that the secondary protein structure in erythrocyte membranes and in myelin corresponds to that of an α-helix or random coil, and that the β-form is virtually absent. These results have been obtained with both dry and wet preparations. In addition, the character of the spectrum is unchanged by removal of lipids from the membranes (Maddy, 1967) or by changing the temperature from $-150°$ to $+150°C$ (Jenkinson et al., 1969).

It has also been reported that the β-form is absent in mitochondrial membranes (Wrigglesworth and Packer, 1969a), but it was later discovered that the ratio between the protein conformations depends on the state of the energetics of the mitochondria (Wallach et al., 1969; Graham and Wallach, 1969). The helix and random coil structure are predominant in mitochondrial membranes in state III (optimal conditions for oxidative phosphorylation: ADP, O_2, and substrate present), but with the transition to state I (O_2 present, substrate and ADP absent) or state IV (O_2 and substrate present, ADP absent) the formation of the β-form is observed (Graham and Wallach, 1969). The results of IR spectroscopy of the membranes of *Mycoplasma laidlawii* give further evidence that membrane protein exists as the α-helix or random structure (Steim et al., 1969), but a high content of protein in the β-form has been found in the membranes of strain A of this organism by IR spectroscopy and optical rotatory dispersion methods (Choules and Bjorklund, 1970).

The fact that the IR spectra in the region of the amide I and amide II bands is unchanged by removal of lipids from the membranes is regarded as evidence of the absence of strong interactions between lipids and proteins of the membranes, although the possibility is not ruled out that stabilization of the conformation in this way may be the result of the removal of bound water.

The IR spectra of membranes of *M. lysodeikticus, E. coli,* and *B. stearothermophilus* do not differ significantly from those described above for erythrocyte membranes and myelin. Thus, the conclusions for the secondary protein structure can be regarded as valid both for the membranes of higher

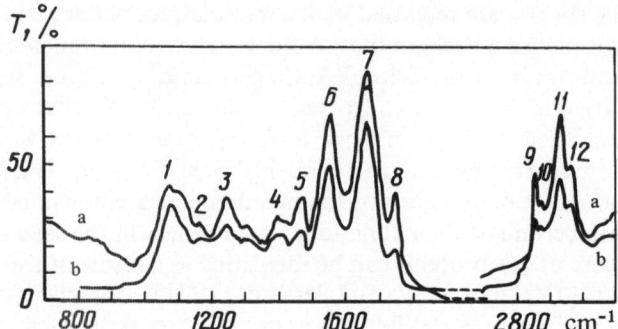

Fig. 11. Infrared spectra of air-dried films obtained from membranes of *Micrococcus lysodeikticus* (a) and *Bacillus stearothermophilus* (b). Membranes obtained by lysis of the cells with lysozyme in a hypotonic medium. Preparation from *B. stearothermophilus* additionally sonicated for 1 min and sedimented by centrifugation for 15 min at 30,000 g. Films applied to germanium disks. Identities of oscillations: (1) P–O–C and PO_4 valency; (2) C–O–C valency (sym.); (3) P=O valency and C–O–C valency (asym.); (4) CH_3 deformation (sym.); (5) CH_2 shear and deformation; (6) Amide II; (7) Amide I; (8) C=O valency; (9) CH_2 valency (sym.); (10) CH_3 valency (sym.); (11) CH_2 valency (asym.); (12) CH_3 valency (asym.).

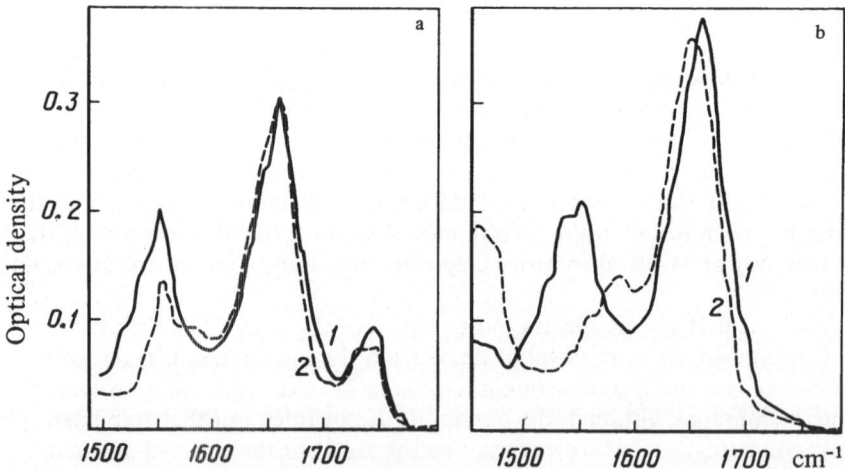

Fig. 12. Deuterium exchange in membranes of *Micrococcus lysodeikticus*. Infrared spectra (1) before; and (2) 25 min after addition of D_2O to dry preparations of intact membranes (a), and of membranes after removal of lipids (b).

organisms and for bacterial membranes (Green and Salton, 1970; Binyukov et al., 1971a; Buyalo et al., 1972). By way of illustration, the IR spectra of dry films obtained from the membranes of certain bacteria within the range from 700 to 4000 cm^{-1} are given in Fig. 11. Absorption spectra in the infrared region for membranes of *M. lysodeikticus* in the intact state and after removal of lipids, before and 25 min after contact with heavy water, are shown in Fig. 12. These spectra are given in optical density units calculated for points on the transmission curves every 5 cm^{-1}, and reduced to the same value of E_{1660} cm^{-1} for the membranes in accordance with the recommendation of Blout et al. (1961).

Three features distinguishing the spectra deserve special attention. First, the position of the maximum and the symmetry of the amide I band are not significantly changed by transfer of the dry film of membranes into D_2O. Presumably the protein molecules in the membrane do not undergo denaturation — at least of the type of helix (or coil) $\rightleftharpoons$ β-form conversion — during drying of the membrane in the course of the thin film's preparation.

Second, if membranes from which the lipids have been removed are transferred into an aqueous medium, a considerable band appears at 1580 cm^{-1}; this is considered to be due to asymmetrical oscillation of the ionized carboxyl group and/or the ND_3 group in arginine (cited by Abaturov, 1970), which are much less marked in preparations of whole membranes.

Third, a mere qualitative estimation of the amide II band for preparations of membranes and membrane proteins in heavy water is sufficient to show the essential role of lipids in slowing the exchange between the protein molecule and heavy water, possibly as a result of an increase in the accessibility of specific sites of the molecule to water.

The screening position of the lipid becomes still more obvious if the curves for the kinetics of deuterium exchange are examined (Fig. 13). They show

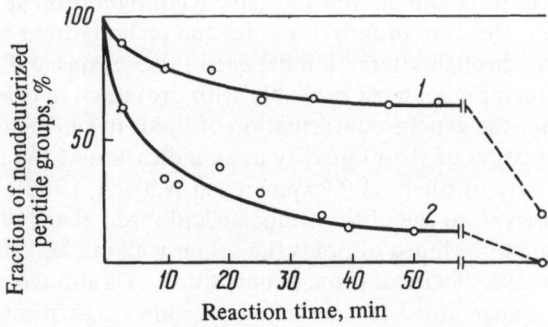

Fig. 13. H $\rightleftharpoons$ D exchange kinetics in preparations of membranes of *Micrococcus lysodeikticus* (1) in the intact state; and (2) after removal of lipids. After the gap on the curves (1 and 2), the preparations were heated for 1 h at 60°C.

that removal of the lipid makes the overwhelming majority of $>$N—H groups in the peptide components of the membrane proteins accessible. This phenomenon, however, can also be explained by the more active influence of the lipid on the conformation of the membrane protein, assuming that the lipid maintains a large part of the polypeptide chain in the α-helical state, and that on removal of the lipid the protein molecule is converted into a random coil. This interpretation is a perfectly natural one, for removal of the lipid inactivates the enzyme systems of membranes, although in this case, the direct destructive action of organic solvents on the protein cannot be ruled out.

The coincidence between the shape and position of the maximum of the amide I band for dry and wet films may imply preservation of the native structure or, on the other hand, irreversible denaturation of the membrane proteins while drying in air.

To assess the possible denaturation changes, malate—2,6-dichlorophenol-indophenol—oxidoreductase activity and the rate of oxygen consumption were determined in membranes of *M. lysodeikticus.* After drying, about 50% of the oxidoreductase and 100% of the respiratory chain activity still remained.

From the results of IR spectroscopy, the state of the membrane lipids could be regarded as fairly orderly, for the lipids extracted from the membranes were found to absorb radiation with a wave number of 720 cm^{-1}, while the original membranes of erythrocytes or *M. lysodeikticus* have no maximum in this region. Absorption at 720 cm^{-1} arises as a result of the rocking mode of the protons in CH$_2$-groups, which is possible when the carbon atoms are in the *trans*-position. The transition of individual components of the carbohydrate chain into the gauche-position leads to a less orderly structure and to disappearance of the maximum at 720 cm^{-1}. Lipids outside the membrane are more orderly than those in the membrane, but it would seem more logical to postulate that the gauche configuration appearing as a result of certain freedom of movement for the carbohydrate chains is fixed in the membrane through interaction between these chains and other components of the membrane, most probably with proteins. In other words, it is considered that the gauche conformation of lipids in the membrane is not so much an indication of their liquidity as an indication of the influence of protein on the state of the lipid (Chapman and Wallach, 1968). A similar conclusion follows from the observations of Jenkinson et al.(1969), who demonstrated the orderliness of lipids (i.e., their well-marked absorption at 720 cm^{-1}) in myelin sheaths at room temperature. It is also worth noting that the absence of change in the IR spectra over a wide range of temperatures (Jenkinson et al., 1969) is also presumptive evidence of the marked stabilizing influence of lipids on the formation of the membrane protein, as has been conclusively shown in the case of blood lipoproteins (Scanu et al., 1969; Day and Levy, 1969).

The results of IR spectroscopy thus suggest that the secondary structure of most of the protein of bacterial membranes is of the α-helical and random coil type. Indirect observations suggest that interaction occurs between proteins and lipids, but no strict proof of this interaction or of its absence is provided by methods available at the present time.

Nuclear Magnetic Resonance (NMR)

This method has only comparatively recently begun to be used for the investigation of biological membranes. For this reason, most results of investigations using NMR apply to membranes of erythrocytes and myelin, the traditional objects in biophysics. Data on bacterial membranes and, in particular, on the membranes of *M. lysodeikticus* are still in the course of publication (Ostrovskii et al., 1972; Ostrovskii and Sofronova, 1972).

The principle underlying the method is as follows. The substance for testing is placed in a strong magnetic field. The field interacts with the nuclei of certain atoms (H^1, P^{31}, etc.) which have an intrinsic magnetic moment μ; it turns them in such a way that some of their magnetic moments are in the direction of the field, others against it. Transition from one position to the other can take place with the expenditure or liberation of energy of $2\mu H_0$, which corresponds to the energy of a quantum of electromagnetic radiation of the order of 50-200 MHz. The intensity of the field or the frequency of the radiation generator can be changed, but absorption of the radiation takes place only when the equation $h\nu = 2\mu H_0$ is satisfied. Until now, the effect of NMR in the study of membranes has been virtually confined to hydrogen atoms (proton magnetic resonance, PMR). When the more general term of NMR is employed, it is PMR which is really implied.

Each group of atoms (CH_2, $-CH_3$, $HC=CH$) has a characteristic maximum of absorption on the NMR curves, the appearance of which depends on the effect of surrounding chemical groups. For this reason, the study of NMR spectra of membrane components — both separately and in the intact membrane — provides an approach to the understanding of their interaction.

When a molecule or group of atoms can move about sufficiently freely, i.e., when the effect of the external environment is small, absorption maxima are sharply pointed (resolved spectrum). If, however, the substance is a component of large and solid structures in which dipole–dipole interaction is considerable and differs for corresponding protons in different parts of the structure, the maxima are flattened and, just as in the case of intact membranes, are converted into inconspicuous eminences (Chapman, 1966). The width of the lines on the spectra of solid and viscous substances is 10^3-10^5 times greater than the width of the absorption lines of the same substances in solution (Chapman et al., 1968).

Definitive NMR spectra of biological membranes can be obtained only after their irradiation with ultrasound, accompanied by fragmentation into

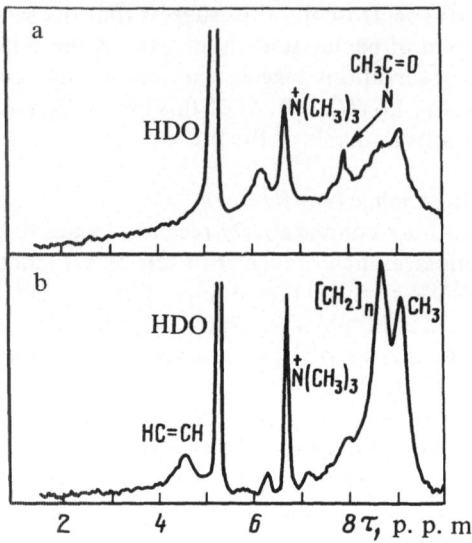

Fig. 14. NMR spectrum of erythrocyte membranes
(a), and of membrane lipids (b), dispersed by
sonication in D_2O. Concentration of material
5%. Records obtained on instrument with storage
element (64 cycles). Chemical shift (τ) expressed
in parts per million relative to tetramethylsilane
(10 p.p.m.) (Chapman et al., 1968).

particles with a diameter of several hundred Angstrom units (Chapman, 1969).
The authors of this method claim that particles of this side are reoriented
quickly enough, so that the negative influence of the magnetic nonhomo-
geneity is reduced to zero, given adequate mobility of individual segments
of the molecules. Other investigators hold different opinions (Sheard, 1969;
Glaser et al., 1970), for they assume that considerable structural changes
take place in a biological membrane during sonication. This does not mean
that the results obtained with ultrasonic fragments are worthless, for such
fragments still retain many of the properties of whole membranes.

Chapman and co-workers (Chapman et al., 1968; Chapman and Kamat,
1968; Jenkinson et al., 1969) showed that certain groups in preparations of
erythrocyte membranes and myelin (i.e., the $-N(CH_3)_3$ group of lecithin
and the cyclic protons of carbohydrates) possess some freedom of movement (Fig. 14). Signals are very weak or completely undetectable from the
protons of HC=CH groups; $-CH_2-CH_3$ groups give wide signals of low intensity. At the same time, $-CH_2-$ and CH_3-groups are well marked in an
emulsion of lipids. These workers conclude that freedom of movement of

the hydrocarbon chains of lipids in a membrane is very limited on account of their interaction with the membrane proteins or cholesterol, or with both these components simultaneously. In fact, in emulsions with lecithin, cholesterol retards the movement and impairs resolution of the signals, but on removal of cholesterol from erythrocyte membranes with ether, the resolution of the NMR spectra is not improved. This suggests that lipid–protein interactions are of predominant importance in these phenomena.

By the action of trifluoroacetic acid on the erythrocyte membrane, the orderly organization of the membrane as a whole is disturbed, and maxima appear on the NMR spectra from amino acids and from hydrocarbon chains. Urea liberates only some amino acids without affecting the mobility of the alkyl groups. The effect of sodium deoxycholate on the membrane is interesting. Evidently, this surface-active substance considerably weakens lipid-protein interactions, for it facilitates the detection of signals from hydrocarbon chains. At the same time, as a result of its inclusion in certain complexes, it does not give a signal itself. Recently, high-resolution NMR spectra of unfragmented erythrocyte membranes have been obtained by raising the temperature of the specimen to 80°C (Glaser et al., 1970). These spectra agreed with those predicted theoretically for the protein components of the membrane, from which it was concluded that lipids and proteins possess an independent spatial arrangement, assuming that only about 25% of the lipids are in contact with protein. This interpretation does not agree with the views of Chapman and co-workers, and further experiments are evidently required in order to establish the contribution of each component of the membrane to the PMR spectra.

The NMR spectra of bacterial membranes (Ostrovskii et al., 1972) are characterized by two wide and strong signals with chemical shift values of 0.8-0.9 and 1.2-1.3 p.p.m., corresponding to the resonance of protons of terminal methyl groups and of $-CH_2$ groups in the hydrocarbon chain. The considerable increase in intensity of both signals when the lipid content of membranes is increased (as a result of the action of the antibiotic chloramphenicol, which leads to disproportion in membrane synthesis) suggests that the spectrum recorded is due mainly to resonance of lipid protons, although some contribution from the protons of membrane proteins cannot be ruled out. Other signals on the NMR spectrum of the membranes of *M. lysodeikticus,* by analogy with data in the literature (Chapman and Morrison, 1966; Chapman and Wallach, 1968), can be allocated as follows: 1.95 p.p.m. $-CH_2-C=C$; 2.4 p.p.m. $-CH_2-COO$; protons at C atoms of carbohydrates, and also protons situated close to phosphate groups give signals in the region of 3.5-3.9 p.p.m.

The other case in which the NMR method has been used to study bacterial membranes is in connection with the study of dissociation of the membranes of *Halobacterium halobium* (Chapman and Salsbury, 1970).

Spectra of the membranes of *H. halobium* in 4 *M* NaCl, made up in D_2O with indistinct and weak absorption bands between 8 and 9 p.p.m., were sharply altered by dilution of the suspension with heavy water to a salt concentration below 0.8 *M*. Under those conditions the membranes dissociated into liproprotein complexes (A. Brown, 1965), of which 80% consisted of protein. Their well resolved spectra are evidence of a marked increase in freedom of movement for some protons in the amino acids.

Some interesting work has been started in Mexico (Cerbon, 1970). This worker used whole cells of *Mycobacterium smegmatis* as the test object and assessed the state of the protons of the hydrocarbon moieties of organic molecules in relation to pH and temperature. Resolution of the NMR spectra was low, but Cerbon found an interesting feature: the character of the NMR signal from the methyl and methylene groups of the lipids (1.3 p.p.m.) was significantly altered, indicating an increase in the mobility of these groups, by addition of alkali to the medium, to pH 8.0. This demonstrates that the state of the lipids of the cell wall and membrane of bacteria depends on the composition of the medium in which they live.

The pattern of proton magnetic resonance in the organic substances of the membranes of erythrocytes, myelin, and certain bacteria thus serve as a pointer to the orderly structure of lipids in bacterial membranes in general, and this orderliness is evidently due to interaction between the lipids and proteins of the membrane.

Effect of Temperature on the State of Membranes. Differential Thermal Analysis

The temperature of the medium in which bacteria live is one of the most powerful natural factors regulating their development. Different species of bacteria respond differently to the temperature conditions of the medium. In nature there are some thermophilic species which live in hot springs at temperatures of 85-89°C (Bauman and Simmonds, 1969), mesophilic species living at 20-40°C, and psychrophilic species which prefer temperatures of between 0 and 20°C. Bacteria which are most fastidious as regards their temperature conditions are those living as parasites, saprophytes, or even symbionts of warm-blooded organisms.

Fluctuations in environmental temperature cause not only a general change in the rate of the metabolic reactions, but also marked functional and structural changes in the cells of all poikilothermic organisms and, in particular, in bacteria (Shaw, 1968; Farrel and Campbell, 1969). For instance, reducing the temperature sharply under experimental conditions from 30°C to 10-15°C, a procedure known as cold shock, leads to conversions. As a result of these conversions on return to the previous conditions, the culture shows a tendency toward synchronous growth (Fitz-James, 1968).

If the temperature falls below 25°C, a new enzyme, desaturase, which catalyzes the formation of unsaturated (double) bonds in the fatty acids of lipids, appears in bacterial cells of the *Bacillus* group (Fulco, 1969, 1970). The composition of the fatty acids is greatly altered in the cells of *B. stearothermophilus* by elevation of the temperature (Yao et al., 1970). At 45°C, for instance, branched-chain C_{15} fatty acids account for 65.5% they account for only 32.6%, as the result of an increase in the proportion of acids with a longer chain. Comparison of thermophilic and mesophilic strains of the genus *Bacillus* shows that the mean length of the fatty acid molecule in thermophiles is 16, and in mesophiles is 15 carbon atoms (Shen et al., 1970). The content of palmitic acid in *E. coli* cells falls from 31.7 to 25.4% if the temperature of the medium is lowered from 35 to 20°C, while the content of octadecenic acid rises from 24.6 to 34.2% (Marr and Ingraham, 1962). The content of saturated fatty acids in the phospholipids of *E. coli* rises if the temperature of growth is increased (Haest et al., 1969; Sinesky, 1971; Cronan and Vagelos, 1972). The possibility cannot be ruled out that the change in composition of the fatty acids which accompanies the change from growth at 20°C to growth at 40°C has some relationship to the increased formation of mesosomes in *E. coli* cells at 40°C (Weigand et al., 1970).

With an increase in the temperature of growth, the content of unsaturated acids decreases in *Mycobacterium butyricum;* branched-chain fatty acids are replaced by normal fatty acids in *Bacillus* sp. (Okuyama et al., 1967; Daron, 1970). In psychrophilic species of bacilli, the lipids consist mainly of unsaturated and branched-chain fatty acids (Kaneda, 1971).

With a comparatively small change in temperature from 37 to 20°C during growth of *B. megaterium,* the sensitivity of the cells to the action of megacin A is reduced (Holland, 1969b). Mutants of *E. coli* have been obtained which are completely insensitive to colicins at a certain temperature (Nomura and Witten, 1967).

Judging by the fact that the composition of fatty acids in membrane lipids depends on the temperature of the medium, presumably a certain degree of orderliness of the lipids is essential for normal function of the membrane. This conclusion, however, is not general in character, for correlation between the degree of saturation of the fatty acids and the environmental temperature has been investigated only for a few bacteria. The absence of such correlation is also known: the polar lipids of cultures of *Mycoplasma laidlawii* A, grown at 25 and 37°C, did not differ in their fatty acid composition (Rottem and Panos, 1969). In the case of *S. aureus* also, the content of monoene acids is virtually unchanged if the temperature of cultivation is reduced from 37 to 25°C (Joyce et al., 1970). However, a specific structure of the cell lipids can be maintained not only by a change in the composition of the fatty acids, but also by a change in the degree of polarity of the lipids (Okuyama, 1969).

Physiological disturbances produced by a change of temperature are connected one way or another with the state of the bacterial membranes. In fact, in the modern view, the site of action of bacteriocins is considered to be the

cell membrane (Nomura, 1969; Smarda and Taubeneck, 1968). Practically
all the cell lipids are located in the membranes. The enzyme desaturase acts
on membrane lipids and, finally, the synchronization of cell growth may be
connected with the straightening out of two cyclic DNA molecules, fixed
at closely situated points of the bacterial membrane, with the aid of that
membrane (Jacob et al., 1963). Protoplasts of *Streptococcus faecalis* and
Sarcina lutea undergo lysis at 60°C (Ray and Brock, 1971).

This coincidence is evidently not fortuitous; it can therefore be supposed
that the action of heat on the bacterial cell is primarily an action on its membrane. This is also shown by the results obtained on model lipid systems
(Luzzati and Husson, 1962) and on membranes of myelin and erythrocytes
(Ladbrooke et al., 1968). The results obtained by differential thermal analysis
(DTA) are particularly interesting (Ladbrooke et al., 1968; Steim et al., 1969),
and these will be discussed in more detail below.

Luzzati and Husson (1962), using an x-ray diffraction method, observed
temperature-dependent phase transitions in various water–lipid mixtures. These
workers postulated that such transitions also take place in natural membranes,
and are one of several methods by which membranes regulate cell metabolism.

Similar structural changes have also been observed during changes in conditions of negative staining of erythrocyte membranes in the course of their
preparation for electron microscopy (Benedetti and Emmelot, 1965). When
stained at 2°C the membranes appear as a random mixture of globules, but at
37°C the globules assume the appearance of hexagonally packed structures.
It is suggested that these structural changes take place also in bacterial membranes, for patterns very similar to the photographs taken of membranes of
higher organisms have been observed on the inner side of the membranes
of *Micrococcus* sp. (cited by Kellenberger and Ryter, 1964). At temperatures of about 70°C, Erokhin and Sinegub (1970a,b) found rearrangment of
the pigments of photosynthesis in the membranes of purple bacteria.

The above experiments with pure lipids (Luzzati, 1962) suggest that
lipids in fact play the dominant role in the structural organization of membranes. However, in more complex systems such as the blood lipoproteins,
an experimental rise of temperature leads to "melting" of the protein α-helix
and to an increase in the relative number of peptide chains in the β-form
(Kobozev and Troitskii, 1967; Dearborn and Wetlaufer, 1969). Dearborn
and Wetlaufer also noted that the higher the lipid content in a lipoprotein
(this may vary from 76 to 85.5%), the higher the content of the β-form in
proteins. This observations, however, could be interpreted somewhat differently by stating that the higher the content of the β-form of a protein, the
more lipid it can bind.

Temperature changes in the blood lipoproteins are reversible, but only up
to a certain limit. The permissible maximum to which different lipoproteins
can be heated lies between 40 and 60°C (Dearborn and Wetlaufer, 1969).

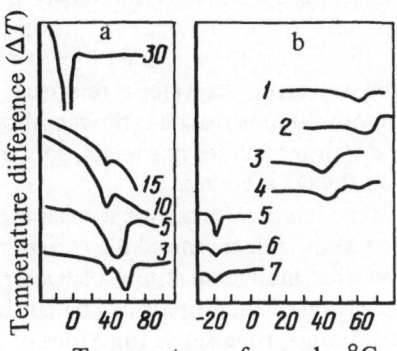

Temperature of sample, °C

Fig. 15. Structural temperature transitions
in biological membranes: a) differential
temperature analysis of erythrocyte membranes using specimens of different water
content; numbers above curves indicate
water content in specimens in % (Ladbrooke et al., 1968). b) Differential thermal analysis of preparations from *Micrococcus laidlawii:* lipids (1) and membranes
(2) from cells grown on medium with
stearate; lipids (3) and membranes (4) from
cells grown in medium without additives;
lipids (5) and membranes (6) from cells (7)
grown on medium with oleate. Specimens
1-4 were investigated as aqueous suspensions; specimens 5-7, as suspensions in 50%
ethylene glycol and 0.15 M NaCl (Steim
et al., 1969).

This phenomenon is probably attributable to denaturation of the protein,
and it indicates the important role of the protein in the conversions studied.

A convenient tool for studying structural transitions in biological membranes is differential thermal analysis (DTA), the possibilities of which were
demonstrated brilliantly by investigations in the laboratories of Chapman
(1966) and Steim (Steim et al., 1969). The principle of the method is very
simple. A small quantity of the substance (for example, 20 mg of the dry
membranes) is placed in the ampule of the analyzer together with a thermocouple, which is connected differentially to another thermocouple in contact
with a standard thermostable specimen. When the temperature of a certain
phase change is reached, the instrument records a "peak" on the curve of
$\Delta T/t$ (Fig. 15a).

Three phenomena discovered are of exceptional interest. First, the existence of phase changes (evidently lipid) in the biological membrane; second,
the fact that the character of these changes is influenced by slight changes in

the technique of preparing the specimen (lyophilization); and, finally, the role of water in this process. In an aqueous suspension of myelin sheaths the only phase change observed is the melting of ice, and no changes in the lipids or proteins can be detected. As water is removed, when a certain level of moisture content (about 20%) is reached, the membrane water ceases to behave as a phase, i.e., it neither freezes nor melts, so that the characteristic lipid extrema appear on the DTA curves.

The action of different temperatures on bacterial membranes has been studied mainly in connection with thermophilia (Loginova et al., 1966). Recently, however, now that individual strains of bacteria with a predetermined fatty acid composition of their lipids can be cultivated, the number of investigations of membrane processes as functions of temperature has increased extraordinarily (see the surveys of Cronan and Vagelos, 1972; Chapman and Dodd, 1971).

The fatty acid composition of lipids of a thermophilic strain of bacteria from the Yellowstone hot springs differs only slightly from that of mesophilic strains (Bauman and Simmonds, 1969). There is only a small increase in the relative proportion of fatty acids with 17 and 19 carbon atoms at the expense of the C_{15} fatty acids. It is accordingly postulated that the membranes of thermophiles are distinguished not by their chemical composition, but rather by their structural organization (Bodman and Welker, 1969), although attempts to find a solid basis for this hypothesis have not yet been fully successful.

In the section of EPR we mentioned our experiments with a paramagnetic probe, which demonstrated a stepwise change in microviscosity in the membranes of M. lysodeikticus at a temperature of about 45°C, possibly indicating the existence of a structural transition in this region. The DTA method reveals no such transition, but an endothermic transition is found at about 20°C due to melting of the lipids (Ashe and Steim, 1971). Although a proper comparison of these results must await fresh experimental evidence, it can be postulated even now that several structural transitions, differing in their energy capacity and localization, can take place in the same membrane. A more detailed study of the behavior of different spin probes has led to the discovery of four temperature-dependent structural transitions in the membranes of M. lysodeikticus, whose parameters are largely determined by the nativeness of the membrane protein (Binyukov et al., 1972).

Some very interesting observations have been made by Kaback and Deuel (1969) on vesicles from membranes of E. coli W6, in which these workers studied active proline transport. During a change in temperature from 27 to 46°C, the permeability of the vesicles increases sharply. This phenomenon agrees with a recent communication by Lucy (1969), who states that one of the three mechanisms of the increase in permeability of artificial lipid membranes under the influence of radiation is local overheating, leading to a local structural reorganization of the membrane.

Using methods of depolarization of the fluorescence of dansylated derivatives of phosphatidylethanolamine, as well as x-ray diffraction, Shechter et al. (1972) recorded three points in membrane vesicles from *E. coli* at which a change takes place in the rate of transport of certain substances through the membrane. The first point (20°C) is ascribed to melting of lipids; the second point (between 30 and 46°C) is linked with an increase in the intensity of hydrophobic interactions and a sharp decrease in thickness of the membrane; the nature of the third point (45-50°C) is not yet clear.

The study of the activity of some bacterial enzymes such as glycoside permease (Fox et al., 1970; Wilson et al., 1970) or respiratory enzymes (Overath et al., 1970), as a function of temperature, has shed some light on the character of interaction between proteins and lipids and their relative arrangement in the membrane. A sharp inflection is observed on the curve of activity of these enzymes. It indicates structural transitions in the bacterial membranes and their importance in the regulation of metabolism. It is interesting to note that inflections on the curve of activity of different membrane enzymes (logarithm of activity versus T^{-1}) do not coincide in the same membrane preparation (Esfahami et al., 1971). Although it was shown even in reconstructed membranes that the temperature range of the inflection is determined by the composition of the fatty acids (Esfahami and Wakil, 1972), these results indicate the important role of protein in these phenomena and show that the distribution of lipids may vary throughout the membrane. The conclusion that the distribution of lipids in the membrane of *E. coli* is nonrandom in character is also reached by Mavis and Vagelos (1972), who showed that the character of the temperature dependence of activity of three membrane enzymes (glycerol-3-phosphate acyltransferase, 1-acylglycerol-3-phosphate acyltransferase, and glycerol-3-phosphate dehydrogenase) differs significantly.

X-ray diffraction studies have shown that a marked structural transition occurs at 35-42°C in wet mycoplasma membranes, both isolated and in intact cells (Engelman, 1970, 1971; Wilkins et al., 1971; Metcalfe et al., 1971a). By changing the composition of the nutrient medium, it is possible to secure an excess of unsaturated or saturated fatty acids in the membrane lipids of *M. laidlawii*, but the range of the temperature transition does not change, further confirming the important role of proteins in this process. The test organism grows badly at 35°C, but the mechanism linking this phenomenon with the structural transition is not quite clear. According to Melchior et al. (1970), the phase transition (20-40°C) recorded by the DTA method in membranes of *M. laidlawii* is due to structural conversions of the lipids, while the absorption of heat between 60 and 80°C is due to denaturation of the protein.

Work of exceptional interest in this connection has been done by Steim et al. (1969), some of whose thermograms are reproduced here (Fig. 15b). These

workers showed that at about 40°C a phase transition of the lipids takes place
in the membranes of *M. laidlawii*, but the temperature range of this transition
varies greatly with the nature of the fatty acids added to the culture fluid.
For instance, the region of phase transition in the membranes can be shifted
from −15 to +60°C by the addition of oleic or stearic acid, respectively, to
the growth medium. For normal development of the microorganism, the
melting point of the lipids in its membranes must be below the temperature
of the medium, i.e., the membrane lipids must be in a "liquefied" state.

The structural transition in the membranes absorbs 25% less energy than
the analogous transition in the corresponding quantity of membrane lipids.
From this, Steim and co-workers conclude that the lipids and proteins of
membranes are virtually not linked at all by hydrophobic bonds, for the
existence of hydrophobic interactions, in their opinion, ought to distort the
picture of the structural transitions in the membrane considerably. It is
interesting that only one phase transition of lipids has been found; the second minimum on the curve is due to denaturation of proteins. This means
that the lipids of membranes are arranged absolutely uniformly.

In summary, it can be stated that bacterial membranes, and, indeed, the
membranes of other organisms may exist in several structural forms which
are brought to light by the study of various parameters as functions of temperature. In the membrane *in vivo,* the various forms evidently coexist and
the components of the membrane are in an unstable state, which facilitates
the functional interconnection between membrane and cytoplasm of the
cell. Phase transitions of lipids are evidently the basis for the transition from
one form to the other (Luzzati, 1962; Reiss-Husson and Luzzati, 1967).
However, the protein and water of the membrane have a very considerable
influence on the character of these transitions,and, indeed, on their possibility (Davis and Inesi, 1971).

It is interesting to note that breakdown of the bacterial membrane by an
electric current takes place when the voltage on the membrane is about 1 V.
This figure is calculated from the sensitivity of protoplasts and spheroplasts
of the bacteria *M. lysodeikticus, B. subtilis, B. megaterium, Sarcina lutea,*
and *E. coli;* pseudomonads; and also erythrocytes to a high-voltage pulsed
current (15-22 kV/cm) (Sale and Hamilton, 1968). Injury to artificial phospholipids, certain plant membranes, and the membranes of nerve cells arises
at a voltage of 0.2 V. This is a figure of the same order, and it suggests that
all biological membranes are constructed in accordance with the same principle. However, the somewhat higher breakdown voltage suggests that certain factors, perhaps the orderliness of the lipids or the character of their
interaction with proteins in bacterial membranes, increase the dielectric
constant of the membranes, a matter of the greatest importance to the understanding of certain membrane functions (see the section on oxidative phosphorylation).

State of the Water

Water is not the only medium in which biological membranes can function, but it is also a component of the membrane which influences many processes and determines its structure (Askogenskaya and Petinov, 1972). Unfortunately, methods are not yet available even for estimating the quantity of water incorporated into membranes. The study of the role of water can be approached only by certain physical experiments accompanied by drying of the membranes (Luzzati and Husson, 1962; Husson and Luzzati, 1963).

X-ray structural analysis, confirmed by DTA (Finean, 1969; Chapman, 1969), has shown that water, in an amount of about 20% of the weight of the membrane, is in a special state in which it does not freeze as the temperature falls. This is known as bound water. Unbound water gives characteristic spots at 3.2-3.4 Å on high-angle x-ray diffraction, so that the moment of disappearance of ordinary water from a specimen of membranes can be detected, and correlation established between the water content and the substantial changes in the character of x-ray diffraction which take place when the moisture content of the material reaches a certain level.

It is considered that when the critical water concentration is reached the lipids of membranes start to undergo conversion into an independent phase, whose state can be studied by various physical methods. Total removal of water from bacterial membranes leads in some cases to a significant change in the character of the protein-protein and protein-lipid interactions. For instance, up to 70% of the firmly bound enzyme $NADH_2$ dehydrogenase dissociates from freeze-dried membranes of *M. laidlawii* simply on washing with buffer solutions, whereas this enzyme cannot be extracted from the original moist membranes even by treatment with digitonin and phospholipase A (Gutman et al., 1968). During fixation of membrane preparation and their passage through solvents, their thickness is reduced by 30% on account of dehydration (Elfvin, 1961, 1963). The hypothesis has been put forward that the action of general anesthetics, hormones, and other membranophilic agents on the cell membranes of animals is affected by their action on the structure of the membrane-bound water (Chapman, 1969).

Whatever the case, it has been precisely established that the presence of water in artificial phospholipid systems has a considerable effect on their structural properties, marked by a significant decrease in the mobility of hydrophobic spin probes (Hsia et al., 1970). It can therefore be expected that a disturbance of the structure of water, the state of which is described as "nonliquid" (Grigera and Cereijido, 1971) or as "special" (Samuilov et al., 1971), recorded by the nuclear spin echo method, can in fact lead to important morphological and functional changes in the membrane. It is assumed that the bound water can line the pores in biological membranes in a layer about 22 Å in thickness, consisting of clusters each combining 162-191 molecules (Schulz and Asunman, 1970).

After studying water-lipid mixtures, Phillips et al. (1970) showed that in
dipalmitoyl lecithin micelles there are 10 molecules of bound water per
molecule of lipid, although under the same conditions there are 12 molecules
of water to 1 molecule of dioleyl lecithin. In this case, the water is undoubtedly located close to the polar groups of the phospholipids, but the localization
of the bound water in natural membranes is not yet known.

Work is presently being conducted on the cultivation of bacteria with the
total replacement of hydrogen by deuterium (Crespi and Ketz, 1969; Den'ko,
1970). The membranes of these bacteria are of exceptional interest as material
for the study of many physical properties. At the same time, the replacement
of ordinary water by heavy water is a promising tool for the investigation of
the role of water in the membrane.

Paramagnetic and Fluorescence Labeling

The method of studying the conformation of the protein molecule by
means of substances which form a bond with the reacting groups of the protein
but leave their physicochemical properties unaffected, has long been known
in chemistry. The manifestation of these properties (paramagnetism, fluorescence) depends on the polarity of the areas of the protein molecule surrounding
the label, and it is thus an indicator of the state of the protein molecule (Likhtenshtein et al., 1968, 1970).

A less specific label, but one specially developed for the investigation of
membranes, is that known as the hydrophobic — a fluorescent substance
or free radical with marked lipophilia, located in the hydrophobic areas of
the membrane. Such a probe can be used, in principle, to examine conformation changes of a protein and also the state of the lipid components of
the membrane, although at present it is very difficult to determine the precise
localization of the probe.

Long-life, free radicals were first synthesized in the USSR by a group of
workers headed by E. G. Rozantsev (1970). These compounds were quickly
used as paramagnetic labels for the study of hemoglobin (Likhtenshtein et
al., 1969), and they have been used quite recently as hydrophobic probes
for the investigation of mitochondrial membranes (Gendel' et al., 1968; Keith
et al., 1968; Koltover et al., 1968). The structural formulas of some paramagnetic probes used in the study of the molecular organization of biological
membranes are given on the following page.

The unpaired electron in these free radicals is recorded by the method of
electron paramagnetic resonance, similar in principle to the NMR method.
The specimen is also placed in a powerful magnetic field, where definite orientation of the spins of the unpaired electrons takes place, and is irradiated
with ultrashort radio waves with a wavelength usually of about 3 cm. The
characteristic absorption of ultrashort-wave energy occurs at the moment of

a) 2,2,6,6-Tetramethylpiperidine-1-oxyl (Hubbel and McConnell, 1968);
b) 2,2,6,6-Tetramethyl-4-capriloyloxylpiperidine (Gendel' et al., 1968);
c) 1,1,3,3,7-Pentamethyl-7,8-benz-1,2,3,4-tetrahydropyrrolo-3,4–β-indole-2-oxyl (Binyukov, 1972);
d) n-Oxyl-4',4'-dimethyloxazolidine-5α-androstan-3-on-17β-ol (Hubbel and McConnell, 1969);
e) Ethylenedi-(1-hydroxy-2,2,5,5-tetramethyl-3-pyrrolidine-3 carboxamide) (Calvin et al., 1969).

resonance, i.e., when

$$hv = g\beta H,$$

where ν represents the frequency of the superhigh-frequency radiation; H, the intensity of the magnetic field; g, the spectroscopic splitting factor; h, Planck's constant; and β, Bohr's magneton (see the surveys by Kalmanson, 1963; Rosantsev, 1970).

The EPR spectrum of one of the paramagnetic probes in ethanol is shown in Fig. 16a.

The character of the spectra depends on the freedom of movement of the probe, i.e., on the isotropy of the medium and on its local viscosity (microviscosity). Microviscosity is estimated experimentally from the time of reorientation of the probe molecules (τ_c) or from the expression $\alpha = \sqrt{I_{+1}/I_{-1}}$, which is proportional to τ_c under certain conditions. (In this case I_{+1} and I_{-1} represent the amplitudes of the first and third components of the spectrum.)

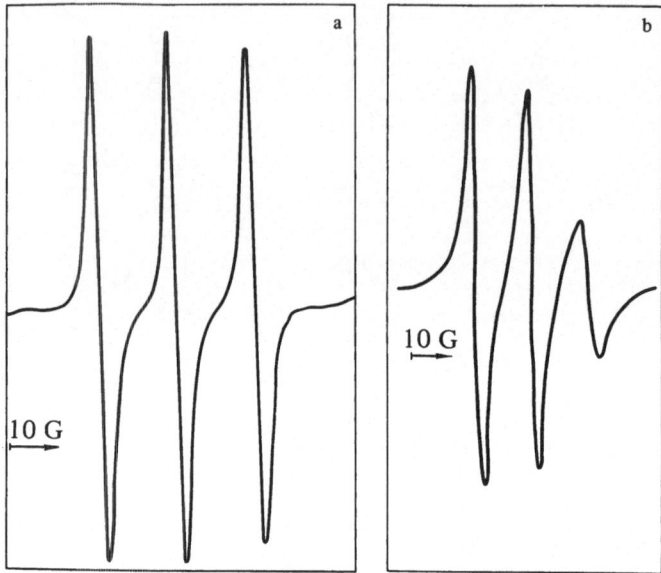

Fig. 16. EPR spectra of a probe radical in ethanol (a) and in membranes of *Micrococcus lysodeikticus* (b).

Jointly with M. G. Gol'dfel'd and E. G. Rosantsev, the writers have attempted by means of the hydrophobic probe to estimate the state of the lipids and proteins in the bacterial membranes of *M. lysodeikticus*. A solution of the probe in ethanol was added to a suspension of membranes or of membrane lipids and proteins separately, and the EPR spectra were photographed as a function of temperature. The experimental results showed that the value of α for the probe in the membrane at temperatures of about 45°C changes stepwise. The closeness of the absolute values of α for a suspension of lipids and membranes indicates that the probe radical in the membrane (Fig. 16b) is evidently in the neighborhood of lipids, but the differences in the character of the $\alpha/t°$ curves are evidence of the fundamental role of protein in maintenance of the internal structure of the membrane subunits. Presumably at the temperature of the change in direction of the curve, conformational changes of the protein molecules or phase transitions of the lipids take place and induce a structural transition in the lipoprotein complexes. Although the EPR spectra of several spin probes are similar in lipids and in the membrane of *M. laidlawii*, and the character of the spectrum is largely determined by the composition of the fatty acids, the presence of protein influences the physical state of the lipids in the membrane (Rottem et al., 1970; Tourtellotte et al., 1970).

The hypothesis of the role of structural transformations of the protein in the molecular organization of membranes rests on a solid basis. As long ago as 1963, P. L. Privalov and D. R. Monaselidze found reversible structural transitions in globular proteins preceding their denaturation, by means of the method of differential thermal analysis. Something similar can be recorded by means of spectropolarimetry (ORD) (Zav'yalov et al., 1970), while recently Binyukov et al. (1971b) found a conformation change in the proteins of bacterial membranes by the use of a spin probe, which has the property of adsorption on protein molecules. The essential role of intact membrane proteins in the structural reorganization of lipids has also been demonstrated by the NMR method for membranes of the sarcoplasmic reticulum (Davi and Inesi, 1971) and by the IRS method for membranes of erythrocytes (Graham and Wallach, 1971).

It is interesting to note that the maximum activity of a membrane enzyme (malic acid dehydrogenase) coincides with the region of inflection of the α/t^{o} curve. Presumably the structural transition observed is the region of greatest physiological activity of the membrane, and at the same time it serves as a means of regulation of the biochemical processes. Very probably, in the living organism, the region of transition is observed at lower temperatures and the transition itself is induced not so much by temperature as by the action of biochemical mediators. This evidently depends on the redox potential of the medium, the ionic strength of the cytoplasm, etc. Temperature-dependent structural changes have been found by the spin-probe method in membranes of *Mycoplasma* (Tourtellotte et al., 1970) and also in membranes of the endoplasmic reticulum (Annaev et al., 1971), while conformational changes depending on the presence of ATP, and thus linked with the state of the membrane system for the transformation of energy, have been found in mitochondria (Kol'tover et al., 1971).

Investigators nowadays have great hopes for the use of biradicals in the study of membranes (Calvin et al., 1969). As a result of exchange interaction of unpaired electrons, the EPR spectra of these compounds become extremely sensitive to the slightest changes in structure of the protein molecule on which they are adsorbed. The use of radicals on the basis of large non-flexible molecules, such as molecules of sterols, is also extremely promising. Some paramagnetic derivatives of the sterols are oriented in a membrane along the hydrocarbon chains of the lipids, and they perform only rotatory movements around the axis of orientation (Hubbel and McConnell, 1969). This feature makes them very sensitive to structural changes in the membrane.

In systems simulating the structure of biological membranes, i.e., in systems of purely lipid micelles with various proteins, it has been shown by spin probe and spin labeling (the paramagnetic molecule linked by a covalent bond with the protein molecule) methods that the combination of protein with lipid, as well as variations in the cationic composition of the medium cause marked changes in the mobility of the various areas of both protein and lipid molecules (Butler et al., 1970; Berger et al., 1970; Drott et al., 1970).

Fluorescent substances were used initially as the tool for localizing the barrier of permeability in erythrocytes and bacterial membranes. Since these dyes do not penetrate inside cells, they can be used to selectively label the membrane proteins on the outer side of the membrane. Maddy (1967) attempted to solve this extremely important problem for the membranes of erythrocytes. Having treated whole cells with a dye (4-acetamido-4'-iso-thiocyanostilbene-2,2'-disulfonic acid) and then having fractionated the proteins, Maddy found that the label was present in virtually all the protein fractions. This means that the proteins of erythrocyte membranes are constantly or periodically in contact with the external medium, assuming that the methods of fractionation of the membrane proteins were sufficiently effective.

New prospects were opened with the use of fluorescent dyes (Gulick-Krzywicki et al., 1970; Romeo et al., 1970; Wallach et al., 1970) whose fluorescence depends on the polarity of the surroundings (Azzi et al., 1969b). For example, it was found that 8-aniline-1-naphthalenesulfonic acid (ANS) has its fluorescence increased by 25 times by adsorption on mitochondrial membranes, and by 35 times if ATP or respiration substrate and O_2 are added to these membranes (Azzi et al., 1969a). The intensity of fluorescence of ANS in the membranes of *Mycobacterium phlei* changes if the membranes are cooled ($-75°C$) and heated ($+50°C$) (Kalra et al., 1972). These observations are definite evidence of the considerable conformational changes in the membrane and of the good prospects for the use of fluorescent dyes as indicators of these conformational changes. Fluorescent labels even have certain advantages over paramagnetic labels, for fluorescence can be recorded continuously, while EPR spectra are recorded periodically.

The study of the intrinsic luminescence of proteins (Vladimirov, 1965) in membranes and also of the spectral properties of natural pigment-protein complexes (Erokhin and Sinegub, 1970a,b) is, of course, of particular interest.

Relative Arrangement of the Components of the Membrane

The relative arrangement of the components of the membrane can be deduced from the results of chemical analyses and direct observations. The most effective methods at the present time are: determination of the sensitivity of the membrane to the action of lytic enzymes; determination of the accessibility of the reacting groups of the protein and lipid to titration by various chemical reagents; the study of the character of immunological reactions; observations in the electron microscope and x-ray structural analysis.

Electron Microscopy and X-Ray Diffraction

Because of the clarity of the pictures obtained by electron microscopy, this method cannot be used despite the great difficulties which may arise in correct interpretation of the photomicrographs. These difficulties have been fully examined elsewhere (Benedetti and Emmelot, 1968; Chapman and Wallach, 1968; Lenard and Singer, 1968, Sjöstrand and Barajas, 1968; Stoeckenius and Engelman, 1969).

Investigations by the method of positive staining using OsO_4 or $KMnO_4$ as fixing agents have shown that bacterial membranes consist of an electron-transparent layer about 30 Å in thickness and two electron-dense layers, each 20-30 Å in thickness, one on each side. The central part of the membrane is considered to be filled with lipids, forming two molecular layers with their polar groups facing outward. The peripheral parts of the membrane (the electron-dense layers) are regarded as layers of protein.

To obtain an image of a bacterial membrane in the electron microscope, vigorous treatments are required: treatment with a powerful oxidizing agent (OsO_4), treatment with an organic solvent (ethanol), and treatment with a beam of electrons *in vacuo*. These procedures could produce substantial changes in the structural organization of the membrane, and the supposed interposition of a double layer of lipid between the two layers of protein must be regarded as hypothetical.

Interesting observations have been made by negative staining of bacterial membranes with phosphotungstic acid (Fig. 5).

In many fragments of the membrane, the mushroom-like outgrowths so characteristic of the internal membranes of mitochondria are found. This structural similarity is evidence that regions performing the functions of mitochondria also exist in the system of bacterial membranes, and the occurrence of regular outgrowths from the membrane suggests that the membrane itself also consists of regularly arranged subunits (Benedetti and Emmelot, 1968).

If membranes of *M. lysodeikticus* are treated with proteolytic enzymes, the mushroom-like outgrowths disappear simultaneously with the ability of the organism to perform oxidative phosphorylation (Fig. 5). Since under these circumstances the work of the respiratory chain of enzymes is almost undisturbed, the mushroom-like outgrowths can be regarded as equivalent to the factors coupling oxidation with the synthesis of high-energy compounds. It can also be regarded that the respiratory enzymes are located inside the membrane itself (Gel'man et al., 1968).

The negative staining method is not without its disadvantages. Some specialists consider that omitting the stage of fixation of the membrane may lead to even more serious artifacts during subsequent drying of the material than if the material had been fixed at once (Borovyagin, 1969; Finean, 1969). However, even if the appearance of mushroom-like outgrowths is an artifact, the asymmetry of the membranes as revealed by negative staining would appear to be perfectly clearly established.

It is interesting to note in this connection that proteolysis disturbs oxidative phosphorylation if the proteases act on the inner side of the membrane, whereas treatment of protoplasts (i.e., the outer side of the membrane) with proteases does not affect the performance of oxidative phosphorylation (Ostrovskii et al., 1963). The enzymic activity of the ATPase of mycoplasma membranes is also exhibited only if the substrate (ATP) is brought up to the inner side of the membrane (Razin, 1969).

New prospects for the study of the molecular organization of membranes were presented by the introduction of the freeze-etching method (Moor et al., 1961; Moor, 1969). The basic operations will be recalled; the section of frozen material is rapidly dried *in vacuo*, when the curved surfaces of the various organelles are exposed; a metal is then sprinkled on the section, followed by carbon, after which the material is removed and the carbon replica from it is examined in the electron microscope.

By this method, regular conical outgrowths corresponding to indentations on the surface of the cell wall have been found on the outer surface of bacterial (*Steptococcus lactis*) membranes. These observations (Hurst and Stubbs, 1969) are reminiscent of the complex union between the membrane and wall of the bacterial cell, which has hitherto been almost completely disregarded. A network of granules measuring 20-60 Å, which was transformed into a paracrystalline state if the medium was deficient in Mg^{++}, was found on the outer surface of membranes of *E. coli* by the freeze-etching method. Randomly arranged granules are also present on the inner surface (Fiil and Branton, 1969; Nanninga, 1970a).

Specialists are by no means unanimous about the results of freeze-etching of membranes. For instance, Moor (1969) considers that in preparation of the section, one of the outer sides of the membranes is exposed. Branton (1966, 1969) considers that the membrane splits in the midline (at the double lipid layer), so that the membrane is seen from within, although the sublimation of water from the perimembranous space also allows the outer surfaces of the membrane to be seen. Evidence of the arrangement of the granules within the membrane, but not on its surface, has also been obtained by freeze-etching experiments with *Bacillus sphaericus* (Sleytr, 1970).

The method based on x-ray diffraction occupies a special place in the structural analysis of biopolymers. This unique method has provided a complete picture of the structure of some proteins (myoglobin, lysozyme, etc.). X-ray structural analysis has been used successfully to determine the molecular organization of biological membranes (Chapman and Wallach, 1968; Finean, 1969). Finean concludes from his study of roentgenograms that the membranes of animal cells contain a layer of lipid 20-30 Å in thickness. The state of this lipid is regarded as disorganized with limited orderliness. Meanwhile, x-ray diffraction studies of membranes of chloroplasts and of the rods of the retina have revealed granularity in the plane of the membrane, which has not been observed in other objects.

The widespread use of x-ray structural analysis is limited by the need for a certain orientation of the test object, but these difficulties will undoubtedly soon be overcome (Lucy, 1969). This method has already made an important contribution to our understanding of the structure of the complex membrane of Gram-negative bacteria (Burge and Draper, 1967). Attempts are being made to analyze the molecular organization of bacterial membranes *in vivo* (Grossbard and Preston, 1957; Burge and Draper, 1967), but it is still too early to draw any definite conclusions as regards these structures.

Roentgenograms of dry powders of the cells of *B. subtilis, B. megaterium, Proteus vulgaris,* and *E. coli* do not differ significantly from one another. The characteristic circular reflections, corresponding to periods of 3.8-3.9 Å, and 4.2 Å, resemble meridian arcs if thin dry bacterial films are studied and the beam strikes the surface of the film at a low angle. These observations suggest that the bacterial membrane contains a layer of lipids with the hydrocarbon chains arranged perpendicularly to the cell surface (Grossbard and Preston, 1957).

Similar conclusions can be drawn from analysis of the roentgenograms of aqueous suspensions of the membranes of *M. laidlawii* (Wilkins et al., 1971). The character of diffraction depends on the composition and phase state of the membrane lipids. The protein evidently does not form a homogeneous layer on the surface of the lipid, but may penetrate into its depth (Engelman, 1971).

Action of Enzymes on Membranes, Acylation, and Immunological Reactions

A traditional approach to the study of the localization of the components of bacterial membranes is to treat them with lytic enzymes (lipase, phospholipases, proteases). Because of the specificity of action of the enzymes under mild conditions of treatment, this approach has proved very productive.

For instance, it has been shown that treatment with lipolytic enzymes inactivates certain membrane enzymes, showing that the ester bonds of the lipids are accessible to the action of external agents. In other words, it shows that the polar groups of the lipids are located on the surface. For example, pancreatic lipase, acting only on the outer side of the protoplast membrane of *M. lysodeikticus* in the presence of 1 M sucrose and 0.005 M $MgSO_4$, inhibits respiration. It has long been known that pancreatic lipase reacts with the membrane of *B. megaterium* in such a way as to alter its permeability very considerably (Spigelman et al., 1958). Phospholipase A from snake venom also disturbs the working of the respiratory chain in a preparation of *M. lysodeikticus* membranes. A decrease in the content of ester groups in the lipids of membranes after treatment with phospholipase and lipase by 10 and 19%, respectively, of the control level has been demonstrated by IR spectroscopy (Oparin et al., 1965b). Lipolysis for 1-3 h did not lead to fragmentation of the membranes, although changes took place in their ultrastructure: perforations appeared, the layers separated, and the mushroomlike outgrowths disappeared (Lukoyanova and Biryuzova, 1965). Destruction of the membrane of *Bacillus cereus* by intracellular phospholipase C in the process of autolysis is accompanied by detachment of a proteolipid (lipid: protein = 1.56:1), as the result of which lipid areas responsible for linking together the lipoprotein components can be found in the membrane (Koga and Kusaka, 1970). Meanwhile it is known that phospholipase C hydrolyzes up to 70% of the phospholipids of the erythrocyte membrane, thus clearly demonstrating that these components of the membrane are superficial in position (Bowman et al., 1971; Coleman et al., 1970).

Proteolytic enzymes have a comparatively weak action on bacterial membranes. By the action of subtilopeptidase A on the membranes of *M. lysodeikticus* not more than 30% of the protein could be digested, the membranes were not fragmented, and only the surface proteins were removed (Lukoyanova et al., 1967; Simakova, 1970). Membranes of the chromatophores of *Chromatium* also were resistant to proteolysis (Bril, 1960). Meanwhile proteolysis of the membrane proteins of the chromatophores of *Rhodospirillum rubrum* is intensified after their solubilization by Triton X-100 (Vernon and Garcia, 1967). Proteolysis of the outer membranes of *E. coli* leads to the breakdown of two of the more than ten proteins. Destruction of the membrane by SDS, followed by its reconstruction, did not alter this selective sensitivity to proteolysis (Bragg and Hou, 1972). Low sensitivity to proteolysis is evidence of protection of the membrane protein by lipid. However, there are exceptions to this rule. A proteolytic enzyme from fungi (pronase) digested almost 80% of the protein on membranes of *M. laidlawii* (Morowitz and Terry, 1969). It has also been reported that proteolysis of a considerable part of the protein did not prevent membranes from acting as osmotic barriers. These results can evidently be regarded as evidence of the inadequate organization of the various membranes.

Membranes and their fragments are known to move toward the anode in an electric field at neutral pH. The effect of surface charges on the state of membranes is so great that in order to stabilize membranes, bivalent cations (Mg^{++}) are added to buffered mixtures, for by interacting with the anionic groups they weaken their repulsion and thus prevent disaggregation of the membranes. The role of the surface charge is seen in a particularly clear form in membranes of the halophilic bacteria, which break up into small fragments merely if the salt (NaCl) concentration is reduced from 4 to 0.8-1 M.

Considerable disaggregation of the membranes of *M. lysodeikticus* takes place on the removal of Mg^{++} ions by means of the chelating agent EDTA. Fragments with a mean Stokes radius (Rs) of about 100 Å are removed from the membranes. These fragments contain $NADH_2$ dehydrogenase, malic acid dehydrogenase, and certain other proteins as well as lipids (Ostrovskii et al., 1969). Since the proteins of the resulting fragments are, on the whole, neutral, presumably the negative charge on their surface is due to ionized phosphate groups of lipids. After treatment of whole cells of *Haemophilus parainfluenzae* with EDTA (Tucker and White, 1970a,c), lipoprotein complexes with a content of cardiolipin and phosphatidylglycerol relatively 2-5 times higher than in the rest of the cell membrane are detached from them.

The stabilizing action of the neutralization of anionic groups on the surface of membranes is so great that the addition of 0.1 M Mg^{++} or acidification of the medium to pH 5 prevents lysis of the protoplasts (or, as the workers concerned called them, "gymnoplasts") in an osmotically unbalanced medium (Van Iterson and Op den Kamp, 1969). In the presence of Mg^{++}, the protoplasts become spherical in shape, while on acidification they remain rod-shaped like the original cells.

An artificial increase in the number of anionic groups on the membrane by chemical addition of acyls of dicarboxylic acids, on the other hand, increases the lability of membranes and may even lead to the solubilization of sparingly soluble membrane proteins (Brown, 1965; McLennan et al., 1965; Smith et al., 1970). The claim is made that teichoic acid in the membranes of Gram-positive bacteria and acid polysaccharides in the membranes of Gram-negative bacteria play the role of accumulators of bivalent ions essential for the normal work of the membrane (Heptinstall et al., 1970).

Although a deficiency of Mg^{++} ions in the cultivation medium causes serious disturbances in bacterial metabolism as well as disturbances in the morphology of the bacterial membranes (Kennell and Kotoulas, 1967; Fiil and Branton, 1969), it is difficult to state whether in fact these ions play such an important role in the maintenance of the structure of the membrane *in vivo*. For instance, spin-probe investigation results show the structure of the lipid areas of the membrane in *M. laidlawii* not to change after removal of Mg^{++} or addition of Mg^{++} to the medium up to a concentration of 0.02 M (Rottem et al., 1970). There is reason to suppose that this function can also be performed by other substances such as polyamines of the spermine and spermidine type (Tabor, 1962; Tabor and Tabor, 1966).

In the attempt to acylate mitochondrial membranes, some interesting observations were made by Lenaz et al. (1969a,b). Intact membranes hardly reacted at all with the acylating agent (succinic anhydride), whereas the defatted proteins of the membranes readily underwent acylation, during which a succinic acid residue was attached to the ϵ-amino groups of lysine and the N-terminal amino groups. The impression is created that the inaccessibility of the reaction groups of the membrane protein is due to their protection by lipid, i.e., to the more superficial position of the lipid in the membrane.

As already mentioned, bacterial membranes are easily succinylated (A. Brown, 1965), although the material treated in this way consisted of membranes of *Halobacterium halobium* with an unusually high protein content (up to 80% of the weight of the membrane), as a result of specific interactions with the cell membrane. In thise case, no protection by lipid was to be expected. In membranes of *M. laidlawii* 50-70% of amino groups, determined by the ninhydrin method, are acylated by succinic anhydride, with resulting breakdown of the membrane into fragments with a sedimentation coefficient of between 0.9 and 523 S (P. Smith et al., 1970).

The use of immunological methods is a very promising approach to the study of molecular anatomy of membranes. It is evident that by the use of antibodies (labeled with ferritin or mercury) against individual membrane proteins, it will be possible to localize each of these proteins in a very short time. However, the immunology of membranes is presently going through the formative period (Salton, 1968; Demus and Mehl, 1970). It has been shown (Argaman and Razin, 1969), for instance, that the fraction of hydrophobic proteins from membranes of *M. laidlawii* exhibit antigenic properties, but the antibodies against these proteins do not react with the original membrane. Leaving aside for the moment any doubts regarding the native state of the proteins after the vigorous procedure of isolation, this fact may signify that the protein occupies a central position in the membrane and is protected by its other components. Kenny and Grayton (1965) further consider that the fixation of complement on the membranes of *Miyagawanella pneumoniae* takes place through interaction with lipids, especially glycolipids; this naturallly assumes their accessibility for reaction with macromolecular substances. It is also interesting to note that antiserum against bacterial membranes reacts with an emulsion of the membrane lipids (Salton, 1968), while antilipid serum also reacts with the original membranes (Razin et al., 1970). However, the lipids of certain species of mycoplasmas (*M. laidlawii*) do not react at all with antimembrane serum, whereas the proteins of these membranes, after removal of lipids by *n*-butanol, show some affinity for the antiserum (Kahane and Razin, 1969b). The important fact is that the immunospecific properties of the membrane proteins were considerably strengthened by their combination with the membrane lipids or even with detergents, thus emphasizing yet again the importance of interaction between the lipids and proteins in the bacterial membrane *in vivo*. Pure preparations of membrane enzymes from *M. lysodeikticus* (ATPase and $NADH_2$ dehydrogenase) showed marked specific binding with antimembrane serum (Fukui et al., 1971).

An original method has recently been suggested for determining the location of cholesterol in membranes and tested on erythrocyte "ghosts" on *M. gallisepticus* and *A. laidlawii* (Pendleton et al., 1972). Cholesterol forms a complex with the bacterial toxin cereolysin, and interaction between this complex and antistreptomycin serum, labeled with ferritin, enables the character of distribution of the cholesterol in the membrane to be determined. The existing data suggest a nonchaotic distribution of cholesterol in the membrane, but in the case of *A. laidlawii* cereolysin did not combine with the cholesterol of the membrane; thus, the results obtained must be interpreted with very great care.

The superficial localization of lipids in the membranes is also shown by experiments in which natural cholesterol in the membranes of erythrocytes was replaced artificially by certain of its derivatives (Bruckdorfer et al., 1969). This replacement was observed on simple mixing of a suspension of membranes with an emulsion of the added lipid. The first condition of such an exchange is evidently the ease with which contact is made between the lipids of the membranes and the lipid micelles outside the membranes, presumably either because of the existence of large "windows" in the surface layer of protein or the superficial position of the lipids themselves.

Something similar takes place with the membranes of mitochondria and microsomes, the lipids of which can be exchanged in the presence of a protein cytoplasmic fraction (Wirtz and Zilversmit, 1969) and also with bacterial membranes (Tsukagoshi and Fox, 1971).

Character of the Linkage between Components of the Membrane

General Remarks

Turning now to discussion of the character of the link between components of the membrane, it must be stated *a priori* that because of the extremely wide variety of chemical groups in bacterial membranes, many different types of interaction must take place.

Also, it becomes perfectly obvious that the nature of the linking forces in a "living" membrane is continuously changing during the functioning of its individual systems. The task of determining the character of interactions in its full range is thus extremely complex. In this section, we shall try to show what types of bond play the dominant role in maintaining the structure of the bacterial membrane as a whole and what interactions are essential for the performance of some of its functions. We shall concentrate attention on the two main components, viz., the proteins and lipids of the membrane, for, as already mentioned, the distribution and role of the other substances (polysaccharides, nucleic acids) have still received only little study.

To understand the stabilizing role of proteins in the membrane, the evidence of protein chemistry must first be considered. Naturally it is unnecessary in this book to describe the vast quantity of material examined in the many excellent surveys (Kauzmann, 1959; Sheraga, 1963; Nemethy, 1967; Ptisyn, 1967; Dickerson and Geis, 1969). To understand the role of proteins in the membrane it is essential to realize that the structure of protein molecules is determined by a combination of covalent, hydrogen, and electrostatic bonds, and of London–Van der Waals forces. Sheraga considers electrostatic and hydrogen bonds together, and defines them as interactions between polar groups. Interactions between nonpolar groups of a protein in water are described as "hydrophobic bonds" (Sheraga, 1963). Considerable importance is now attached to these bonds in the stabilization of the structure of globular proteins.

The main partners of proteins in the membrane, i.e., lipids, carry both polar and nonpolar groups. Because of these structural features in their molecule, they have a well-marked capacity for orderly arrangement and for the formation of structures.

It is generally held that electrostatic forces, London– van der Waals forces, and hydrophobic interactions all participate in the stabilization of biological membranes (Salem, 1964; Kavanau, 1965; Van Deenen, 1965; Chapman, 1969; Wallach, 1969).

Wallach (1969) attaches particular importance to three types of inter-actions between components of the membrane: 1) Van der Waals interactions; 2) interactions of nonpolar groups with water arranged in an orderly manner on hydrophobic areas, which is accompanied by adhesion of these areas, the relief of water, an increase of entropy, and consequently, a decrease in the free energy of the system (hydrophobic interactions); and 3) an increase in the strength of hydrogen and ionic bonds as a result of the removal of water and creation of a medium with low dielectric constant on the appearance of nonpolar associations. Hydrophobic interactions are usually regarded as the chief type of linkage between protein and lipid in biological membranes. These linkages are variously insensitive to the ionic strength of the medium; they are strengthened by a rise of temperature. Their strength is affected by the length and structure of they hydrocarbon chain of the fatty acid of the lipid.

The most general information about the forces stabilizing membranes have been obtained by the action of substances disturbing particular bonds on membranes. These include treatment of the membranes with solutions of high ionic strength, chelating agents, urea, solvents, detergents, and what are known as "chaotropic agents."

As can be seen from the above account, the NMR method has proved to be a very promising tool for the investigation of interactions between components of membranes, and has enabled the linkage proteins and lipids to be studied in the intact membrane. Measurements of the optical rotatory dispersion and circular dichroism have been less successful in this respect (Chapman, 1969; Wallach, 1969).

Important results shedding light on the interactions between the components of biological membranes have been obtained by modification of the membranes with proteolytic and lipolytic enzymes.

Action of Organic Solvents, Ionic Strength, and H^+ and OH^- Ions on Membranes

In an attempt to picture the mechanism of interaction between proteins and lipids in a membrane, we can draw in rather greater detail some actual possibilities (Fig. 17), as has been done by Green and Fleischer (1964a,b). We know that up to 95% of the lipids can be extracted from biological membranes by solvents and detergents. The extraction of so much of the lipids is readily explained by possibilities a and b (Fig. 17) but it would hardly be possible in the presence of strong polar (especially ionic) interactions (Fig. 17c,d) (Wallach, 1969). The conclusion that covalent bonds play an

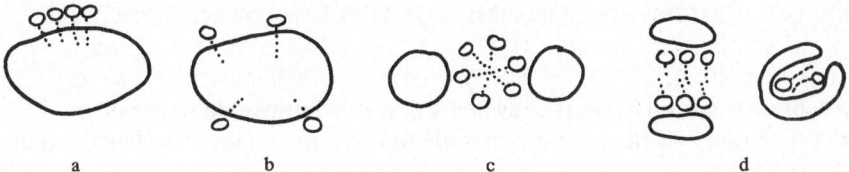

Fig. 17. Possible relative arrangements and interactions of protein and lipid. Explanation in text.

unimportant role in protein-lipid interactions is thus perfectly justified (Wallach and Gordon, 1968). It must be mentioned at this point, however, that lipids are easily extracted from all biological membranes, and in particular from myelin sheaths, for which most of the evidence supports linkages of type d (Fig. 17). The solubility of membrane lipids in organic solvents thus points indirectly to hydrophobic interactions between these lipids and proteins.

Electrostatic forces may develop between the polar groups of protein molecules and phospholipids. Phosphate groups of the phospholipids, the quaternary nitrogen atom of phosphatidyl choline, the carboxyl and amino groups of phosphatidyl serine, the amino group of phosphatidyl ethanolamine, and the free amino, carboxyl, imino, and hydroxyl groups of the proteins molecules can participate in these interactions. Van Deenen (1966, 1968) states that ionized groups of phosphatidyl choline can link up with two oppositely charged groups of the protein molecule; the resulting complex is stabilized by Van der Waals forces between the hydrocarbon chains. Ionic lipid–protein interactions can also occur on account of acid phospholipids: phosphatidyl glycerol, diphosphatidyl glycerol, and sulfatides.

Electrostatic interactions are illustrated by model experiments which have shown the formation of a complex between cytochrome c (animal) and diphosphatidyl glycerol or phosphatidyl glycerol (Green and Fleischer, 1963; Das et al., 1965; Gulik-Krzywicki et al., 1969). Protein–protein interactions can also be electrostatic in character, as has been shown experimentally by the formation of complexes of cytochrome c with negatively charged cytochrome oxidase (Smith and Minnaert, 1965). When the role of electrostatic interactions in the stabilization of bacterial membranes is examined, it must be remembered that as a rule bacterial membranes contain no phosphatidyl choline; the ratio of phosphatidyl ethanolamine to phosphatidyl glycerol varies within wide limits; and there are many bacteria whose membranes contain only phosphatidyl glycerol and diphosphatidyl glycerol.

On the surface of all biological membranes the majority of ionogenic groups are anionic. It is therefore not surprising that on acidification of the medium to pH 5 or below the membranes aggregate. Conversely, if the pH is increased, deaggregation of the membranes is intensified, although without any significant effect on lipid-protein interactions. It must be mentioned at this point that solvents of membrane proteins based on acetic acid, such as glacial acetic acid: phenol: H_2O — $1:2:1$ (Takayama et al., 1966; Rottem and Razin, 1967), are used in analytical biochemistry. However, these solvents are such vigorous systems that it is impossible to speak of the specificity of their action on ionic linkages in membranes (Bagdasarian et al., 1964).

Some interesting results for the distribution of anionic groups on the surface of membranes of *M. lysodeikticus* have been described by Italian workers (Cutinelli et al., 1969).

Membranes kept for 4 h at 25°C in HCl (pH 4), washed with water and freeze-dried, were shown to be capable of binding H^+ ions to the extent of 0.2 meq/g dry weight through phosphate groups, 0.28 μeq/g dry weight through α-carboxyl groups, and 0.25 meq/g dry weight through γ- and β-carboxyl groups. More than 25% of the cation-exchange groups thus consist of phosphate groups, evidently belonging to lipids. Our very rough calculations show that, given a binding power of membranes of 0.73 meq/g dry weight, the density of the cations must be about 5 per 100 $Å^2$, i.e., 1 ion in an area of $4 Å \times 5 Å$. Treatment with acid and alkali during titration of the membranes must, of course, introduce distortions into its structure, and all the figures cited above must be taken as extremely approximate, but this approach, which has also been developed by A. Brown (1965), may prove extremely useful in the elucidation of the mechanism of transporting substances through the membrane.

It has long been known that alkaline solutions cause considerable degradation of bacterial membranes (Edebo, 1961a,b; Bakerman and Wasemiller, 1967), but the connection between these phenomena and membrane structure is still not quite clear. This confronts us with the problem of accessibility of individual areas of the membrane and of individual reaction groups to external agents.

It is considered that an increase in pH increases deaggregation of membranes on account of an increase in dissociation and mutual repulsion of anionic groups. Since even at neutral pH values, groups located on the surface are dissociated this may be a question of penetration of hydroxyl ions into the interior of the membrane, leading to disturbance of ionic, hydrogen, and possibly other bonds. However, over a very wide range of pH values (1.4-11.7), disintegration of the bacterial membrane does not go beyond the formation of lipoprotein complexes (A. Brown, 1965). The penetration of hydroxyl ions into the membrane is facilitated if the salt bridges are first broken by removal of cations. After prolonged dialysis, 60% of the membranes of *B. megaterium* dissolved in a weak NaOH solution at pH 8.0 (Mizushima et al., 1966a,b; Yamaguchi et al., 1967). About 30% of the protein is removed from the membranes of *M. lysodeikticus* at pH 10 (Table 6).

TABLE 6. Role of Electrostatic Interactions in Stabilization of the
Membrane of *Micrococcus lysodeikticus*

Conditions of treatment	Protein, % of control	Cytochromes $a + b_{556} + b_{560} + c_{550}$, μmoles $\times 10^{-3}$		ATPase, μatoms P_{inorg} removed	
		Control	Experiment	Control	Experiment
H_2O, 15 min, 0°	77	–	–	38	25.4
H_2O, 15 h, 0°	77	104	103	–	–
EDTA (0.01 M) 15 min, 0°C	73	–	–	38	22.6
EDTA (0.01 M), 60 min, 0°C	73	99	99	–	–
KCl (4 M), 30 min, 25°	100	104	104	–	–
pH 9, 30 min, 25°	83	153	137	–	–
pH 10, 30 min, 25°	67	153	135	–	–

Mutual repulsion of the anionic groups of bacterial membranes under
natural conditions is evidently controlled not by the concentration of hydrogen
(and hydroxyl) ions, for the pH of the cytoplasm and medium are usually
close to neutral, but more probably by the existence of other cations (bivalent metals, polyamines) which block the negatively charged groups, and
possibly link together the separate complexes and membrane fragments by
metallic or hydrocarbon bridges. Polyamines of the spermine and spermidine
type have been found in some bacteria (Tabor, 1962; Quigley and Cohen,
1969; Bachrach, 1970), as well as large quantities of bivalent cations. Proof
of their participation in stabilizing the structure of bacterial membranes
in vivo is still required, although there are reports of the stabilizing action
of spermine and spermidine ont he membranes of some bacteria (Tabor et
al., 1961; Tabor, 1962; Grassowicz and Ariel, 1963; Rodwell, 1965; Inouye
and Pardee, 1970a,b). The mechanism of this phenomenon has not been identified, but there is reason to assume that interaction between the polyamine
and the acid groups of the membrane, and a decrease in its hydrophilic properties take place (A. Brown, 1964; Harold, 1964). The existence of other very
strong linkages stabilizing the membrane with the aid of the wall's components
likewise cannot be ruled out. In the living cell, in fact, the membrane on one
side is pressed against the cell wall while on the other side, it is in contact with
the substances of the cytoplasm. The character of the connection between
the membrane and the wall is still a matter of conjecture. The cell wall of
Gram-positive bacteria can be removed fairly easily by the action of lysozyme.
In general, when the cells of these bacteria are destroyed the membranes
are easily separated from the wall fraction. The separation of membranes
of Gram-negative organisms from the cell wall, however, is a difficult task.

As our knowledge of the functions of the cell wall and its connection
with the membrane increases, it will evidently be possible to progress from

the examination of these cell components separately to the examination of the cell envelope (membrane plus wall) as a whole. However, at the present time, the study of membranes at the molecular level seems to be particularly useful and necessary.

Let us turn now to the hypothetical role of cations as stabilizers of bacterial membranes. Experiments using model bilayer membranes (Shah, 1970) have shown that the addition of Ca^{++} ions causes a decrease in area of a cardiolipin film by 15-20% of its original value. This must lead to an increase in the packing density of the hydrocarbon parts of the lipid. Considerable disturbances in the structure of the bacterial membrane can therefore be expected if bivalent cations are removed. In fact, treatment with chelating agents (EDTA, etc.), which selectively bind ions of the bivalent metals, leads to considerable disturbances in the membrane, despite the fact that according to Razin (1972), about 30% of the Mg^{++} present in the membranes before treatment with EDTA still remains bound to them after treatment. Treatment with 10^{-2} M EDTA, for instance, leads to partial destruction of the membranes of $M.$ $lysodeikticus$ (Table 6) (Lukoyanova et al., 1961, 1972; Simakova et al., 1969) and of the membranes of $Haemophilus$ $parainfluenzae$ (Tucker and White, 1970a,b,c). Membranes of $Mycoplasma$ are stabilized by Mg^{++} and Ca^{++}, and are broken up by the action of EDTA (Rodwell, 1965). EDTA, which "undoes" the magnesium cross-linkages, also enables lysozyme to penetrate through the external membrane of $E.$ $coli$ to peptidoglycan. The combined action of EDTA and lysozyme is used to disturb the mechanical strength of membranes of Gram-negative bacteria and other microorganisms for the purpose of obtaining spheroplasts and lysates of the cells (Shibuya et al., 1967; Miura and Mizushima, 1968). Lysis of some Gram-negative bacteria ($Pseudomonas$ $aeruginosa$ 64, $Alcaligenes$ $faecalis,$ $Vibrio$ $succinogenes$) can be obtained with EDTA (Eagon and Carson, 1965; Gray and Wilkinson, 1965a,b; Wolin, 1966). Rupture of the membrane of $E.$ $coli,$ when damaged by "ghosts" of phage T2, is prevented by Mg^{++} ions (Gershanovich et al., 1966).

It might be supposed that ionic bonds in a membrane, if accessible to water, would be considerably weakened by solutions with high ionic strength. However, there are no reports in the literature of any considerable destruction of bacterial membranes by solutions of high ionic strength. At most, slight destruction of the membrane is produced with detachment of weakly bound lipoproteins or of individual proteins.

Our own results obtained by the treatment of membranes of $M.$ $lysodeikticus$ with these agents (Table 6) will serve as an example. In this case, slight destruction of the membrane is observed only in solutions of alkaline pH. Under these conditions, the cytochrome proteins do not leave the membrane. According to Broberg and Smith (1967), 10% NaCl likewise does not extract cytochrome from the membranes of $B.$ $megaterium.$ Treatment of membranes with solutions of high ionic strength in some cases actually causes them to aggregate. For example, with an increase in the salt concentration in aqueous suspensions of membrane fragments of *M. lysodeikticus,* the degree of aggregation of these particles is gradually increased. Ultimately they are completely salted out by ammonium sulfate in concentrations of, for example, about 20-40% of saturation (Zhukova et al., 1966). There is no separation of lipids which, under those conditions, should rush to the surface of the suspension.

Besides the general effect of destruction of the bacterial membrane on rupture of the salt bridges, certain specific features can be detected in the linkage of individual proteins. For instance, cytochromes are not extracted at all by treatment of the membrane with chelating agents (Table 6). Meanwhile, as the result of treatment with EDTA ($0.01 M$), respiration in the membranes of *M. lysodeikticus* is inhibited by 40-60%, and under these conditions the lipoprotein complexes containing malic acid dehydrogenase and $NADH_2$ dehydrogenase leave the membrane (Cohn, 1956, 1958; Gel'man et al., 1963b). Meanwhile ATPase, bound in the membranes of *M. lysodeikticus* by salt bridges, is extracted by water or EDTA, and moreover in the form of an individual protein, not a lipoprotein (Simakova et al., 1968; Munoz et al., 1968). ATPase and nectine are also removed from the membranes of *S. faecalis* by disturbance of electrostatic interactions (Abrams, 1965; Baron and Abrams, 1971).

Some bacteria, because of their specific conditions of existence, may even require an increased salt concentration. These are the so-called halophiles (*Halobacterium halobium, Halobacterium cutirubrum,* etc.). An increased salt concentration ($4 M$ NaCl) is an essential condition for stability of the membranes of the halophiles, for lowering the salt concentration, even to $1 M,$ causes the membrane to break up into lipoprotein particles (A. Brown, 1965). In that case, the high ionic strength weakens negative interactions (repulsion mainly of carboxyl groups). Yet it does not lead to separation of the lipids and proteins. Besides screening the negative groups of the membrane, cations (Na^+, K^+, Li^+, NH_4^+) in halophilic bacteria may also stabilize hydrophobic interactions, as has been shown by the study of the $NADH_2$ dehydrogenase activity of membranes of halophilic bacteria as a function of the concentration and nature of the salts. Activity of the enzyme is determined principally by the nature of the anions, which are arranged in the following descending order: $Cl^- > H_2PO_4^- > Br^- > NO_3^- > ClO_4^- > SCN^-$ (Lanyi and Stevenson, 1970).

The concentration of cations essential to bind proteins in the membrane differs for different components, including flavoproteins, cytochrome b, and cytochrome oxidase (Lanyi, 1971).

In general, a decrease in the cation concentration in the membrane leads to different consequences in different bacteria. This serves as evidence of the specificity of organization of the membranes, probably in connection with differences in their lipid and protein composition. Whereas treatment of *M. lysodeikticus* with water leads to removal of a comparatively small proportion of the protein (Table 6), under the same conditions the whole membrane of halophilic bacteria breaks up into lipoprotein fragments (Onishi and Kushner, 1966; Stoeckenius and Rowen, 1967; De Voe and Oginsky, 1969a,b). However, in neither case are the lipids and proteins separated from one another. Perhaps the lipoprotein complexes are so constructed that the polar groups of their components face outward, and the membrane can be assembled only if the electrostatic repulsion developing between these two groups is overcome by screening with cations or by the creation of Mg^{++} or Ca^{++} bridges, while the complexes themselves are stabilized by different forces (most probably hydrophobic). The bond linking the protein and lipids within lipoproteins is stable over a wide range of pH values and ionic strengths. Even a considerable increase in the strength of repulsion of the anionic groups induced artificially by blocking the ϵ-amino groups of lysine, and the terminal amino groups chemically be means of succinic anhydride, does not lead to dissociation of the lipoproteins. For instance, although an increase in the number of COO-groups in the membranes of *Pseudomonas* NCMB 845 by succinylation was accompanied by an increase in the hydrophilic properties, and more concentrated solutions of salts were required to produce stabilization of the membrane, no liberation of lipids or proteins was observed (A. Brown, 1964).

Succinylation of the membrane proteins of *M. laidlawii,* while not leading to liberation of proteins and lipids, was accompanied by the formation of a series of fragments which differed in their lipid : protein ratio (P. Smith et ai., 1970).

Manipulations of the ionic strength by a change in pH, by chelating agents, and by concentrated salt solutions thus do not lead to dissociation of the membrane into protein and lipid. From this it can be deduced that other types of interaction between these substances are predominant. It could admittedly be assumed that, as has already been pointed out, a significant proportion of the ionic bonds lies in nonpolar areas of the membrane and is inaccessible to the action of aqueous solutions. However, this hypothesis (Stoeckenius and Engelman, 1969) seems somewhat artificial in character. A more likely suggestion is that powerful hydrophobic interactions are superposed upon electrostatic bonds, and hold together the components of the membrane despite local changes in the pattern of the charges on the membrane. The essential features are that fragmentation of membranes by detergents is potentiated in the presence of salts and that ionic detergents fragment membranes more strongly than nonionic detergents.

The role of hydrogen bonds in the stabilization of bacterial membranes is unimportant. For that reason, urea does not bring about any significant destruction of the membrane; components of the membrane such as cytochrome proteins are not in general affected by treatment with urea (Simakova et al., 1969; Lukoyanova et al., 1972). The relative contribution of hydrogen bonds to maintenance of the structure of mitochondrial membranes is also small (Fleischer et al., 1962).

The study of the forces stabilizing bacterial membranes, as well as other biological membranes, thus consists essentially of the study of hydrophobic interactions. No direct methods of measuring the strength of these interactions yet exist. For a general assessment of these interactions, it is therefore necessary to examine the effects of solvents and detergents on the membrane and its components.

Investigations have shown that the membranes of *M. lysodeikticus* and of mycoplasmas retain their original appearance under the electron microscope after extraction of lipids (Grula et al., 1966; Terry, 1967); a similar phenomenon has been described in the case of mitochondrial membranes (Fleischer et al., 1967). Removal of 80% of the lipids from membranes of *M. lysodeikticus* by treatment with 1% deoxycholate does not cause the membranes to disintegrate (see Fig. 5c).

The fact that the membrane retains its structure after removal of the lipids can be interpreted as evidence of the existence of powerful interactions between proteins (Salton et al., 1968; Lukoyanova et al., 1972). Internal membranes not containing lipids extractable by a mixture of chloroform and methanol have been found in halophiles (Stoeckenius and Kunau, 1968), suggesting a dominant role for the protein in the structure of the biological membrane, whatever the true nature of the one light and two dark bands on photomicrographs of the membrane.

After removal of the lipids, the membrane proteins can be dissoved only in vigorous solubilizing systems. This is certain evidence of the existence of hydrophobic interactions between the protein molecules after removal of the lipid, although the evidence that this is so in native membranes is not conclusive.

Detergents as Tools for the Investigation of Membranes

It has been known for a long time that many surface-active agents (detergents) possess a bactericidal action. Many attempts have been made to use this phenomenon therapeutically, but because their action is nonspecific, the detergents proved to be toxic to the experimental animals also. Special investigations have shown that detergents act on the cytoplasmic membrane of the cell, increasing its permeability. For example, Tween-80 leads to disappearance of UV-absorbing substances from cells of *Pseudomonas aeruginosa,* whatever the temperature (Brown and Winsley, 1969).

In recent years, surface-active agents have been widely used in investigations of the molecular organization of biological membranes, for in a certain concentration these agents can solubilize the components of membranes.

Work on the use of detergents and salts of bile acids was stimulated, in particular, by the successful experiments of Green and his collaborators on the solubilization of mitochondrial membranes, which led to the isolation of lipoprotein complexes which are components of the mitochondrial respiratory chain. Besides bile acids, there are other natural detergents such as the lipopeptide from the cells of B. subtilis (Tsukagoshi et al., 1970).

Several monographs have been written on the composition and properties of detergents (Schwartz and Perry, 1953; Schonfeldt, 1965; Shinoda et al., 1963). Because of the structural properties of a molecule with hydrophilic and hydrophobic areas, detergents can arrange themselves on the surfaces of a phase boundary and reduce surface tension. If their concentration exceeds the so-called critical concentration of micelle formation (CCM), molecules of the detergent aggregate to form micelles. The CCM value for ionic detergents is relatively high; for example, for sodium dodecylsulfate, widely used in biochemistry, it is 0.081 mole/liter (micellar weight 17,800, aggregation number 62). The CCM of nonionic detergents is much lower; for Triton X-100 or Tween-20, for example, it is 0.0009 and 0.0003 mole/liter, respectively. The micellar weight of Triton X-100 varies from 53,000 to 208,000, depending on the composition of the polymer homologues. In work on the fragmentation of membranes, nonionic detergents are usually used in a concentration two or three orders of magnitude higher than the CCM, i.e., in a micellar rather than a molecular form. Ionic detergents such as dodecylsulfate are used in concentrations close to the CCM, although more often than not they exceed the critical value, i.e., these detergents also are used in the micellar form. The necessary concentration is determined empirically in each case, depending on the rate and completeness of solubilization of the membrane material. Like synthetic detergents, the natural detergents cholate and deoxycholate form micelles by fusion of hydrophobic areas of their molecules (Small et al., 1969). The action of detergents on biological membranes is usually described as solubilization, a term taken from colloid chemistry.*

The intimate mechanism of interaction between micelles of a detergent and the membrane is unknown, for the structure of membranes, and in particular of bacterial membranes, is not yet deciphered. In the future, therefore, we shall speak mainly of the end result of collision between detergent micelles and membranes, and not of their interaction. The basic similarity between the action of ionic (including cholate and deoxycholate) and nonionic detergents indicates that the hydrophobic part of the molecule, and not the charged, hydrophilic groups of the detergent, is the main agent producing fragmentation of the membrane. It thus follows that when the micelle comes into contact with the membrane surface, it probably changes its configuration,

---

*Solubilization is defined as the preparation of a thermodynamically stable isotropic solution of a subtance normally insoluble or very slightly soluble in a given solvent by the introduction of an additional amphiphilic component or components (Elworthy et al., 1968).

exposing hydrophobic groups and perhaps even breaking up into molecules. In this form the detergent interacts with the membrane chiefly with the non-polar, hydrophobic part of the molecule, in the course of which the detergent molecule probably sinks into the hydrophobic parts of the membrane. Penetration of detergents (sodium dodecylsulfate and deoxycholate) into the erythrocyte membrane (Chapman, 1969) induces the appearance of an NMR signal from protons of the hydrocarbon chains of the lipids. This indicates that detergents induce disorganization of the lipids and endow the fatty acid residues with considerable freedom of movement. In addition, sodium dodecylsulfate increases the mobility of certain groups in the protein, which also is recorded by the NMR method. Obviously some features which distinguish the action of ionic and nonionic detergents (the size of the membrane fragments, the activity of its enzymes) can be explained by interaction between the charged groups of detergent and the corresponding groups in the molecules of the membrane proteins and lipids.

Under the action of detergents, the membrane breaks up into fragments of varying size which are solubilized by virtue of the detergent they contain. The nature of the bond between the detergent and the lipoproteins of the membrane has not yet been explained. All that is known is that the detergent can be removed by dialysis or gel-filtration, and that under certain conditions the detergent can cause separation of the lipids of the membrane from its proteins. The degree of destruction of the membrane depends on the length of the hydrocarbon chain of the alkyl sulfates (Salton, 1968). Our own observations show that destruction of the membrane depends on the structure of the detergent molecule. Cholate and deoxycholate, which differ from each other in a single polar group, have a different action on membranes. Tween-80 and Triton X-100, both derivatives of polyhydroxyethylene, also differ in their action on membranes (Lukoyanova et al., 1972). Data on the effect of detergents on the content of protein and phospholipids in particles and supernatant (144,000 g, 60 min), obtained by treatment of membranes of *M. lysodeikticus,* are given below (in per cent of their content in the original membranes):

		Particles	Supernatant
Tween 80, 1%	Protein	61	31
	Phospholipid	89	10
Cholate, 1%	Protein	51	50
	Phospholipid	88	12
Deoxycholate, 1%	Protein	18	80
	Phospholipid	9	90
Triton X-100, 1%	Protein	16	81
	Phospholipid	8	91
Dodecylsulfate, 0.4%	Protein	0	100
	Phospholipid	—	—

Details of the action of detergents on membranes and isolated proteins are given by several authors (Elworthy et al., 1968; Gel'man, 1969; Harold, 1970; Tanford, 1970; Mosolov, 1971). We can now go on to consider the action of the various detergents on bacterial membranes.

An anionic detergent which is frequently used is sodium dodecylsulfate (SDS) $CH_3(CH_2)_{10}CH_2-O-SO_3=Na$. Salton and Netschey (1965) treated membranes of M. *lysodeikticus* with 1% SDS and obtained fragments with a sedimentation coefficient of 1.7 S. Similar fragments have also been isolated from membranes of mycoplasmas (Razin et al., 1965). The results of these experiments aroused considerable interest, and it was even postulated as a result that the biological membrane consists of a system of homogeneous lipoprotein subunits, stabilized chiefly by hydrophobic interactions (Green and Perdue, 1966; Green et al., 1967). The further discovery was made, however, that the size of the membrane fragments depends on the structure of the detergent molecule and may vary within wide limits (Salton and Netschey, 1965). The character of fragmentation is determined not only by the structure, but also by the concentration of the detergent. For instance, with an increase in the concentration of SDS, the sedimentation coefficient of particles from membranes of mycoplasmas decreased (Engelman et al., 1967). Analysis of the solubilized material in a sucrose gradient showed that with an increase in the SDS concentration from 3.0 to 7.0 mM, the fragments are reduced in size, but in a concentration of 10 mM, the homogeneous fragments (about 4 S) are in fact a mixture of lipid and protein.

The problem of whether the fragments of the membrane are lipoprotein complexes or whether a mixture of lipid-detergent and protein-detergent complexes, capable of interaction, is formed on solubilization has also been studied on membranes of *Mycoplasma* by Rodwell et al. (1967). These experiments also showed that during fractionation of membranes solubilized by SDS in a sucrose gradient, free proteins and lipids appear. It was later shown that if material from *Mycoplasma* membranes is solubilized with SDS, much of the protein can be separated from the lipid by electrophoresis in polyacrylamide gel (Rottem et al., 1968). These workers emphasize that the symmetrical 3.3 S peak observed on analytical centrifugation of the solubilized material gives no indication of the true state of the preparation. Furthermore, a symmetrical peak (2.8 S) could be obtained by centrifugation of proteins liberated from lipid, although the protein fraction consists of at least ten different proteins. In the analytical centrifuge a preparation of membrane lipids in the presence of SDS gave two peaks: 2.0 and 6.3. When the protein was mixed with lipid (2:1), the original peak of the solubilized membranes (3.5 S) appeared. The experiments of Engelman and of Rodwell and coworkers thus raised the question of whether the methods of sedimentation analysis can suitably be applied to the products of solubilization of membranes by detergents.

A special examination of the disintegration of membranes of *M. lysodeikticus* into protein and lipid has been made by Butler et al. (1967). They found no free lipids after treatment of membranes with detergent, but on electrophoresis in polyacrylamide gel the lipids moved as a separate band. In the study of other biological membranes, such as by solubilization of *B. subtilis* membranes with 0.2% SDS, it was found that the fragments do not dissociate into lipid and protein, and do not aggregate with an increase in particle size when the detergent concentration is reduced (Bishop et al., 1967b). Careful investigation of the fragmentation of erythrocyte membranes by SDS likewise does not justify the conclusion that the membrane breaks up into lipids and proteins (Bakerman and Wasemiller, 1967).

Meanwhile, treatment of membrane proteins, and of hydrophobic proteins in general, with SDS followed by electrophoresis is polyacrylamide gel has gained a firm foothold in analytical biochemistry as one of the most effective and reliable methods of fractionating proteins and of rapidly determining their molecular weight (Shapiro, et al., 1967). Depending on the purpose of the experiment and the character of the membrane, its contact with detergent may thus end either in a slight degree of fragmentation or in its profound destruction with denaturation of the protein components.

In some cases, admittedly, the individual specific properties of membrane proteins have been preserved even after total disintegration of the membrane. Proteins obtained from the membranes of *M. laidlawii* by treatment with 0.5% SDS have largely lost their specific antigenic qualitites (Argaman and Razin, 1969), but some of the proteins still preserve their antigenic specificity and even their $NADH_2$ dehydrogenase enzyme activity.

Nonionic detergents have a milder action on the enzyme proteins of membranes, and for this reason they are more frequently used in membrane fragmentation work (Newton, 1960). Some investigations have been carried out with Triton X-100 and its related nonidet P-40, some with Tween-80; Triton X-100 is polyhydroxyethylene-isooctylphenyl ester $C_8H_{17}C_6H_4O(CH_2CH_2O)H_{10}$ (molecular weight about 680). The nonidet P-40 is also a compound of the polyhydroxyethylated alkylphenol type. Tween-80 is a hydroxyethylated anhydrosorbitol mono-oleate.

The process of fragmentation of *E. coli* membranes with a 1% solution of the nonidet P-40 has been described in detail by Shibuya et al. (1967). The protein and lipid peaks (relative to P^{32} and S^{35}) of the solubilized material did not coincide in a sucrose gradient, although on analytical centrifugation the material gave one peak. It thus follows that sedimentation cannot be a criterion of the homogeneity of the fragments for solubilization by nonionic detergents. A similar conclusion can be drawn from the results on solubilization of the membranes of *M. lysodeikticus* by P-40 (Salton and Netschey, 1965; Salton et al., 1967).

Fragmentation by detergents not only shows the important role of hydrophobic interactions in the maintenenace of membrane structure as a whole, but in some cases it also demonstrates the role of these forces in the organization of individual areas of the membrane. Treatment of membranes of *B. megaterium* with cadmium laurylsarcosinate yielded nine lipoprotein fractions which differed from each other in their protein:lipid ratio and in their qualitative phospholipid composition (Daniels, 1971). When membranes of *M. lysodeikticus* were treated with P-40, a fraction rich in protein (protein:phospholipid = 7:1) and containing cytochromes, and a fraction poor in protein (protein: phospholipid = 0:6) were found (Salton et al., 1967). Similar results were obtained in our own experiments to study the action of Triton X-100 and deoxycholate on the membranes of *M. lysodeikticus* (Tikhonova et al., 1970). The relative ease of solubilization of lipids by detergents, while the protein aggregates remain intact, is explained by the probable existence of powerful interprotein hydrophobic interactions in specific areas of the membrane containing proteins and cytochromes.

As an illustration, a photograph of the membranes of *M. lysodeikticus* after treatment with deoxycholate is shown (Fig. 5b). In these membranes the lipid:protein ratio is 0:2, compared with the normal 0:45. Despite this, the membrane structure is preserved; the concentration of cytochromes in these membranes is increased by 4 or 5 times (Lukoyanova et al., 1972; Simakova, 1970). Further evidence of the resistance of the membrane proteins to detergents is given by the results of solubilization of *Mycoplasma* membranes by Triton X-100 (Razin and Barash, 1969). It is curious that different membrane proteins differ in their solubilization, evidently because of differences in the character of their packing in the membrane and the absence of standard subunits.

The effectiveness of fragmentation of the membrane and of solubilization of its components by detergent also depends on the state of the charged groups. For instance, Tween-80 extracts nearly all the malate–$NADH_2$ dehydrogenase complex from membranes of *M. lysodeikticus* if they are first treated with EDTA. Only 20% of the complex can be removed by the direct action of Tween-80 (Tsfasman et al., 1972a,b).

In several investigations in which Triton X-100 has been used to fragment the chromatophores of photosynthesizing bacteria, the action of the same detergent in different concentrations on physiologically similar structures has been compared. Treatment of the chromatophores of *Chromatium* with 4% Triton X-100 led to the formation of light and heavy fragments which differed in the structure and content of the forms of bacteriochlorophyl (Garcia et al., 1966). In later work, these authors compared the structural and photochemical properties of particles obtained by the action of Triton X-100 on the chromatophores of *Rhodopseudomonas palustris* and *Rhodopseudomonas* HTC-133 (Garcia et al., 1968a,b). They found that treatment of the chromatophores of *R. palustris* with 0.6% Triton X-100 solution disturbs the organization of the various forms of bacteriochlorophyll; rupture of the chromatophore membranes, however, arises only after prolonged treatment with 4% detergent. In the character of their fragmentation by the action of Triton X-100, the chromatophores of *R. palustris* are similar to those of *Chromatium*. Surprising as it may seem, treatment of the chromatophores of *Rhodopseudomonas* HTC-133 with Triton X-100 (Garcia et al., 1968b) gave completely different results from those obtained with *R. palustris* and *Chromatium*. A 0.6% solution of Triton X-100 did not affect the arrangement of the components of the photosynthetic apparatus in the membrane. Although 4% Triton ruptured the membrane with the formation of light and heavy components, these did not correspond in their properties to the analogous fragments from *Chromatium* and *R. palustris*. The result of treatment with the detergent is thus determined by several factors: the organization of the membrane, the nature of the detergent, and the conditions of the treatment (pH, ionic strength, concentrations of protein and detergent, duration of treatment, temperature). Whereas according to some observations Triton X-100 can disintegrate the membrane of *Mycoplasma* down to proteins and lipids, the same detergent can extract a still active photosynthetic center from the chromatophore membrane (Reed, 1969) and a complex of cytochromes b_{560}, c_{550}, and a_{601} from the membranes of *M. lysodeikticus* (Tikhonova et al., 1970).

The use of bile salts has played an important role in the study of membranes. Examples of the specific action of deoxycholate on individual areas of the bacterial membrane and of the isolation of heterogeneous fragments are given by solubilization of the membranes of *A. vinelandii* (Jones and Redfearn, 1967a,b), and also the isolation of the malate–$NADH_2$ dehydrogenase–cytochrome b_{556} block and the cytochrome b_{560}, c_{550}, a_{601} block from the membranes of *M. lysodeikticus* by the action of cholate (Tikhonova et al., 1970; Gel'man et al., 1970). The specific action of bile salts on membranes, accompanied by excision of certain fragments, is difficult to explain in terms of the results obtained by Salton and Schmitt (1967), who showed that the lipid and protein are so weakly bound in the products of solubilization of the bacterial membrane by deoxycholate that they can easily be separated by disk electrophoresis in polyacrylamide gel. A similar conclusion can be drawn from the experiments of Gibson (1965), who studied the effect of deoxycholate on the chromatophores of *Rhodopseudomonas spheroides*. As a result of treatment of the chromatophores, up to 70% of the lipids but less than 20% of the protein and carbohydrates can be liberated.

Treatment with deoxycholate, cholate, or a combination of the two substances in the presence of ammonium sulfate has been used to isolate four different fragments of the membrane: $NADH_2$ coenzyme Q reductase, succinate coenzyme Q reductase, coenzyme QH_2 cytochrome c reductase, and cytochrome oxidase (Hatefi et al., 1962; Green and Fleischer, 1964). Even a cursory comparison of the various papers published by Green's group will show that by varying the concentration and nature of the detergent, the concentration of the salts, and the duration of their action, it is possible to obtain not only different enzyme complexes, but also particles containing all enzymes of the respiratory chain of mitochondria membranes at the same time (Blair et al., 1963). The cause of the various effects produced by the detergent during destruction of the same membrane must evidently lie in the different conditions of treatment.

There have been few investigations into the action of different detergents on the same membrane. According to Salton and co-workers, the nonidet P-40 solubilized the membrane of M. lysodeikticus by 96%, and the fragments formed had a sedimentation coefficient of 3.1 S; sodium dodecylsulfate gave 75% solubilization and a sedimentation coefficient of 1.7 S; dodecyl-trimethylammonium bromide gave 40% solubilization and a sedimentation coefficient of 0.7 S (Salton and Netschey, 1965; Salton et al., 1967; Salton, 1968). Butanol extracted 13%, digitonin, 11%; lipase treatment, 18%, and Triton X-100 extracted 45% of the hydrogenase from the chromatophores of Chromatium (Feigenblum and Krasna, 1970). The relationship between the size and properties of the fragments, and the structure of the detergent molecule has been demonstrated by comparing the action of various detergents on membranes of erythrocytes (Bakerman and Wasemiller, 1967), of M. laid-lawii (P. Smith et al., 1970), and of the chromatophores of R. rubrum (Biedermann, 1971).

All detergents evidently fragment the membrane to some extent by disturbing hydrophobic interactions, but the character of the fragmentation is largely determined by the structure of the detergent. For this reason, it is difficult to draw any conclusions regarding the organization of the membrane by comparing the action of various detergents on it. Meanwhile, treatment of membranes by a series of detergents does allow the behavior of small areas, marked by enzymes of the respiratory chain, to be studied. Some information in this respect is given by fragmentation of the membrane of M. lysodeikticus by a series of detergents (Tikhonova et al., 1970; Gel'man et al., 1970). Although the degree of fragmentation of the membrane is determined by the nature of the detergent, the forces of hydrophobic bonding of the membrane components are unequal, as is clear from the example of the respiratory chain. Over the extent of the respiratory chain, hydrophobic interactions are weakest in the area between cytochromes b_{556} and b_{560}. That is why, during fragmentation of the membrane by different detergents (cholate, deoxycholate, Triton), the respiratory chain breaks up into two complexes: $NADH_2$ dehydrogenase–cytochrome b_{556}, and cytochromes b_{560}, c_{550}, and a_{601}.

Work on reconstruction of membranes from their fragments has developed hand in hand with work on fragmentation of the membrane by detergents and bile salts. The first of these investigations were regarded as the pathway to obtaining membranes outside the living cell. However, as it was gradually discovered that the process of fragmentation of membranes by the action of detergents depends on many factors (including the nature of the detergent, pH, salt composition of the medium, duration of action, etc.) and that the size and composition of the fragments are not constant, but that fragments or individual proteins and lipids of the membrane may aggregate together with only a negligible change in the conditions, the question arose whether reaggregation of the membrane from the fragments must be regarded as an artifact due to the hydrophobic character of the components of the membrane. Reaggregation of membranes from fragments has been studied in several objects, including *Mycoplasma* membranes, *M. lysodeikticus* membranes, and mitochondrial membranes.

Reaggregation of a membrane from its fragments was first observed in material obtained from membranes of *Mycoplasma* by solubilization with sodium dodecylsulfate (Razin et al., 1965). Dialysis against a buffer containing Mg^{++} led to the formation of a precipitate consisting of membranes morphologically identical to the original membranes and differing from them only in a higher lipid content (up to 60%). On treatment with sodium dodecylsulfate, the reaggregated membranes were again solubilized to 3.3 S fragments. Next followed a series of investigations aimed at studying the mechanism and biological significance of artificial reaggregation of membranes of *Mycoplasma*, L-forms of *Streptobacillus moniliformis, Halobacterium salinarum*, and other organisms (Terry et al., 1967; Engelman et al., 1967; Rodwell et al., 1967; Engelman and Morowitz, 1968; Rottem et al., 1968; Razin et al., 1969; Razin and Barash, 1969; Stevenson and Brown, 1969). These investigations are fully disucssed in the surveys of Gel'man (1969), Poglazov (1970), and Razin (1972); for this reason their results will be considered only very briefly. The components of different membranes, after solubilization with 0.01 M sodium dodecylsulfate, react readily with the formation of a membrane-like structure in the presence of Mg^{++} ions. The lipid : protein ratio in the aggregate, as well as the velocity of the process, depend on the Mg^{++} concentration. Evidently, the role of Mg^{++} in this process is essentially one of neutralizing the negative charges of the proteins and lipids which prevent reaggregation through electrostatic repulsion. Bivalent cations probably form salt bridges which stabilize the membrane. The possibility cannot be ruled out that Mg^{++} links charged molecules of dodecylsulfate, thereby producing aggregation of the proteins or lipoproteins bound with them (Smith et al., 1970).

Grula et al. (1967) have made a detailed study of the reaggregation of membranes from fragments obtained by solubilization of membranes of *M. lysodeikticus* with 0.03 M sodium dodecylsulfate. For aggregation to take place, the detergent had to be removed by dialysis against magnesium ions (0.01 M $MgCl_2$). Morphological similarity between the original and the reaggregated membranes has been reported (Butler et al., 1967). Some very interesting results helping to explain the mechanism of aggregation of membranes and the role of lipids in this process are given in a paper by Grula et al. (1967). Practically all the lipids can be extracted from the membranes of *M. lysodeikticus* by treating them with alkaline acetone in the cold, without producing any disaggregation of the membranes. Fragmentation of these lipid-free preparations of membranes can then be brought about by treating them with sodium dodecylsulfate. During dialysis in the presence of Mg^{++}, the protein complexes reaggregate with the formation of typical membranes. A high yield of reaggregation is obtained if phospholipids of the membranes and carotenoids are added together with Mg^{++} during dialysis. The process of reaggregation was found to be nonspecific. It can take place from proteins alone, from proteins and lipids present in the original membrane, and also from proteins and lipids obtained from other bacteria and differing from the lipids of *M. lysodeikticus* by a high content of phosphatidyl ethanolamine. Proteins and lipids of membranes of different mycoplasmas, having no common antigens and differing in their buoyant density, can be hybridized. Aggregates containing both types of proteins and lipids can be obtained (Razin and Kahane, 1969; Cole et al., 1971). A soluble protein, penicillinase, can even be included in the process of reaggregation into a membrane (Rottem et al., 1971).

A "classical" three-layered membrane is observed in the electron microscope when reaggregated material is examined after fixation in the usual way. However, the study of intact and reaggregated membranes of *M. laidlawii* by the freeze-etching method has revealed significant differences between them: the aggregated membranes have no globular outgrowths (Tillack et al., 1970). Native and reaggregated membranes of *M. laidlawii* give a thermal phase transition at the same temperature. However, the fatty acid residues in the reaggregated membrane are less orderly than in the native membrane (x-ray diffraction). Differences in the structure of these types of membranes can be revealed more clearly by the study of their benzyl alcohol NMR spectra and ANS fluorescence (Metcalfe et al., 1971a,b).

Detection of the specificity of assembly is difficult in membranes of *Mycoplasma*, because they do not contain multicomponent systems which could act as criteria of the native state of the newly assembled material. Specificity of reaggregation in bacterial membranes is characterized by suppression of $NADH_2$ oxidase activity on reaggregation of solubilized membranes from *B. megaterium* in the presence of Mg^{++} (Eisenberg et al., 1970a,b; Yu and Wolin, 1970, 1972) and in analogous experiments with solubilized membranes of *M. lysodeikticus* (Eisenberg, 1971) and *B. stearothermophilus* (Kiszkiss and Downey, 1972). On reaggregation of chromatophore membranes solubilized by dodecylsulfate, photophosphorylation is restored by 60% (Takacs and Holt, 1971). The important conclusion which can be drawn from the results of these experiments on reaggregation of bacterial membranes is that lipids and proteins, as well as individual lipoporotein complexes, can interact to form a new membrane even in the absence of a template.

Specificity of reaggregation of membranes has been shown in the case of interaction between fragments of mitochondrial membranes after removal of the bile acids (McConnell et al., 1966; Green et al., 1967). On reaggregation, the activity of the whole chain of electron transport was restored. Green and co-workers interpret the phenomenon of reaggregation of membranes as the result of interaction between native subunits liberated from the mitochondrial membrane and characteristic of many different membranes. However, the views expressed by Green and co-workers cannot have such a general significance, for investigations of the fragmentation of bacterial membranes have shown conclusively that although reaggregation is a spontaneous process which accompanied the removal of detergents or deoxycholate, it can nevertheless take place from fragments of different sizes, from proteins and lipids, and even from proteins alone. According to Green and co-workers, the internal membrane of mitochondria is built from repeating lipoprotein subunits, uniform in size but differing in enzyme composition.

The system of subunits is stabilized chiefly by hydrophobic interactions in which a special role belongs to phospholipids situated in certain areas on the outer surface of the subunits (Green et al., 1967). According to the opposite view, phospholipids do not play an important role in the organization of the membrane, for reaggregation can also take place from membrane proteins alone, and these must therefore constitute the structural basis of the membrane (Grula et al., 1967). A compromise solution to the problem of the arrangement of protein and lipid can be found in the conclusions drawn by Gibson (1965). In his view, the aggregate of pigments and catalytic proteins (the chromatophore) has a lipid covering which stabilizes the whole system of chromatophores into a single membrane. The possibility cannot be ruled out that the lipid component in biological membranes possesses a certain freedom of movement over the lipoprotein particle, and this property probably operates during the formation of membranes (Ostrovskii et al., 1968).

There is no doubt that hydrophobic bonds play a very important role in linking together the components of membranes, for a membrane can be broken up into small fragments only with the aid of detergents. This is evidently the only reliable conclusion which can be drawn at the present time by comparing the results of various investigations into the fragmentation of membranes by detergents. These investigations have as yet produced no answer to the question of how lipids and proteins are arranged in the membrane, how the lipoporoteins are packed, and whether uniform subunits exist or not. Detergents have recently been used in membrane biochemistry in a completely different role. Insoluble salts of the detergent Sarkosyl have been used with great success for locating bonds between bacterial membranes and the DNA molecule, and for isolating this complex from a cell digest (Tremblay et al., 1969). In these experiments the protoplasts or spheroplasts of *B. mega-terium, B. subtilis, M. lysodeikticus,* and *E. coli* were mixed with Na-Sarkosyl and centrifuged in a sucrose concentration gradient in the presence of Mg^{++} ions. Under these conditions, the Mg-Sarkosyl formed crystals; as a result of hydrophobic interactions, fragments of the membrane adhered to these crystals. In turn, ribosomes and DNA molecules were attached to them. In this way, a fraction (M-band) containing 90-99% of the DNA, 33% of the RNA, 15% of the protein, and 10-30% of the phospholipids of the cell was obtained.

The experiments of Tremblay and co-workers thus offered tempting prospects for the investigation of those areas of the membrane which are responsible for the attachment of DNA molecules and which possibly participate in initiating replication and distribution the genetic material between the daughter cells.

Complexes

Bacterial membranes are somewhat labile structural formations. Simply mixing a suspension of membranes with buffer causes them to fragment, and the process can be intensified by the addition of agents inducing repulsion or weakening attraction between the components of the membrane.

The existence of powerful lipid—protein interactions in membranes is not in dispute. It is also evident that fragments of the membrane obtained by different methods (lipoprotein complexes) are the basic form in which proteins and lipids are linked together. These general principles are not in themselves sufficient to allow a decision whether lipoproteins of different size and composition are united into larger repeating subunits in a membrane of approximately equal size, or whether comparatively small lipoprotein complexes form a mosaic type of system in which the arrangement of the components may vary.

That the membrane is constructed from large subunits is usually proved by the fact that upon fragmentation the membrane breaks up into large fragments which preserve their functional activity. For example, upon sonication, the membranes of *Sarcina lutea* break up into fragments with a sedimentation coefficient of 4.2 S (Salton, 1968). Upon electrophoresis in sucrose (0-50%, 0.03 M borate buffer, pH 8.7), these particles form a single band. However, they were too large for electrophoresis in polyacrylamide gel, which could have demonstrated their homogeneity. Mitochondrial membranes, if treated by sonication, break up into fragments 80-90 Å in diameter (Tzagoloff et al., 1968), which these workers regard as the natural structural elements of the membrane.

The view that the membrane is a mosaic of lipoproteins is equally well justified. Because of the lability of such a system, any lipoprotein may lie next to various other lipoproteins or proteins. It will thus have a different molecular environment in different parts of the membrane. The molecular environment of a lipoprotein naturally determines the strength of its bonding in the membrane. Consequently, it will be expected that some molecules of a given lipoprotein will be solubilized by one agent, and others by other agents. The writers have observed this phenomenon in a study of the complex of malate and $NADH_2$ dehydrogenases from membranes of *M. lysodeikticus*. The complex was partially detached from the membrane by deformation in the course of osmotic shock during lysis of the cell (20%) and was extracted to that extent by EDTA, while the part that remained in the membrane (about 40%) was solubilized with Tween-80. The complex of these dehydrogenases is not an exception: all other proteins of the membrane can also be extracted consecutively by deformation of the membrane, removal of Mg^{++} ions, and weakening of the hydrophobic interactions by detergent. This has been shown by gel-electrophoresis of various preparations of membranes by Takayama's method (Tsfasman et al., 1972a,b).

As shown by the results of analytical centrifugation (sedimentation coefficient 12.8 S) and of electrophoresis (its mobility in a buffer, pH 7.4, is 1.54×10^{-4} $cm^2V^{-1}sec^{-1}$), after several stages of purification (filtration through Sephadex G-200, centrifugation in a sucrose density gradient, etc.), the dehydrogenase complex solubilized by EDTA is an almost homogeneous preparation. Further investigation of this enzyme complex showed, however, that besides malate dehydrogenase and $NADH_2$ dehydrogenase, it also contains several other proteins and is a heterogeneous mixture of particles with a Stokes radius (R_s) of about 100 Å (shown by the results of filtration through Sag-4 agarose). After negative staining, bands of the membrane forming closed rings or spherical particles can be seen in the electron microscope. The number of basic and acidic amino acids in the protein of this complex is approximately equal, so that the marked acidic properties of the complex may be explained by additional acid groups, possibly phosphate groups of lipids. About 55% of the amino acids are nonpolar (Zhukova et al., 1966; Zhukova, 1969). By treatment with EDTA, Nachbar and Salton (1970) also obtained fairly large fragments of the membrane of *M. lysodeikticus* containing $NADH_2$ dehydrogenase, but without malate dehydrogenase.

The distribution of the same enzymes among particles of different sizes suggests that these enzymes in different parts of the membranes of *M. lyso-deikticus* are surrounded by different proteins and lipids, and that the magnesium "bridges," which are broken by EDTA during extraction of the complex, are arranged irregularly. In other words, the complexes which the writers isolated are not uniform structural units. The existence of such units as blocks measuring $114 \times 114 \times 50$ Å has been postulated and, to some extent, proved for mitochondrial membranes (Green et al., 1967).

In contrast to the fragments extractable from membranes by EDTA, in which malate dehydrogenase and $NADH_2$ dehydrogenase coexist, there is another series of proteins (a dehydrogenase complex) which has been obtained by extraction with Tween-80 from membranes treated with EDTA. It contains two dominant protein components (mol. wt. about 70,000) and several minor components. This is thus a small lipoprotein whose extractability from the membrane is determined by its molecular environment (Tsfasman et al., 1972b). Similar uniform lipoprotein complexes also exist in other bacterial membranes. Cells of mutant M 46 of $R.$ $rubrum,$ if grown under certain conditions, secrete a lipopolysaccharide–protein complex with molecular weight of about 1.5×10^6 into the medium (Schick and Drews, 1969). This complex (sedimentation coefficient 22 S) contains 21.5% bacteriopheophytin, 61% protein, and 17.5% carbohydrates. After negative staining with phosphotungstic acid, it appears as long filaments 40 ± 5 Å in diameter, which break up into smaller subunits ($K_s = 1.85$ S) on treatment with detergents. The filament structure is evidently an artifact, for the complex can move in a 2.6% polyacrylamide gel with a pore diameter of about 50 Å. The authors cited claim that the complex has lost its ability to be incorporated into the membrane as a result of a mistake during the synthesis of its protein. The mechanism of excretion of this huge particle from the bacterial cell is purely a matter of conjecture.

Arguments for and against the existence of biological membranes as an aggregation of lipoprotein complexes are not confined to the facts already discussed (see Stoeckenius and Engelman, 1969); we shall return to this problem in the next chapter when describing cytochrome–dehydrogenase complexes. The only other point we shall make here is that the velocity of isotope exchange in the protein and lipid molecules varies in animal membranes (Omura et al., 1967). This means that even if lipoprotein complexes are components in the molecular organization of membranes, they can hardly play the role of unified construction blocks, for in that case the breakdown and synthesis of the proteins and lipids of the membranes would take place simultaneously. Generally, protein synthesis in mycoplasmas can be inhibited by chloramphenicol, whereas lipid synthesis still continues. As a result, the lipid: protein ratio in the membrane is altered (Razin, 1969). The protein:lipid ratio can also vary considerably (from 4.5 to 1.7) in the membranes of *Streptococcus pyogenes* on conversion into the L-form (Panos, 1968).

What is the nature of the bond between lipids and proteins within the lipoprotein complexes from bacterial membranes? Nothing more can be said at the present time regarding the structure of the complexes than to discuss the molecular anatomy of the membrane as a whole. The amino acid composition of the total protein and the behavior of dehydrogenase complexes from membranes of *M. lysodeikticus* during fractionation on ion-exchange resins suggest that the phospholipids of the complex are located superficially, with their polar groups facing outward. They possess some freedom of movement over the particle (Ostrovskii, 1968). McClare (1967), who analyzed the fragmentation of *H. halobium* membranes by the action of various agents,

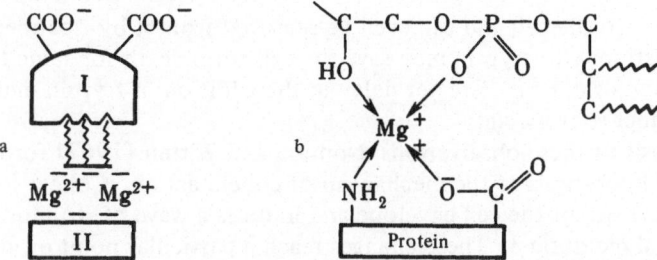

Fig. 18. Hypothetical structure of lipoprotein complexes in membranes of the halophile *Halobacterium halobium* (a), and role of Mg^{++} ions in lipid—protein interactions (b) (McClare, 1967). Broken line represents hydrocarbon moieties of lipids; I and II represent protein molecules.

considers that the components of the complex are linked together by both hydrophobic and ionic bonds (Fig. 18a), and that the strength of the ionic bond between lipids and protein through Mg^{++} is determined by additional coordination bonds (Fig. 18b).

Bacteriocins and Bacterial Membranes

Many authorities consider that the key to understanding the structure of bacterial membranes can be obtained by more careful and intimate study of the action of bacteriocins and bacteriophage envelopes on the membrane.

Bacteriocins are a varied group of proteins with molecular weight of 30,000-100,000 secreted by bacterial cells (Maeda and Nomura, 1966; Nomura, 1969; Holland, 1969a; Kunugita and Matsuhashi, 1970). They are essentially antibiotics for certain closely related strains (Kudlai and Likhoded, 1966; Garyaev, 1970). The most widely studied group of bacteriocins is that of the colicins, produced by *E. coli*. Colicins attack cells by a "single hit" mechanism, i.e., the character of the titration curve of a sensitive culture with the colicin suggests that one molecule of colicin is sufficient to cause the death of one cell. In fact, for a cell to be infected, many molecules of colicin (about 100) have to be adsorbed, but only one of them is effective.

The cause of death of bacterial cells after adsorption of colicin varies (Reeves, 1968). For instance, colicin E_1 uncouples oxidation and phosphorylation without inhibiting respiration or producing any significant disturbance of membrane permeability (Hirata et al., 1969; Fields and Luria, 1969). E_2 causes degradation of DNA; E_3 inhibits protein synthesis without affecting the synthesis of nucleic acids (Nomura, 1969). The most important factor in the action of the colicins, and one which is common to all of them, is that they exert their action through the membrane. Special experiments by Maeda and Nomura (1966) showed that colicin, labeled with tritium and C^{14}, does not penetrate into the cell and can even be removed from it by treatment with trypsin. Digestion with trypsin may save the cell from death for some time after adsorption of colicin. Greater delay in the addition of trypsin makes the colicin effect irreversible.

On the basis of these observations, Nomura and Witten (1967) formulated the following hypothesis of the mechanism of colicin action. Colicin settles on the acceptor site of the cell envelope and induces a wave of structural changes in the bacterial membrane. These changes reach a particular point on the membrane, from which the effect is transmitted inside the cell through interaction with nucleic acid or with intracellular proteins.

Recently, some details have been added to this general picture. It is considered that when the colicin is adsorbed on the surface of the cell, it forms a complex (K-I) with a certain substance in the membrane or rigid layer. After this formation, K-I is converted into a new complex (K-II), which activates the lytic enzymes located on the inside of the membrane. The probability of K-I's conversion into K-II is slight and depends on many factors, including the protein composition of the membrane. Consequently, a change in the assortment of membrane proteins as the result of mutation may lead to a change in sensitivity to colicins (Holland et al., 1971; Rolfe and Onodera, 1971).

Special experiments have shown that colicins of the E group (E_1, E_2, E_3, etc.) compete for the same acceptor, but they transmit their lethal impulse from the point of absorption of the membrane in different ways (Nagel de Zwaig and Luria, 1967; Nomura and Witten, 1967). This was demonstrated with the aid of mutants sensitive to some colicins and insensitive to others of the same group (Nagel de Zwaig and Luria, 1967). It is considered that the number of adsorption points on the bacterial cell varies from several dozen to thousands. For example, E. coli can adsorb up to 2000-3000 molecules of E_2 (Maeda and Nomura, 1966), but the mechanism of adsorption is still a matter for discussion. Smarda and Taubeneck (1968) claim that adsorption of colicin takes place directly on the cytoplasmic membrane, for in their experiments colicin acted so quickly on the L-forms of E. coli and Proteus mirabilis that the addition of trypsin very soon after the bacteriocin could no longer save the cell from death. In some cases, the colicin acted as if it were added at the same time as the trypsin. Since, as the results of chemical analysis showed, the L-forms used did not contain components of the cell wall and did not react with bacteriophages, the hypothesis of the localization of bacteriocin acceptors on the membrane proposed by these workers seems perfectly justified. Other evidence of the direct action of bacteriocins on the cell membrane is given by the discovery of sensititivy to the action of colicins E_1 and K in membranes obtained from a resistant strain of E. coli (Bhattachariya et al., 1970).

It is interesting to note that incubation of E. coli CA-42 cells for 30 min with 2,4-dinitrophenol (2×10^{-3} M) before treatment with colicin delays the action of colicin E_2 by at least 2 h. During this time, trypsin can remove the colicin adsorbed on the cell surface (Reynolds and Reeves, 1969). The compound 2,4-dinitrophenol is an inhibitor of oxidative phosphorylation and a disorganizer of biological membranes. This confirms once again the view (Changeux et al., 1967; Nomura, 1969) that the lethal impulse is transmitted by bacteriocins in the form of a wave of conformational changes of the components of the membrane. Changes of this type have been recorded as an increase in ANS fluorescence in the membranes of *E. coli* treated with colicin E_1 (Cramer and Phillips, 1970).

Although the action of bacteriocins on bacterial cells has not yet departed from the realm of conjecture, it does nevertheless suggest that the bacterial membrane is a structural and functional mosaic, whose state may largely depend on the adsorption of single molecules with regulatory functions.

While the ultimate effect of bacteriocins on the cell is lethal, the initial effect of certain colicins, such as that of colicin E_2 on *Salmonella typhimurium,* is manifested as the stabilization of the bacterial membranes, which can be recorded as a decrease in the intensity of lysis of the cells in a hypotonic medium (Nose et al., 1970). Meanwhile, colicin E_2 in the presence of ATP causes dissociation of the DNA + RNA + membrane complex, whereas without ATP, the nucleic acids are not removed from the membrane (Beppa and Arima, 1970). Feingold (1970) considers that the primary process in the action of colicin E_1 on the cell is to induce or catalyze the oxidation of the membrane lipids, thereby causing profound conformational changes in the membrane.

Several hypotheses have been put forward with regard to colicin E_3. Boon (1972) showed that colicin E_3 causes degradation of ribosomes *in vitro.* He considers that E_3 may force some of the molecules through the membrane and act directly on the ribosomes, although other investigators consider that E_3 merely induces the appearance of a substance in the cytoplasm which leads to destruction of the ribosomes (Ohsumi and Maeda, 1972).

Sensitivity to bacteriocins can be altered reversibly in certain strains (conditional mutants) by lowering or raising the temperature of the medium (Nomura and Witten, 1967). For instance, strain *E. coli* ER 437, which is sensitive to colicin E_2 at 30°C, becomes resistant after incubation for 30 min at 40°C if chloramphenicol is added to the ordinary medium, but remains sensitive after incubation for the same time in buffer. These workers infer that energy must be supplied to the membrane for the transition from sensitivity to resistance to take place, and that this process is unconnected with protein synthesis.

A similar phenomenon has been observed with another strain, ER 438, which is resistant to colicins at 37°C but sensitive to them at 40°C. The conversion from one state into the other could easily be observed in the presence of chloramphenicol, suggesting that certain internal reversible structural changes take place in the membrane which are independent of the synthesis or death of the protein in the cell itself.

Changes in the sensitivity of mutant cells to individual colicins are accompanied by the acquisition of certain new properties by the membranes of these cells. For instance, mutants tol II and tol III (Nagel de Zwaig and Luria, 1967) die (undergo lysis) in a medium containing 0.5% deoxycholate or 0.001 M versene, although the original strain and the colicinogenic strain (immune to colicin) can be grown in the presence of these substances.

The task of elucidating the mechanism of action of the bacteriocins would be made much easier if studies could be conducted on subcellular fractions, for example, in fragments of bacterial membranes. However, it has not yet been found possible to transfer the study to the subcellular level. Holland, for instance, showed that megacin A disturbs the permeability of the protoplasts of sensitive cells so that they gradually undergo lysis, but it has no appreciable lytic action on a preparation of the membranes (Holland, 1969b). As has already been mentioned, colicin E_1 strengthens ANS fluorescence in the membranes of $E.$ $coli$ (Cramer and Phillips, 1970), but colicin K (Hirata et al., 1969) uncouples oxidation and phosphorylation in some cells of $E.$ $coli,$ although it does not inhibit oxidative phosphorylation in a preparation of membranes. All these facts indicate that if the bacterial cell is destroyed under "mild" conditions, substantial changes can be introduced into the structure of its membrane.

How Can the Bacterial Membrane Be Represented?

If we approach the molecular organization of bacterial membranes from the standpoint of the theory of the origin of life on earth developed by Oparin (1960), we can assume, as do Stoeckenius and Engelman (1969), that the primary membrane of the protobionts (the first inhabitants of the earth) was a double hydrocarbon film. Stoeckenius and Engelman consider that the basis of this membrane, consisting of a bimolecular layer of lipids with the polar groups facing outward, has persisted without much change until the present day, and that in all living organisms its appearance is very close to Robertson's modification of the scheme of Davson and Danielli (Fig. 19a).

We consider, however, that biological membranes, as they have developed, have not remained exactly as they were in their first stage but have undergone a series of consecutive qualitative changes connected with the formation of closer contact between the lipids and proteins of the cell. In some contemporary bacterial cells there are droplets of lipid substances, but it is unlikely that lipid molecules exist in a state of true solution in the cytoplasm. Lipid molecules are synthesized by a wide range of enzyme proteins, and until the termination of their stay in the cell they are evidently in the form of complexes with protein. In other words, we consider that the

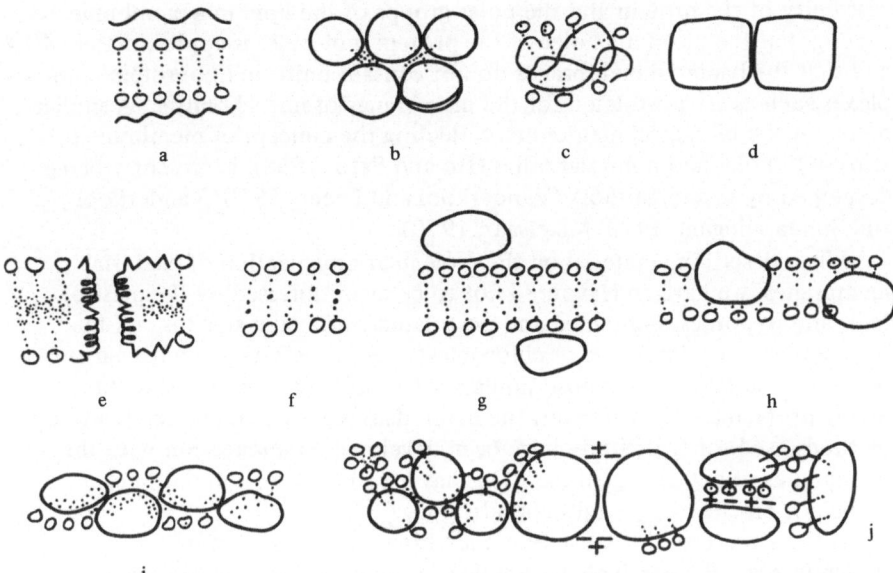

Fig. 19. Structural models of biological membranes: a) The Danielli–Davson–Robertson model: proteins of different types, conjecturally in the β-form, cover a bimolecular layer of lipid with its polar groups facing outward (Danielli and Davson, 1935; Robertson, 1959, 1960). b) The model of Frey-Wyssling and Mühlethaler (1968): double layer of globular proteins (diameter of the molecule about 45 Å), linked by hydrophobic amino-acid residues; the lipid lies between the protein molecules. c) Benson's (1966) model. d) The Green–Perdue–Sjöstrand model: complexes consisting of several enzymes and noncatalytic proteins; the molecules of the lipid are situated on the outer surfaces of the complexes with their polar groups facing outward (Green and Perdue, 1966; Sjöstrand, 1963). e) The Lenard–Singer–Wallach–Zahler model (Lenard and Singer, 1966; Wallach and Zahler, 1966): the protein molecules are so arranged that their helical segments are surrounded by hydrophobic groups, while their random segments are situated more superficially. f-i) Hypothetical path of development of the biological membrane. j) Model of a bacterial membrane: The membrane consists of qualitatively different areas linked together mainly by forces of hydrophobic interaction; a,b,d,e proteins shown as continuous wavy or broken lines. In schemes b,g,h,i,j, proteins shown by large circles or ellipses; small circles represent polar groups of lipids; broken lines represent hydrophobic parts of molecules; + and − signs indicate ionic interactions, including magneisum "bridges."

most natural state for the lipid in a living cell is in the form of a complex with protein, stabilized by hydrophobic interactions.

To speculate for a moment, let us try to represent the path of development of biological membranes from lipid vesicles on a lifeless earth to contemporary bacteria (Fig. 19f-i). We suggested the variant 19i in 1968 at a symposium in Erevan as the most convenient method of explaining the simultaneous accessibility of the protein and the polar groups of the lipid in a membrane, assuming that the mean diameter of the protein molecules is approximately 40 Å and that the bacterial membranes do not contain uniform lipoprotein complexes such as are postulated for the membranes of mitochondria. A similar model of the biological membrane, reflecting the concept of membrane proteins with polar and nonpolar poles (Ito and Sato, 1968), is presently being developed by several authors (Vanderkooi and Green, 1970; Vanderkooi and Sundaralingam, 1970; Mikelsaar, 1970).

When describing material on the molecular organization of bacterial membranes, we have endeavoured not to be overimpressed by the mass of facts and hypotheses obtained and postulated for myelin sheaths and the membranes of erythrocytes or chloroplasts, i.e., for structures with narrowly specialized functions, whose similarity to bacterial membranes is still highly problematical. However, the main ideas regarding membranes and the methods used for their study have been developed in connection with the membranes of higher organisms. This has led to the creation of a special branch of biochemistry: mitochondriology.

If the structure of the bacterial membrane is represented by the scheme shown in Fig. 19j, which allows ample scope for criticism, we are forced to label the individual components with names taken from model schemes of various biological membranes (see the survey by Singer, 1971).

The following assumptions were made in the preparation of this scheme:

1. The structural basis of membranes consists of about 30 proteins in number, no one component is quantitatively predominant, and their molecular weights vary from 10,000 to 200,000.
2. The polypeptide chains of the protein molecules in the membrane form random coils, in some places forming α-helices.
3. Protein molecules are linked with one another and with lipids mainly by hydrophobic bonds, although ionic and hydrogen bonds are also important.
4. The lipid, which is located near the hydrophobic sites on the protein, may form large aggregates and, perhaps, a continuous network in the plane of the membrane.
5. Components of the membrane are mobile, and their motion may lead to a change in the charge and lipophilicity of individual areas of the membrane as they perform their functions.

It is relevant to mention, as Luria (1969) points out, that the bacterial membrane can undergo considerable functional changes by acquiring new enzyme systems or losing a large proportion of certain components. We have already pointed out that because of their unusual polyfunctional character, bacterial membranes are evidently very heterogeneous from the structural point of view. An example of the narrow specialization of individual areas of the bacterial membrane is provided by the structure known as the proximal disk (Vaituzis and Doetsch, 1969), the basis of the flagellum or organ of movement of the cell. It is believed that the disk performs its rotary motion through the asynchronous contraction and relaxation of its polar lamellae, which function like a transmission wheel. The chemical composition of the proximal disk is not yet known; still less is known of its molecular organization.

Any scheme which purports to explain all the properties of a biological membrane at one and the same time must therefore be treated with considerable reservations.

A critical experiment which would explain the true structure of the bacterial membrane has not yet been devised. In the writers' opinion, one of the most promising lines of membrane research is the study of the primary structure of membrane proteins. In conjunction with existing physicochemical data on the state of the proteins, knowledge of their primary structure would give a more accurate interpretation of the conformation of the protein molecules responsible for the most important functions of membranes.

Physical methods of evaluating the state of the components of a membrane, such as spin-labeling, infrared spectroscopy, and differential temperature analysis, are also unquestionably very promising. Specific enzymic and chemical modification of membrane proteins, with the aid of fluorescent or paramagnetic agents, for example, is particularly interesting. This approach has already been used to study membranes (Maddy, 1967), and further developments may yield significant results.

CHAPTER IV

Bacterial Membranes as Polyfunctional Systems

As they learned more about the physiology of bacteria, research workers were struck by the remarkable variety of functions performed by the membranes. Electron transfer in the respiratory chain, transformation of energy during photosynthesis, nitrogen fixation, the transport of ions and metabolites, and the synthesis of components of the cell wall, as well as of other substances, all take place in the membranes. The membrane apparatus of the bacterial cell is also concerned with such complex physiological processes as division, sporulation, and cyst formation. All functions which are distributed in plant and animal cells among specialized membrane structures (Nass, 1969; Broda, 1970) are connected in bacteria with the cytoplasmic membrane and the system of internal membranes. The main difficulty hindering our understanding of the way in which bacterial membranes function stems from the absence of morphological differentiation and the impossibility of ascribing a particular function to a particular structure of the membrane. However, this does not mean that membranes themselves do not include areas which have undergone specialization for the performance of particular functions, although there is yet very little evidence in this respect. At present, we must regard the functions of the bacterial cell membranes as a whole. It is not yet possible to determine experimentally the precise localization of these functions.

Electron Transport

One of the most important functions of bacterial membranes is the organization of energy transformation processes. Enzymes of the respiratory chain are installed in the membranes of aerobic and facultatively anærobic, heterotrophic bacteria. Systems of electron-transporting enzymes for photosynthesis and chemosynthesis are installed in the membranes of autotrophs. In this sense, the membrane apparatus of bacteria can be compared with mitochondria or chloroplasts. The localization of electron-transporting enzyme systems in bacterial membranes is now universally accepted. The evidence on which this conclusion is based can be found in surveys; no further discussion of this subject is required (Gel'man et al., 1966; Ryter, 1969; Van Iterson, 1969a,b,c). The question of the distribution of electron transport enzymes between mesosomes and the cytoplasmic membrane is much less clear. We have already dwelt on this point when examining the biochemical data obtained by investigation of the mesosomes and cytoplasmic membrane separately (see page 17). The evidence of cytochemistry on this matter is extremely contradictory. A detailed analysis of this complex material will be found in the survey by Ryter (1969); it will not be repeated here. Ryter concludes that both mesosomes and cytoplasmic membrane carry electron-transporting enzymes, but that there is a difference between them because triphenyltetrazolium chloride is reduced only in mesosomes, tellurite is reduced only in the cytoplasmic membrane, and tetra-nitro-BT is reduced in both structures. The unequal reduction of redox indicators suggests either that there are differences in the composition of the respiratory chains or that their organization is different in the cytoplasmic membrane and mesosomes.

Any examination of the function of bacterial membranes cannot ignore the question of the function of the fungus-like appendages covering their surface (see page 83, Fig. 5). Similar structural elements are also found in mitochondrial membranes where, according to Racker (1967), they contain ATPase, an enzyme of oxidative phosphorylation. The same view regarding the functions of the fungus-like appendages of bacterial membranes (*Micrococcus lysodeikticus*) is held by Munoz and Salton and their collaborators (Munoz et al., 1968, 1969). A different view is held by Simakova and co-workers, who found that ATPase is localized in the stroma of the membrane (Simakova et al., 1969; Simakova, 1970). The possibility cannot be ruled out that other phosphorylation factors are localized in the appendages (Ostrovskii, 1964; Ishikawa, 1970). The localization of oxidative phosphorylation will be examined later in this chapter (see page 197).

Transport of Materials

One of the most enigmatic processes of the activity of living matter, the transport of materials into the cell, is entirely a function of the membranes; the permeability of the bacterial cell is determined by that of the cytoplasmic membrane (Mitchell, 1963, 1970; Marquis and Gerhardt, 1964; Kaback and Stadtman, 1966; Kaback, 1970a,b). It is possible that invaginations of the cytoplasmic membrane in the direction of the cell wall also perform an excretory function (Kushnarev, 1966c). Nothing is yet known of the role of mesosomes in the transport of materials, but such a role cannot be ruled out.

Substantial progress has recently been made in the study of the transport of organic substances in bacterial cells (see the surveys by Kaback, 1970a,b,

1972; Antonov, 1970). The situation is less satisfactory with regard to elucidation of the mechanism of permeability to inorganic ions (Harold, 1970).

In some bacteria, sugars are transported by phosphorylation in the membrane, with the participation of a system of P-enol-pyruvate-P-transferase enzymes. This process is carried out by a group of enzymes (enzyme I and enzyme II); the enzymes are installed in the membrane and require phosphatidyl glycerol for their action. Enzymes of the II group have been solubilized from the membrane of *Escherichia coli* and purified (Kundig and Roseman, 1969, 1971; Gale, 1971). Enzyme II B is homogeneous, its molecular weight is 35,000, and it readily aggregates. Enzyme II A, in fact, contains three different proteins: specific for glucose, mannose, and fructose. The whole set of enzymes of the P-enol-pyruvate-P-transferase system transports sugars through the membrane by translocation of groups during vectorial phosphorylation of the sugars in the membrane. Dephosphorylation takes place on the inner side of the membrane or in regions of the cytoplasm lying next to it. The action of the enzyme system is determined by the conformation of protein components bound in the membrane (enzymes of group II) (Kaback, 1969, 1970a,b). Transport of sugars with the aid of the P-enol-transferase system has also been found in obligate anaerobes (*Clostridium thermoaceticum*), aerobes (*Bacillus subtilis*), and facultative anaerobes (*Staphylococcus aureus, E. coli, Salmonella typhimurium*) (Kaback, 1970b). This process is evidently independent of the presence of a respiratory chain.

The role of specific proteins mounted in the membrane is also revealed in the study of the transport of lactose and of β-galactosides in general (Kepes, 1971). A protein (M-protein) responsible for lactose transport has been found in the membranes of *E. coli*. The protein is firmly bound in the membrane and is solubilized only by detergents. The M-protein accounts for up to 3% of the total proteins of the membrane. Its molecular weight is about 30,000, although this figure may depend on the use of a detergent for its isolation and purification (Jones and Kennedy, 1969; Kennedy, 1969). The mechanism of binding lactose or β-methylgalactosides with the M-protein is unknown. Perhaps the conformation of the M-protein is changed on account of the energy of lactate oxidation in the respiratory chain, and this process is coupled with the transfer of β-methylgalactoside (Barnes and Kaback, 1970). There is every reason to suppose that the transport of glycine and proline, which has been studied in isolated bacterial membranes, involves the participation of firmly bound proteins, just as occurs in the transport of sugars (Kaback and Stadtman, 1968; Kaback and Deuel, 1969).

It is more difficult to explain the role of the membrane in the transport of other substances which are bound by proteins that are solubilized by osmotic shock of the cell. This category includes proteins binding phosphate, sulfate, leucine, arginine, arabinose, and galactose (see the surveys by Kaback, 1970a; Antonov, 1970). The phosphate-binding protein leaves the cells of or spheroplasts of *E. coli* during cold osmotic shock. It is evidently bound with the membrane, but the nature of the bond has not been determined (Medveczky and Rosenberg, 1970). The sulfate-binding protein has been purified to homogeneity. It has a molecular weight of 32,000 and 1 mole of the protein binds 1 mole of sulfate (Pardee, 1966, 1967; Langridge et al., 1970). This protein passes into solution on the formation of spheroplasts of *S. typhimurium.* The solubility of the protein and its freedom from lipids suggest that if it is bound with the membrane, the bond is very weak. Thus, the way in which the sulfate–protein complex transports sulfate through the membrane has not been fully explained. Essentially the same problem arises with the transporting role of leucine-binding protein of *E. coli,* which has been obtained in the crystalline form. Its properties have been described in detail by Penrose et al. (1968). From their recent experiments, Penrose et al. (1970) conclude that the protein binds leucine without undergoing any change in its conformation, and the protein–amino acid complex passes in this form to the inside of the membrane, where it breaks up.

The transport of widely different materials through the bacterial membrane is thus accomplished by specific proteins which form labile compounds with the substance to be carried. Despite this considerable progress in the understanding of the transport of materials through the bacterial membrane, the very important question of how energy is transported remains unexplained (Harold, 1970). This mechanism must evidently bring the membrane as a whole, or its individual parts, into a state in which the carrier proteins acquire the necessary conformation for binding the substrate on one side of the membrane and transferring it to the other side, followed by breakdown of the complex formed by the protein and the transported material (Barnes and Kaback, 1971; Kaback and Barnes, 1971; Kerwar et al., 1972; Short et al., 1972).

A feature which distinguishes bacterial transport as a whole is the ability of bacteria belonging to widely different physiological groups, with different types of energy metabolism, to absorb the same substances, e.g., amino acids, sugars, K^+, Na^+, Mg^{++}, SO_4^-, NO_3^-, etc. The obligate anaerobes of fermentation, whose membranes do not generally contain electron-transporting enzymes, do in fact absorb substances from the medium just as readily as bacteria whose membrane apparatus is saturated with electron-transporting systems, and in which the high-energy compounds are generated in the membrane itself. Two assumptions must therefore be made regarding the transport of energy in bacteria: 1) the mechanism of transport of materials and inorganic ions in all bacteria is the same regardless of the source of energy; and 2) the mechanism of transport in bacteria whose membranes do not possess an electron-transfer chain differs from that in bacteria in whose membranes either oxidative or photosynthetic phosphorylation takes place. To judge from the interesting experiments of Harold and co-workers (Harold and Baarda, 1969; Pavlasova and Harold, 1969), the first alternative seems more probable. These experiments showed that the transport of ions and metabolites in the cells of *Streptococ-*

cus faecalis and in anaerobic *E. coli* cells is suppressed by typical uncouplers of oxidative phosphorylation. The fact that such a phenomenon can take place in bacteria which have no respiratory chain or which have switched over to anaerobiosis is evidence that the mechanism of transporting ions and metabolites is the same in all bacterial cells, regardless of how energy is supplied to the membrane.

There can be no doubt that the transport of cations is a fundamental property of all bacterial membranes. In order to understand its mechanism, like that of the transport of metabolites, it must be remembered that the process takes place in membranes with different levels of energetics. Presumably the mechanism of this process is a very old one and has persisted in all bacteria regardless of their further development. More precisely, the mechanisms by which inorganic cations penetrate the cells of anaerobic bacteria (whose membranes have no respiratory chain) must also exist in bacteria that do possess electron-transfer systems. We shall not probe deeper into this extremely complex problem, which is being investigated particularly intensively in work on the membranes of mitochondria and chloroplasts (Skulachev, 1969, 1972; Lieberman and Skulachev, 1970).

It is still difficult to say whether the transport mechanisms of the cytoplasmic membranes of bacteria are repeated in the organelles of higher organelles of higher organisms, or whether they have developed completely different transport systems. The only fact no longer in doubt is that the membranes of bacteria belonging to different physiological groups contain ATPase. This enzyme has been found in the membranes of *S. faecalis, Streptococcus pyogenes, Bacillus megaterium, M. lysodeikticus, S. aureus, Mycoplasma* sp., and *Lactobacillus casei* (Abrams et al., 1960; Weibull et al., 1962; Ishikawa and Lehninger, 1962; Sokawa, 1965; Pollack et al., 1965; Simakova et al., 1968; Gross and Coles, 1968; Munoz et al., 1969; Barker and Thorne, 1970). ATPase is the most constant component of bacterial membranes; this naturally suggests that its physiological role must be exceptional. In membranes of bacteria without respiratory chains, ATPase probably participates in the transport of Na^+ and K^+, while in cells containing a respiratory chain, the action of this enzyme is possibly coupled with the respiratory chain. The role of ATPase in cation transport is illustrated by the increased electrical conduction of the lipid bilayer occurring through interaction with the ATPase from membranes of *S. faecalis*. Conduction is stimulated by Mg^{++}, Na^+, K^+, and ATP (Redwood et al., 1969).

Results showing the formation of complexes of cyclodepsipeptides (valinomycin and enniatin) and K^+, and showing the influence of these substances on K^+ transport are of great importance to understanding the mechanism of permeability of bacterial membranes (Ovchinnikov et al., 1969; Shkrob et al., 1969; Shemyakin et al., 1969). Model experiments to simulate this process have led to the suggestion that the membrane itself may contain

protein molecules which, when in a certain conformation, can bind with cations; being buried in the hydrophobic regions of the membrane, these proteins
protect the cations from water. The similarity between the temperature dependence of phase changes and the kinetics of carbohydrate transport in the membrane of *E. coli* also indicates that the transport systems are located in the
region of fluid, mobile areas of the membrane (Wilson and Fox, 1971b).

Cell Division and Spore Formation

The membrane apparatus of bacteria participates in cell division, and in
the formation of spores and cysts. It is postulated that stretching the membrane during growth of the cell gives the nucleoid the signal to divide. The
distribution of "roles" between the cytoplasmic membrane and the mesosomes in this communication system is not clear.

In a series of investigations, Ryter and collaborators found communica-
tions between the nucleoid and the cytoplasmic membrane of *B. subtilis*
through the mesosomes — unique signposts to cell division (Ryter and
Jacob, 1963; Ryter and Landman, 1964; Jacob et al., 1966). From a comparison of the structure of mesosomes at different stages of development of
the *Lactobacillus plantarum* cell, Kakefuda et al. (1967) also conclude that
mesosomes play some part in cell division. Cells of *E. coli* mutants resistant
to irradiation contain many mesosomes in the lag phase which undergo
degradation in the logarithmic phase. The intensified development of the
mesosomes is evidently connected with division of the genomes in the lag
phase (Pontefract and Thatcher, 1970). The close connection between the
mesosomes and loci of new cell wall formation in streptococci also indicates
a role of the mesosomes in cell division (Cole, 1965). However, division can
take place in cells without mesosomes, as has been shown for *B. subtilis* grown
on gelatin, and for reversive forms of protoplasts and young cells arising during
germination of spores (Ryter and Landman, 1968). For example, in the absence of mesosomes in the protoplasts of Gram-positive bacteria, a direct connection is found between the nucleoid and the cytoplasmic membrane (Ryter
and Jacob, 1966; Fuhs, 1966). In Gram-negative bacteria (*E. coli*), which
have no mesosomes, the nuclear material is connected with the cytoplasmic
membrane (Ryter and Jacob, 1966; Lark, 1966). However, in strains of
E. coli which do have mesosomes, their development does not correlate with
DNA synthesis or with division of the nuclear material (Altenburg and Suit,
1970). Poor development of the mesosomes in young cells of *Clostridium
sporogenes* also points to the essential role of mesosomes in the processes
of cell division (Hoeniger and Headley, 1969).

Recently segments with attached DNA have been isolated from the membranes of several bacteria by means of the detergent Sarkosyl (Tremblay et
al., 1969; Rubenstein et al., 1970; Shull et al., 1971). A complex of DNA with
membrane lipoproteins has been found in lysates of *B. subtilis* cells (Ivarie

and Pene, 1970). The change in the protein composition of membranes of
E. coli mutants with disturbed initiation of DNA synthesis is evidence of a
direct connection between the state of the membranes and replication (Shapiro
et al., 1970). A similar conclusion can also be drawn from the fact that one
protein is deficient in membranes of an *E. coli* mutant which cannot synthesize
DNA (Inouye and Guthrie, 1969).

Participation of membranes in spore formation has been demonstrated
cytologically for several bacteria: *B. subtilis, B. megaterium,* and *Bacillus
cereus* (Fitz-James, 1960; Van Iterson, 1961; Kawata et al., 1963; Ryter,
1965; Ellar and Lundgren, 1966; Freer and Levinson, 1967; Ellar et al.,
1968). During cyst formation the number of mesosomes in the cells of
Sporocytophaga myxococcoides increases (Holt and Leadbetter, 1967).

Although bacteria have no clearly defined endoplasmic reticulum, there
is nevertheless evidence to show that ribosomes can be connected with the
internal membrane structures and that protein synthesis takes place faster
in such ribosomes than in free ribosomes. Observations of this type have
been made for the ribosomes of *S. faecalis, E. coli, B. megaterium, Bacillus
anitratum, M. lysodeikticus,* and *Bacillus coagulans* (Schlessinger, 1963; Abrams
et al., 1964; Tani and Hendler, 1964; Godson and Butler, 1964; Hallberg and
Hauge, 1965; Shul'ga and Tongur, 1966; Zaitseva et al., 1970; Scharff et al.,
1972).

Attachment of ribosomes to membranes has been observed in *B. cereus,
B. megaterium,* and *E. coli* (Pfister and Lundgren, 1964; Schlessinger et al.,
1965; Hendler and Nanninga, 1970; Cundliffe, 1970). According to Aronson,
ribosomes are attached to membranes through a polypeptide formed in them
(Aronson, 1966). In the survey by Burd (1967), the connection between
polysomes and membranes in the period of active protein synthesis is emphasized. Meanwhile, Partterson et al. (1970) assert that binding of ribosomes
to the membrane in *E. coli* is found only if the cells have been "opened" by
lysozyme. Some interesting observations on the link between synthesis of
ribosomal RNA and the cytoplasmic membrane have been made by Haywood
(1971).

The Membrane and Biosynthesis

Bacterial membranes carry not only electron-transfer enzymes and mechanisms of the transport of materials, but also a number of enzymes of biosyntheses which assist in the formation of components of the membrane
itself and, in particular, of the cell wall. The fact that many biosynthetic processes take place in the membrane is determined, on the one hand, by the role
of the membrane lipids as activators of the corresponding enzymes and intermediate products and, on the other hand, by the closeness of the membrane
to the cell wall, so that the transfer of the products formed in the hydrophobic
stroma of the membrane to the cell wall is facilitated. In addition, the bac-

terial membrane also contains enzymes concerned with metabolism of the membrane lipids.

The synthesis of several components of the cell wall (peptidoglycan, teichoic acids, polysaccharides, O-antigen) in various bacteria takes place with the participation of undecaprenol monophosphate, which carries residues of sugars and related substances from nucleotide precursors to the chains of polymers in the cell wall.

Enzymes participating in the synthesis of peptidoglycan, polysaccharides, and lipopolysaccharides are found in the membranes of various bacteria (see the surveys of Ellar, 1970; Ghuysen et al., 1969; Nikaido, 1968). We shall give only a few examples. The membranes of $M.$ $lysodeikticus$ contain isoprenol phosphokinase, responsible for one stage in the complex process of synthesis of the peptidoglycan of the cell wall (Higashi and Strominger, 1970). C_{55}-isoprenol (bactoprenol) has been found in the membranes of $L.$ $casei$ (Barker and Thorne, 1970), where it probably participates in the synthesis of components of the cell wall (Lennarz and Sher, 1972). The mechanism of synthesis of teichoic acid and peptidoglycan in $Bacillus$ $licheniformis$ has been studied in detail by Watkinson et al. (1971). The enzymes UDP-glucose dehydrogenase and UDO-galactose epimerase, which participate in the synthesis of intermediate products of the cell wall polysaccharides, are present in the membranes of $Klebsiella$ $aerogenes$ (Sutherland and Norval, 1970). Another enzyme concerned in the formation of the lipopolysaccharides of the bacterial cell wall is UDP-galactose: lipopolysaccharide α,3-galactosyl transferase, present in the membranes of $E.$ $coli$ and $S.$ $typhimurium$. This enzyme is activated by phosphatidyl ethanolamine (Rothfield et al., 1969, 1972).

Enzymes responsible for the conversion of phosphatidic acid into phosphatidyl ethanolamine and phosphatidyl glycerol are localized in the membrane of $Bacillus$ sp. (Patterson and Lennarz, 1969, 1971). Biosynthesis of glucosyl diglycerides takes place in the membrane of $Mycoplasma$ $laidlawii$ (P. Smith, 1969). All the enzymes for the synthesis of phospholipids are present in the membrane of $E.$ $coli$ and $S.$ $typhimurium$ (Chang and Kennedy, 1967; White et al., 1971; Bell et al., 1971); an enzyme participating in the formation of diphosphatidyl glycerol from phosphatidyl glycerol has been found in the membrane of $M.$ $lysodeikticus$ and $S.$ $aureus$ (De Siervo and Salton, 1971; Short and White, 1972). Mannosyl-1-phosphoryl-undecaprenol synthetase has been found in the membrane of $M.$ $lysodeikticus$ (Lahav et al., 1969). The synthesis of mannosyl diglyceride takes place in the membrane of $M.$ $lysodeikticus$ (Lennarz and Talamo, 1966).

The membranes of $Mycobacterium$ $phlei$ contain phospholipase A (Ono and Nojima, 1969). Those of $Haemophilus$ $parainfluenzae$ contain phospholipase D (Ono and White, 1970), while the membranes of $M.$ $laidlawii$ contain lysophospholipase (Van Golde et al., 1971). This group of enzymes probably participates in the lipid metabolism of the membranes. A protein transferring acyl residues is located in the membrane of *E. coli* (Van den Bosch et al., 1970).

Other enzymes of biosynthesis have been found in bacterial membranes. The membranes of *E. coli,* for instance, contain a firmly bound thiamine kinase (Miyata et al., 1967); the membranes of *S..typhimurium* contain a group of enzymes synthesizing isoleucine (Cronewett and Wagner, 1965); toxin formation in *Corynebacterium diphtheriae* is connected with the membranes (Uchida and Yoneda, 1967). Enzymes concerned with the synthesis of the benzoquinone ring are bound with the membrane (Rudney, 1969), and octaprenol transferase also is located in the membranes (Young et al., 1972).

In the cells of *Azotobacter vinelandii,* nitrogen fixation takes place in the membranes (Yakovlev et al., 1965; Lyubimov, 1969). Hydrogenase is firmly bound with the membrane (Yakovlev and Mitsova, 1970). The development of the membrane systems of *A. vinelandii* depends on the form of nitrogen in the growth medium. Cells fixing nitrogen have a well developed system of internal membranes; cells grown on a medium with NH_4Cl do not contain such a system (Oppenheimer and Marcus, 1970). Despite this difference, some investigators have succeeded in isolating a nitrogen-fixing enzyme system composed of macromolecular protein complexes; these complexes were evidently separated from the membrane (Bulen et al., 1964; L'vov et al., 1970; Gvozdev et al., 1971).

Some functions are found only in the membranes of certain species of bacteria. Examples are the arrangement of the basal particles of the flagellae of some bacilli (Abram et al., 1965, 1966; Mirsky, 1970) and the presence of penicillinase in the membrane of *B. licheniformis* (Sargent and Lampen, 1970).

Every enzyme system, or even every individual enzyme, such as penicillinase, must evidently be so fixed in the membrane that it can bring about interaction between the components of the system and between the system and the substrate, so that the supply of substrates and the removal of the reaction products into the cell or surrounding medium from the membrane are regulated. The study of the organization of enzyme systems in the membrane is thus the ultimate key to understanding the organization of the membrane as a whole. However, at present we have information (and that far from complete) about only one enzyme system — the electron transfer system; virtually nothing is known about the organization of the other enzyme systems listed above or about individual enzymes. It may therefore be useful to examine in some detail the process of electron transfer in heterotrophic and chemoautotrophic bacteria (the respiratory chains) in order to obtain a closer insight into the molecular organization of the system in connection with the organization of the membrane as a whole.

The Respiratory Chain and Its Organization in the Bacterial Membrane

The General Scheme of the Respiratory Chain

Most bacteria have a respiratory chain,* i.e., a system of spatially organized dehydrogenases and cytochromes which carry out the transfer of electrons from the substrate. The main difference between aerobic bacteria and facultative anaerobes is in the terminal electron acceptor: in aerobic bacteria this is oxygen, while in facultative anaerobes it may be either oxygen or an acceptor of the nitrate or sulfate type.

The butyric acid and lactic acid fermentation bacteria have no cytochromes and, consequently, no respiratory chain. The photosynthesizing bacteria have highly organized electron transfer systems, but the study of their respiratory chain is complicated by the impossibility of separating the respiratory chain from the chain of photophosphorylation. Exceptions are found in some species which, when grown in darkness and under aerobic conditions, develop a normal respiratory chain.

The factual evidence of electron transfer enzymes in bacteria is collected in a series of surveys (Dolin, 1963; L. Smith, 1963, 1968; Newton and Kamen, 1963; Kiesow, 1967; Peck, 1968; Singer, 1968a; Kotel'nikova, 1969). In view of the great variety of composition of the respiratory chains, oxidation substrates, and terminal acceptors, it is difficult to combine them all into a single scheme. A general scheme of the respiratory chains proposed by L. Smith (1968) is given at the top of page 130.

At this scheme shows, electrons can join the respiratory chain through various dehydrogenases which are firmly bound to it. In bacteria the oxidation of substrates thus does not necessarily involve their dehydrogenation by

*The term respiratory chain is a synonym for "electron transfer chain."

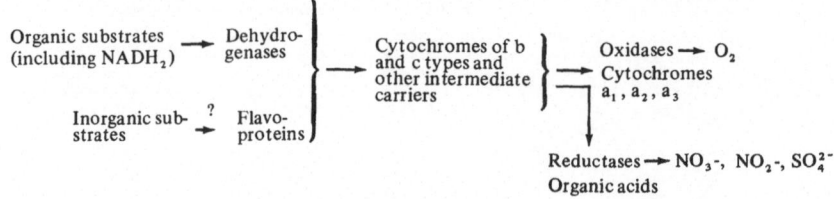

cytoplasmic enzymes with the formation of $NADH_2$ as the first step, and it may begin with enzymes which transfer the electron directly to the respiratory chain. In many chemosynthesizing bacteria, the enzymes responsible for the initial oxidation of the inorganic substrate (hydroxylamine, Fe^{++}, nitrite, H_2, etc.) are also directly connected with the respiratory chain. Membrane dehydrogenases transfer electrons to cytochromes through the intermediary of ubiquinones and menaquinones, just as takes place in mitochondria. Cytochromes and terminal oxidases in aerobic bacteria also exhibit considerable variety.

In most facultative anaerobes, the respiratory chain is similar in its assortment of components to that of the aerobic bacteria, but nitrates, sulfates, and oxidizing agents arising during nitrogen fixation act as the terminal electron acceptor, although oxygen also can be used.

To understand how the respiratory apparatus of bacteria functions, it is important to recall that the membrane systems containing electron transfer enzymes are buried in the cytoplasm and are not separated from it by a special membrane corresponding to the outer membrane of the mitochondria. In bacteria, therefore, direct interaction can take place between the respiratory chain and co-enzymes, substrates, ADP, and inorganic phosphorus of the cytoplasm. The possibility cannot be ruled out that free interaction of electron-transfer chains with the cytoplasm, and also with the external medium (through the cytoplasmic membrane), leaves its imprint on the regulatory mechanisms of oxidation. In most bacteria no respiratory control has been found, let alone higher types of regulation, including hormonal. Meanwhile most bacteria, if presented with anaerobic conditions, "know" how to switch their respiratory chains to other terminal acceptors than oxygen.

Bacteria adapt themselves just as easily to different substrates by forming initial enzymes of the respiratory chain.

Like the analogous systems of mitochondria, the respiratory chains of bacteria are mounted in the membranes. For this reason, all data on the properties, arrangement, and interactions of the chain enzymes are important to the understanding of the molecular organization of the membrane as a whole. In that case, the respiratory chain is the "marker" of the membrane. The study of organization of the respiratory chain in bacterial membranes is still only in its initial stages. Not only is the spatial arrangement of the components in the membrane still unknown, but so also is their mode of interaction with each other. To some extent, our ignorance of the arrangement and functioning of the respiratory chain has led to differences of opinion on the mechanism of electron transfer along the chain. The problem of how electrons are transferred from carrier to carrier, or the significance of the structural organization of carriers in the form of complexes, in the membrane, will evidently be solved only when the molecular organization of the membrane as a whole and of the respiratory chain in particular has been elucidated.

Dehydrogenase
The following dehydrogenases have been found in respiratory chains located in bacterial membranes:

Enzyme	Bacterium	Source
Fragments of Membranes		
Formate dehydrogenase	*Escherichia coli, Pseudomonas fluorescens*	Schachman et al., 1952
Malate dehydrogenase	*Mycobacterium avium,*	Kusunose and Kusunose, 1959
Malate dehydrogenase	*Mycobacterium stegmatis, Mycobacterium butyricum, Mycobacterium avium*	Kimura and Tobari, 1963; Tobari, 1964
NADH$_2$ dehydrogenase	*Mycobacterium flavum*	Biggins and Postgate, 1971
Lactate dehydrogenases	*Mycobacterium flavum*	Biggins and Postgate, 1971
Pipecolate dehydrogenase	*Pseudomonas putida*	Baginsky and Rodwell, 1966
Dihydroorotate dehydrogenase	*Pseudomonas* sp.	Miller and Kerr, 1967
Lactose dehydrogenase	*Pseudomonas graveolens*	Nishizuka and Hayashi, 1962
Aldose dehydrogenase	*Rhodopseudomonas spheroides*	Niederpruem and Doudoroff, 1965
Aldose dehydrogenase	*Pseudomonas fragii*	Weimberg, 1963
D-Alloxyproline dehydrogenase	*Pseudomonas striata*	Adams and Newberry, 1961
Aldehyde dehydrogenase	*Pseudomonas aeruginosa*	Heideman and Azoulay, 1963
Malate dehydrogenase	*Pseudomonas* sp., *Pseudomonas* NCY139869	Kornberg and Phizackerley, 1961; Francis et al., 1963; Hopper et al., 1970
Malate dehydrogenase	*Acetobacter xylinum*	Benziman and Perez, 1965; Benziman and Karniely, 1968
Glucose dehydrogenase	*Acetobacter suboxydans, Pseudomonas fluorescens*	Hauge, 1961a,b, 1964; Hauge and Hallberg, 1964; Hauge and Mürer, 1964
Dehydrogenases of primary and secondary alcohols, polyalcohols, pentoses, aldoses	*Gluconobacter suboxydans*	De Ley and Kersters, 1964; Kersters et al., 1965
Lactate dehydrogenases	*Acetobacter peroxydans*	De Ley, Shell, 1959; Ywasaki, 1960
Glucose dehydrogenase, ethanol dehydrogenase	*Acetobacter* sp.	King and Cheldelin 1957, 1958
Alcohol-cytochrome c_{553}-reductase, aldehyde dehydrogenase	*Acetobacter* sp.	Nakayama, 1961; Nakayama and De Ley, 1965
D-Fructose dehydrogenase	*Gluconobacter arinus*	Yamada et al., 1966

Enzyme	Bacterium	Source
D-, L-Lactate dehydrogenase	*Gluconobacter arinus*	Kline and Mahler, 1965
D-, L-Lactate dehydrogenase	*Escherichia coli*	Bennett et al., 1966
NADH₂ dehydrogenase		Bragg and Hou, 1967a
D-Fructose dehydrogenase		Kerr and Miller, 1968
D-, L-Lactate dehydrogenase	*Azotobacter vinelandii*	Jones and Redfearn, 1966; Jurtshuk and Harper, 1968; Naik and Nicholas, 1966a,b
Formate-cytochrome c oxidoreductase	*Azotobacter vinelandii*	Ackrell et al., 1972
Dihydroorotate-ubiquinone reductase	*Erwinia amylovorum*	Sutton and Starr, 1960
D-, L-Lactate dehydrogenase, malate dehydrogenase, NADH₂-benzylviologen reductase	*Nokardia*	Sih and Bennett, 1962

Membranes

Enzyme	Bacterium	Source
Glucose-6-phosphate dehydrogenase, 6-phosphogluconate dehydrogenase	*Pseudomonas natriegenes*	Walter and Eagon, 1964
Glucose-6-phosphate dehydrogenase, 6-phosphogluconate dehydrogenase, malate dehydrogenase	*Bacterium megaterium*	Weibull et al., 1959
NADH₂ dehydrogenase	*Bacterium megaterium*	Mizushima, 1968
NADH₂ dehydrogenase	*Mycoplasma laidlawii*	Stopkie and Weber, 1967
Steroid-1-dehydrogenase	*Pseudomonas aeruginosa*	Norton et al., 1963
Lactate dehydrogenase, D-alanine dehydrogenase, formate dehydrogenase	*Haemophilus parainfluenzae*	White, 1964, 1966
Malate dehydrogenase NADH₂ dehydrogenase D-Lactate dehydrogenase	*Micrococcus lysodeikticus*	Cohn, 1956, 1958 Gel'man et al., 1960a,b, 1962, Owen and Freer, 1970
L-Lactate dehydrogenase	*Vibrio parahemolyticus*	Unemoto et al., 1965
Succinate dehydrogenase	*Desulfovibrio vulgaris*	Yagi, 1969
NADH₂ dehydrogenase	*Halobacterium cutirubrum*	Lanyi and Stevenson, 1970
D-Lactate dehydrogenase, trimethylamine-N-oxidoreductase	*Halobium salinarium*	Cheah, 1970c
D-Lactate dehydrogenase, malate dehydrogenase, NADH₂ dehydrogenase	*Chlorobium* (cytoplasmic membrane)	Cruden and Stanier, 1970

The prosthetic groups of these enzymes have received very little study, and only in individual cases has the presence of flavin nucleotides been established.

From the point of view of the molecular organization of the bacterial membrane, the arrangement of this group of enzymes is important. Details on the character of the binding of the dehydrogenases in the membrane and on their isolation and purification are interesting in this respect.

Dehydrogenases differ in the strength of their binding with the membrane. Some enzymes are firmly bound with the membrane, others very weakly bound. The latter can be separated from the membrane during lysis or mechanical disintegration of the cell, or they can be extracted from membranes by disturbing their electrostatic interactions (washing with water, treatment with alkali, treatment with alkali, treatment with chelating agents). The firmly bound dehydrogenases can be separated from the membranes only by solubilization with detergents. The subdivision into strongly and weakly bound dehydrogenases is conventional, for there are some enzymes that can be extracted from membranes partially by disturbance of their electrostatic interactions and partially by detergents, i.e., by disturbance of the stronger, hydrophobic interactions.

All difficulties connected with the isolation of membrane proteins apply fully to membrane dehydrogenases also. If the membranes are treated with detergents or with agents disturbing the electrostatic bonds, destruction of the membranes usually takes place, with liberation of fragments of different sizes and composition, some of which contain the desired dehydrogenase. The fragments can be further fractionated by gel filtration, ion-exchange, chromatography or by centrifugation in a sucrose concentration gradient. Despite this complex isolation technique, even now it is impossible to be sure that fragments carrying a particular dehydrogenase are obtained in the "individual" form and not contaminated by other fragments of the membrane. It is not easy to obtain dehydrogenases in the individual form: they are usually accompanied by one or more cytochromes and lipids, and also, possibly, by other proteins whose presence usually is not determined.

The process of cell destruction can itself be accompanied by detachment of membrane enzymes both in the individual form and in the composition of membrane fragments. The so-called soluble $NADH_2$ dehydrogenases of *Mycobacterium tuberculosis* (Bogin et al., 1969), judging by the character of their allosteric inhibition, are identical with the $NADH_2$ dehydrogenase that remains bound with the respiratory chain (Worcel and Goldman, 1967).

In some cases it is possible to detach dehydrogenases from membranes simply by treatment with ultrasound. After such treatment, malate dehydrogenase is found to require phospholipids and menaquinone (Phizackerley and Francis, 1966). The malate-vitamin K reductase of *Mycobacterium phlei* is also weakly bound with the membrane (Asano and Brodie, 1963, 1964; Asano et al., 1965). The enzyme is found in the soluble fraction of the cell after sonication. Evidence that the enzyme is bound with the membrane *in vivo* is given by its activation by phospholipids.

The enzyme Fe^{++} cytochrome c reductase has been found in ultrasonic homogenates of *Ferrobacillus ferrooxidans* cells. In all probability, this dehydrogenase, weakly bound in the membrane, is detached during destruction of the cell (Yates and Nason, 1966). Much of the membrane $NADH_2$ dehydrogenase is extracted by distilled water from membranes of *E. coli* after their lyophilization (Gutman et al., 1968).

D-Alloxyproline dehydrogenase from particles of *Pseudomonas striata*, obtained by centrifugation at 100,000 g, can also be extracted by EDTA (Adams and Newberry, 1961; Yoneya and Adams, 1961). The hydroxylamine-cytochrome c reductase of *Nitrosomonas europea* and *Nitrosomonas oceanus* are separated from the membrane by lysis of the cell as the result of mechanical destruction (Hooper and Nason, 1965). Succinate dehydrogenase is separated from the membranes of *M. phlei* by treatment with alkali (Kalra et al., 1971).

It is possible that when their electrostatic interactions are disturbed, these dehydrogenases are detached from the membrane not in an individual form, but as small lipoprotein complexes, as fragments of the membrane containing several proteins and not sedimented during high-speed centrifugation. A membrane fragment containing malate and $NADH_2$ dehydrogenases has been isolated from the membranes of *M. lysodeikticus* by treatment with EDTA, followed by purification of the extract by ultracentrifugation and gel filtration of the supernatant.

The membrane fragments are shaped like flattened disks with a mean diameter of 200 Å (Ostrovskii et al., 1968a). The complex contains several proteins (Tsfasman et al., 1972a) and lipids essential for activity of the enzymes (Oparin et al., 1965a). The amino acid composition of membrane fragments carrying dehydrogenases has been determined (Zhukova et al., 1966).

A hydroxylamine oxidase, oxidizing hydroxylamine in the presence of artificial electron acceptors, has been isolated from the supernatant obtained by centrifuging the EDTA-lysozyme digest of *N. europea* cells at 144,000 g. The bond between the enzyme and the terminal oxidase is evidently broken during lysis.

The enzyme forms part of a membrane fragment which contains cytochromes b_{557} and c_{549}. Electron-microscopic data show that the fragment is a spherical particle about 160 Å in diameter; its molecular weight is 200,000 and its sedimentation coefficient 10^5 (Rees, 1968).

Membrane fragments containing dehydrogenases can be separated from the membrane not only be treatment with EDTA, but also by rendering the medium alkaline. Mizushima (1968) obtained an $NADH_2$ dehydrogenase with a sedimentation coefficient of 28 S by this method from the membranes of *Bacillus megaterium*. The preparation was not retained on a Sephadex G-200 column, i.e, it evidently had a large particle size. It contained up to 10% of phospholipids and FAD (1×10^{-8} mole/mg protein). It is interesting to note that recombination of the membrane fragment carrying $NADH_2$

dehydrogenase with the membrane could be achieved only after treatment of the "alkaline" membranes with buffer at pH 4.0 in the presence of Mg^{++} ions.

Isolation and purification of firmly bound dehydrogenases can be difficult because either detergents or solvents have to be used for their solubilization and this sometimes leads to inactivation of the enzymes. This happened, for example, during solubilization of lactate and succinate dehydrogenases from the membranes of *E. coli* and *Salmonella typhosa* (Kidwai and Murti, 1965; Nasir and Murti, 1965). Solubilization by detergents is sometimes accompanied by activation of the enzyme. For instance, the activity of $NADH_2$ dehydrogenase was increased by 40 times on its extraction from the membrane of *B. megaterium* KM with 0.3% deoxycholate (Eisenberg et al., 1970a).

Some firmly bound dehydrogenases have nevertheless been isolated, although not in an individual state but as components of membrane fragments or lipoprotein complexes.

Malate dehydrogenase firmly bound with the respiratory chain is present in *Mycobacterium avium* (Kimura and Tobari, 1963). To extract the enzyme, the membrane fraction was first treated with acetone, and then with Tris-buffer and ammonium sulfate. The resulting preparation possessed high activity but it was not homogeneous. The enzyme was a component of a protein–lipid complex. The complex could not be split either by phospholipase A, or by lipase, or by detergents. The best acceptor for this enzyme is phenazine methosulfate. Its activity is stimulated by 2.2×10^{-6} M FAD. Tobari succeeded in isolating an apomalate dehydrogenase from the acetone powder and in purifying it by salting out and fractionation on DEAE-cellulose. Activity appeared when the apoenzyme was added to FAD and lipid from cells of *M. avium*. The active principle in the lipid was evidently diphosphatidyl glycerol (Tobari, 1964).

When membranes of *E. coli* B were treated with a mixture of deoxycholate and EDTA at pH 7.0, a fragment was isolated and its size could be judged indirectly from the fact that it passed through a Sephadex G-200 column without retention. This membrane fragment contained dihydroorotate-ubiquinone reductase, lipids, ubiquinone, FMN, nonheme Fe, and cytochrome b_{560} (Kerr and Miller, 1968).

The membranes of *Gluconobacter suboxidans* are rich in dehydrogenases (De Ley and Kersters, 1964; Kersters et al., 1965). The substrates of these dehydrogenases are primary and secondary alcohols, polyalcohols, pentoses, and hexoses. The enzymes can be solubilized by combined treatment of the membrane with Triton X-100 and EDTA, and dehydrogenases are solubilized in a complex containing cytochrome c_{553}. Membrane fragments containing D- and L-lactate dehydrogenases are distributed in a sucrose gradient among several fractions, depending on their particle size. Both enzymes were solubilized from membranes by dodecylsulfate (Bennett et al., 1966). The succinate dehydrogenase from *Propionibacterium pentosaceum* has been shown to pass into the soluble state only after treatment with butanol. In the purified preparation, the enzyme is found together with the flavin cytochrome b (Singer and Lara, 1958). Fructose dehydrogenase was isolated from ultrasonic fragments of *Gluconobacter arinus* by treatment with deoxycholate and butanol (Yamada et al., 1967). By detergent treatment followed by proteolysis, formate: cytochrome c oxidoreductase with a molecular weight of 60,500 was isolated from particles of *Desulfovibrio vulgaris* obtained at 80,000 g (Yagi, 1969).

In some cases, detergents are used in conjunction with salts to solubilize dehydrogenases. For example, by treating particles from *E. coli* with deoxycholate and ammonium sulfate, and $NADH_2$ dehydrogenase was isolated (Bragg and Hou, 1967b,c; Bragg, 1965).

It is evident that no demarcation line can be drawn between the weakly bound and firmly bound dehydrogenases of bacterial membranes. According to our observations 30% of the malate and $NADH_2$ dehydrogenase complex is removed from membranes of *M. lysodeikticus* by treatment with EDTA, while the rest is extracted by detergents (Ostrovskii et al., 1968; Tsfasman et al., 1972a).

Different dehydrogenases may be bound differently in the same membrane. For example, $NADH_2$ dehydrogenase is bound much less strongly than succinate dehydrogenase in the membranes of *B. megaterium,* as is shown by extraction under alkaline conditions (Mizushima et al., 1966b). The formate, succinate, and $NADH_2$ dehydrogenases of the membranes of *Haemophilus parainfluenzae* are more firmly bound with the respiratory chain than D- and L-lactate dehydrogenases (White, 1965b).

Since some dehydrogenases are weakly bound with the membrane, oxidation of the corresponding substrates is sensitive to deformations of the membrane. According to L. Smith (1962), osmotic shock of the spheroplasts of *B. subtilis* was accompanied by a decrease in the oxidation of endogenous substrates and amino acids, and an increase in $NADH_2$ oxidase activity. The respiratory chain as such is apparently not affected by deformation of the spheroplast membrane, for the $NADH_2$ oxidase activity was unchanged, but the weakly bound dehydrogenases were evidently detached.

To judge from the character of binding of the dehydrogenases, the organization of bacterial membranes is specific. In fact, malate dehydrogenases and $NADH_2$ dehydrogenases of different bacteria differ in the character of their attachment to the membrane, as their extractability shows.

It has been shown that the natural acceptors for some bacterial malate dehydrogenases are menaquinone derivatives. The malate dehydrogenases from *M. avium* (Tobari and Kimura, 1966) and the malate dehydrogenase from *Acetobacter xylinum* (Benziman and Karmely, 1968) may be mentioned. An essential condition for activity of the malate dehydrogenase from *Pseudomonas ovalis* Chester was the addition of phospholipids from the membranes,

FAD, and 2-methyl-1,4-naphthoquinone (Francis and Phizackerley, 1965).
The malate-vitamin K reductase from *M. phlei* also is activated by phospholipids (Asano et al., 1965). These results agree with the observations of
Tobari, who showed that a lipid is essential for the activity of the malate
dehydrogenase from the membranes of *M. avium*. Inactivation of glucose
dehydrogenase by lipolysis of the membranes of *Bacillus anitratum* with
phospholipase A was described by Hauge and Hallberg (1964). Interaction
between the dehydrogenase and lipids of the membrane carrying the respiratory chain provides the necessary conformation for enzyme activity.

The nearest neighbors of the dehydrogenases on the respiratory chain
are the cytochromes, especially cytochromes of the b group. This conclusion
can be drawn from investigations of the kinetics of action of the respiratory
chain enzymes during electron transfer to oxygen or to other acceptors, as
well as from the composition of fragments containing dehydrogenases and a
given cytochrome.

Dehydrogenases interact with the cytochrome components of the respiratory chain so long as the spatial connection between them is preserved
in the membrane (White, 1962; P. Smith, 1964). Any action which disturbs
this connection and converts the dehydrogenase into the soluble state (ultrasound, lysis, and osmotic shock) will prevent oxidation of the corresponding
substrate.

The properties of the isolated dehydrogenase-cytochrome complexes
will be examined below (see page 168).

Ubiquinones and Menaquinones

Several types of lipid-soluble quinones are found in bacterial membranes:

Ubiquinone type   Rhodoquinone

Menaquinone type   Chlorobiumquinone

The quinones of membranes have a long polyprenyl (terpenoid) chain
connected with the quinoid ring (Pennock, 1967; Crane, 1968). Together

with the lipids, quinones are among the few components of the membranes whose chemistry has been well studied ("Nomenclature of Quinones with Isoprenoid Side Chain," 1966; "Biochemistry of Quinones," R. Morton, Ed., 1965).

Ubiquinones are derivatives of 2,3-dimethoxy-5-methylbenzoquinone, in which a polyprenyl chain of varied length occupies position 6. Ubiquinones with a number of prenyl units (n) ranging from 4 to 10 (UQ-4 to UQ-10) have been found in bacteria. The side chains have all their prenyl residues in the *trans*-position; there is information in the literature on natural *cis*-isomers of the ubiquinones. Rhodoquinone, a compound related to ubiquinone, has been isolated from *Rhodospirillum rubrum* (Glover and Threlfall, 1962). An amino group is present in position 3 of the benzoquinone ring instead of the methoxy group in ubiquinone.

Of the two vitamin K's (K_1, phylloquinone; K_2, menaquinone), only the latter is found in bacterial membranes. The menaquinones are regarded as derivatives of menadione, differing from it by having a radical in position 3 of the naphthoquinone ring. In the case of menadione, R=H.

In the menaquinones of bacteria, the number of prenyl residues (n) varies from 6 to 9. The length of the menaquinone side chain is indicated after its name, e.g., menaquinone-8 (MQ-8) or octaprenyl-menaquinone. Menaquinones in which the double bond in one of the prenyl residues (the second prenyl residue from the ring) is saturated have been isolated from two species (*Corynebacterium diphtheriae* and *M. phlei*). These compounds are described as II dihydromenaquinone-8 (2H) and II dihydromenaquinone-9 (2H) [MQ-9 (II-H)] (Gale et al., 1963; Scholes and King, 1965b; Azerad et al., 1967; Campbell and Bentley, 1968). It has been shown that MQ-9 (II-H) from *M. phlei* consists of two isomers which differ in the geometry of the double bond in the prenyl unit next to the ring. The main component is the *trans*-isomer and the minor component the *cis*-isomer; they are present in the ratio of 36:1 (Dunphy et al., 1968).

The demethylmenaquinones differ from menaquinones in the absence of the methyl group in the second position, which is replaced by hydrogen. Demethylmenaquinone-9 has been isolated from *S. faecalis, E. coli,* and *H. parainfluenzae* (Baum and Dolin, 1964; Lester et al., 1965).

The photosynthetic sulfur bacterium *Chlorobium thiosulphatophilum* produces an unusual menaquinone, namely chlorobiumquinone, which has a C_{34} side-chain with double bonds conjugated with the double bonds of the naphthoquinone ring (Frydman and Rapoport, 1963; Powls and Redfearn, 1969).

The data in the literature on the distribution and content of quinones in several taxonomic groups of bacteria, including the acetic acid bacteria, micrococci, enterobacteria, pseudomonads, and mycobacteria, have been summarized in tabular form (Lester and Crane, 1959; Pandya et al., 1961; Bishop et al., 1962; Bishop and King, 1962; Scholes and King, 1963; Crane, 1965; Rebel and Mandel, 1965; White, 1965a; Jones and Redfearn, 1966; Gel'man et al.,

1966; Jeffries et al., 1967; Azerad et al., 1967; Cawthorne et al., 1967; Yamada et al., 1968, 1969; Whistance et al., 1969; Morton, 1971). Gram-negative bacteria have been shown to contain several different types of quinones at the same time. Ubiquinones with 4, 5, 6, 7, and 8 isoprenyl units, menaquinone-8, demethylmenaquinone-8 and several homologues of polyprenyl phenols have been found in *E. coli* W and *Proteus vulgaris* (Whistance et al., 1969). As a rule, menaquinones are found in Gram-positive bacteria; several isoprenyl homologues may be present simultaneously. Staphylococci have been shown to contain MQ-7, MQ-8, and MQ-9; MQ-6, MQ-7, and MQ-8 have been found in micrococci (Jeffries et al., 1967, 1968).

Not all the quinones present evidently participate in the oxidative metabolism of the bacteria; some of them are intermediate products of biosynthesis or catabolism, and they may perhaps perform some other role as yet unknown.

One approach to the study of the function of the ubiquinones and menaquinones in bacteria is by analysis of their distribution in connection with the conditions of aeration during cultivation. For example, the menaquinone concentration in cells of *Staphylococcus albus* grown anaerobically is 140 times smaller than in cells grown in aerobic culture (Bishop et al., 1962). The menaquinone-9 concentration was almost doubled in *Staphylococcus aureus* when transferred from anaerobic to aerobic growth; meanwhile the C-35 isoprenyl homologue was replaced by the C-45 (Frerman and White, 1967). Growth of *E. coli* under aerobic conditions has been shown to favor the formation of ubiquinones, whereas under anaerobic conditions menaquinone biosynthesis is predominant (Polglase et al., 1966). Similar results have been obtained for three other species of bacteria. In *Pseudomonas fluorescens* grown aerobically, the UQ-9 concentration was doubled (Kuligowska and Erecinska, 1967). The synthesis of ubiquinone and cytochromes in *Azotobacter vinelandii* is intensified under conditions of nitrogen fixation (Knowles and Redfearn, 1969). Data for the ubiquinone, menaquinone, and demethylmenaquinone concentrations are given below (in nmoles/g dry weight) in bacteria grown under aerobic and anaerobic conditions (Whistance and Threlfall, 1968):

Conditions of growth	UQ-8	MQ + DMQ
Escherichia freundii		
Aerobic	0.43	0.24
Anaerobic	0.28	0.35
Proteus mirabilis		
Aerobic	0.76	0.56
Anaerobic	0.24	0.89
Aeromonas punctata		
Aerobic	0.42	0.14
Anaerobic	0.22	0.13

The increase in synthesis of quinones under aerobic conditions confirms their role in respiration. However, other functions of the quinones in the membrane cannot be ruled out, for inhibition of quinone synthesis by 50% in *S. aureus* by diphenylamine does not affect the formation of the respiratory chain (Hammond and White, 1970). As a rule, quinones are absent in anaerobic fermentation bacteria (*Clostridium sporogenes, Lactobacillus casei*). Naphthoquinones in *S. faecalis* are an exception; their role has not been discovered (Baum and Dolin, 1963, 1965). Menaquinone-6 has been found in obligate anaerobes, namely, in the sulfate-reducing bacteria *Desulfovibrio vulgaris* and *Desulfovibrio gigas* (Weber et al., 1970), where in conjunction with cytochrome c_3 it forms a shortened electron transfer chain.

The distribution and role of ubiquinones and menaquinones in the photosynthesizing bacteria have been examined elsewhere many times (Sugimura and Okabe, 1962; Osnitskaya et al., 1964; Carr, 1964; A. Green and Mascarenhas, 1964; Carr and Exell, 1965; Takamiya et al., 1967; Maroc et al., 1968; Okayama et al., 1968; Powls and Redfearn, 1969).

The chief features of the biosynthesis of ubiquinones and menaquinones in bacterial systems can now be considered explained (Glover, 1965; Friis et al., 1966; Campbell et al., 1967; R. Jones, 1967; Threlfall, 1967; Cox et al., 1969; Hammond and White, 1969; Rudney, 1969; Whistance et al., 1969, 1970; Bezborodov and Chermenskaya, 1970; Guerin et al., 1970; Snyder and Rapoport, 1970).

The concentration of ubiquinones in bacterial cells ranges from 0.10 to 4 μmoles/g body weight; the concentration of menaquinones varies within almost the same limits (Whistance et al., 1969). These figures are comparable with the ubiquinone concentration in mitochondria: 2.8 μmoles/g dry weight (Pumphrey and Redfearn, 1960; Ernster et al., 1969).

The presence of ubiquinones and menaquinones exclusively in the membrane fraction of bacteria, where the cytochromes also are located, is confirmed by all the existing evidence (Brodie, 1969; Brodie et al., 1970). For instance, whereas the cells of *A. vinelandii* contain 2.6 μmoles UQ-8 per gram dry weight, a preparation of particles transferring electrons contains 6 μmoles ubiquinone (Lester and Crane, 1959). The quinone concentration in electron-transporting particles and in the membranes is given in Table 7.

The specific concentration of ubiquinones in the membranes of the chromatophores of *R. rubrum* rises to 36 μmoles/g dry weight, compared with their concentration in the cells which is 1.3 μmoles/g (Crane, 1965; Carr and Exell, 1965). Of the total menaquinone of *C. diphtheriae* cells, 53% is located in the fraction of particles responsible for electron transfer (Scholes and King, 1963).

After fractionation of the electron-transporting particles, the supernatant of *M. phlei* contained ten times less menaquinone than the particles; after similar fractionation, quinones were absent in the supernatant of *E. coli*

TABLE 7. Functions of Quinones in Electron-Transporting Chains of Bacteria

Object (particle or membrane)	Process (activity)	Quinone	Source
Acetobacter xylinum	Oxidation of $NADH_2$ and malate	Ubiquinone-10	Benziman and Goldhamer, 1968
Hemophilus parainfluenzae	Oxidation of succinate, $NADH_2$, and lactate	Demethylmenaquinone	White, 1965a,b
Escherichia coli	Oxidation of $NADH_2$ $NADH_2$-cytochrome b_1-reductase	Ubiquinone-8	Jones, 1967; Cox et al., 1970
Escherichia coli B.	Oxidation of lactate Oxidation of dihydroorotic acid	Ubiquinone-6	Bragg, 1971 Kerr and Miller, 1968
Aerobacter aerogenes	Oxidation of $NADH_2$	Ubiquinone-8	Knook and Planta, 1971
Halobacterium	Oxidation of $NADH_2$	Menaquinone-8	Marquez and Brodie, 1970
Mycobacterium phlei	Oxidation of $NADH_2$ and malate	Dihydromenaquinone-9; cis-MQ-9(2H)	Dunphy et al. 1968; Mi Mari and Rapoport, 1968; Brodie, 1969; Phillips et al., 1970
Corynebacterium diphtheriae	Oxidative phosphorylation with $NADH_2$	trans-MQ-9(2H)	Krogstad and Howland, 1966
Bacillus megaterium	Oxidation of succinate Oxidation of $NADH_2$, malate, α-glycerophosphate	Menaquinone-9 Menaquinone	Kröger and Dadak, 1969
Proteus rettgeri	Oxidation of $NADH_2$ and succinate Oxidation of formate	Ubiquinone } Menaquinone	Kröger et al., 1971
Hydrogenomonas H20	Oxidation and phosphorylation with hydrogen	Ubiquinone-8 (octaprenyl ubiquinone)	Bongers, 1967
Desulfovibrio vulgaris Desulfovibrio gigas	Reduction of sulfate	Menaquinone-6	Weber et al., 1970
Chlorobium thiosulpha tophilum	Photo-oxidation of sulfide	Chlorobiumquinone	Powls and Redfearn, 1969
Chloropseudomonas ethylicum and Chloropseudomonas thiosulphatophilum	Cyclic photophosphorylation Photosynthetic transfer of electrons	Menaquinone-7	Powls and Redfearn, 1969
Rhodospirillum rubrum g-9	Photophosphorylation of $NADH_2$; cytochrome c_2-reductase	Ubiquinone-10	Okayama et al., 1968; Yamamoto et al., 1970

(Brodie, 1961). Similar results on the localization of menaquinones in particles or membranes responsible for electron transfer have been obtained by fractionation of the respiratory systems of several bacteria, including *E. coli, M. lysodeikticus, A. xylinum, B. megaterium, Micrococcus denitrificans, Pseudomonas pyocyanea, B. subtilis, Proteus* P18, *Thiobacillus thiooxidans,* and several nonsulfur purple bacteria (Bishop et al., 1962; Pandya and King. 1962; Carr and Exell, 1965; Rebel et al., 1964; Adair, 1968; Benziman and Goldhamer, 1968; Kröger and Dadak, 1969; Scholes and Smith, 1968b; Cox et al., 1969). Even in the cytochrome-free anaerobic bacterium *S. faecalis,* the whole of the menaquinone is localized in the membranes (Baum and Dolin, 1965).

Quinones are present in the membranes and chromatophores of bacteria in a considerable molar excess (3-5 or, in some cases, as much as 25 times) compared with the concentration of cytochromes (Knowles and Redfearn, 1966; Crane, 1968; L. Smith, 1968; Powls and Redfearn, 1969; Benziman and Goldhamer, 1969; Kröger and Dadak, 1969; Cox et al., 1970). A similar ratio between ubiquinone and cytochromes is found in mitochondria from animal tissues (Green and Brierley, 1965).

The most interesting problem, but the one whose study has made least progress, is that of the binding of the quinones in the membranes. Quinones are known to be bound firmly with the membrane, for they are extracted together with lipids from the membranes only by solvents, and they are extracted more completely by a mixture of polar and nonpolar solvents than by nonpolar solvents alone. For instance, a 3 : 1 mixture of isooctane and propanol or a mixture of methanol and acetone extracts quinones more completely than isooctane itself (Crane, 1968; Benziman and Goldhamer, 1968; Kröger and Dadak, 1969). These results evidently indicate that the attachment of quinones in the membrane is of a dual character (hydrophobic and hydrophilic), on account of the chemistry of their molecules. The similarity between the extractability of lipids and quinones suggests that quinones are located in lipid, especially phospholipid, areas of the membrane. On introduction of quinones into membranes, short-chain isoprenyl homologues are introduced more easily than long-chain homologues. Since short-chain quinones are readily soluble in more polar solvents than the long-chain kind, the impression is obtained that polar areas which facilitate their introduction exist on the surface of the membranes.

Quinones are compounds of low molecular weight, and their concentration in cell membranes is low; evidently, their effect on the packing of the macromolecular membrane components cannot be great. Meanwhile, the state of the quinones in the membrane is to some extent determined by the packing of the lipids.

Urban and Klingenberg (1969) compared the redox potential of ubiquinone when packed in mitochondrial electron-transporting particles with the potential of isolated ubiquinone. They found that the redox potential of ubiquinone in the particles is + 650 mV at pH 7.0; in the isolated state, it is 104 mV. These workers explain this difference by the way in which the ubiquinone is packed in the membrane. The water-soluble reactive benzoquinone ring of ubiquinone faces the aqueous phase and projects from the lipid phase of the membrane; the isoprenoid chain of the molecule is buried in the lipid phase.

The information on the function of quinones in bacterial cells, given below, is interesting from the standpoint of the arrangement of the components of electron-transporting chains in the membrane. For a long time, even the fact that quinones participate in the electron transfer chain was disputed. After almost a decade of argument, all modern schemes of the electron-transfer chain assume to some extent or other the participation of ubiquinone in the respiratory chain of mitochondria (Green and Brierley, 1965; Chance, 1965; Kröger and Klingenberg, 1967; Ernster et al., 1969; Crane, 1968; Skulachev, 1969; Kotel'nikova, 1969). Ubiquinones participate both in the direct and reverse transfer of electrons along the respiratory chain (Lieberman and Baker, 1965; Asano et al., 1967a,b). The point of localization of ubiquinone in the electron-transfer chain (ETC), which is closely connected with yet another difficult problem in mitochondrial energetics, the participation of cytochrome b in the ETC, is being studied at the present time. Most workers place ubiquinone between succinate and $NADH_2$ dehydrogenases on one side, and cytochrome b on the other side (Crane, 1968; Ernster et al., 1969).

Some workers, who question the role of cytochrome b in the ETC, locate the ubiquinones between the flavoproteins and cytochrome c (Kröger and Klingenberg, 1967; Palmer et al., 1968).

In an attempt to explain the large molar excess of ubiquinone in the mitochondrial membrane by comparison with the cytochrome content, Kröger and Klingenberg put forward an interesting hypothesis that ubiquinone acts as a common pool of reducing equivalents reaching the cytochrome chain from various flavoprotein dehydrogenases (Kröger and Klingenberg, 1967). They have also extended this idea to bacterial respiratory chains (Kröger and Dadak, 1969). If this view proves to be correct, it will follow that each membrane cytochrome is apparently surrounded by a lipid drop with a quinone introduced into it.

The participation of quinones in bacterial electron transfer chains is being studied intensively (Brodie, 1965; Crane, 1968; Brodie et al., 1970; Gel'man et al., 1966). Most of the papers on which this section is based were published in 1965-1970.

As a result of the great variety of electron-transport processes in bacteria (including nitrogen fixation, photosynthesis, reduction of sulfate, nitrate, etc.), the manifold functions of the bacterial quinones are coming to light. Admittedly, although these processes are evidently so varied, they are still nevertheless membrane-bound, electron-transporting processes; it is only the donors or acceptors of the electrons which change. If several quinones exist in one species of bacteria, they participate in alternative pathways (for example, in *E. coli*).

Methods of investigating the role of quinones in the electron transfer chain of bacteria and mitochondria are the same. They include measurement of the kinetics of reduction and oxidation of endogenous quinones, and extraction of quinones by solvents or irradiation at 360 nm, which selectively destroys the menaquinones. Subsequent addition of one or another quinone leads to the reconstruction of oxidation (or of oxidation and phosphorylation). True restoration of activity takes place only in the presence of the specific quinones. In some cases, it has been shown that quinones which are analogues of the true components of the ETC can restore oxidation only. As a rule, they do so by creating bypass electron-transport reactions.

Another method of studying the role of quinones is by the use of inhibitors which, in their chemical nature, are analogues of the natural ubiquinones and menaquinones. These include, for example, dicoumarol, lapachol, and piericidin (Harold, 1970).

In many bacterial ETC's, quinones function between flavoprotein dehydrogenases and cytochrome b, or between nonhemin iron and cytochrome b (Downey, 1964; Itagaki, 1964; Francis and Phizackerley, 1965; Knowles and Redfearn, 1966; R. Jones, 1967). However, several species of bacteria have been found in which the quinone lies between cytochromes b and c. For example, this is true of the succinate-oxidase chain of *Corynebacterium diphtheriae*, which includes menaquinone-9 (Krogstad and Howland, 1966).

Research on mutants which have lost the ability to synthesize quinone is a promising method of discovering the place of quinones in the respiratory chain. For instance, the disturbance of oxidation in a mutant of *C. diphtheriae* is connected with a disturbance of the synthesis of menaquinone (Krogstad and Howland, 1966). Mutant strain *E. coli* 156:53 D2, which cannot synthesize ubiquinone, showed very weak $NADH_2$ oxidase and $NADH_2$ cytochrome b_1 reductase activities; the succinate oxidase complex was less affected. These workers consider that the ubiquinone is located between $NADH_2$ dehydrogenase and cytochrome b_1 (R. Jones, 1967). An ubiquinone-deficient mutant of *E. coli* showed severe impairment of $NADH_2$ and lactate oxidase activities (Cox et al., 1970). The role of demethylmenaquinone-7 in the respiratory chain of *H. parainfluenzae* has been studied in detail by White (1965a). The rate of oxidation and reduction of the quinone correlates with the rate of oxidation and reduction of the other components of the respiratory chain. During inhibition with cyanide, the quinone located between the dehydrogenases and cytochromes receives the whole stream of electrons. Extraction with aqueous acetone inhibits the action of the respiratory chain, which is restored by the addition of demethylmenaquinone-7. The quinone is firmly bound in the membrane, and it interacts only with dehydrogenases bound in the membrane. Removal of the dehydrogenases by treatment of the membranes with ultrasound is accompanied by inhibition of the respiratory chain (White, 1965b).

Comparison of the kinetics of reduction and oxidation of ubiquinone-10 with the general ETC activity in A. $xylinum$ showed that ubiquinone participates in $NADH_2$ and malate oxidase systems, and is located between the dehydrogenases and the cytochrome chain (Benziman and Goldhamer, 1968).

From the kinetic data, the results of inhibitor analysis, experiments on extraction of quinone with pentane, and reconstruction of various oxidases on the addition of menaquinone, it has been concluded that the quinone in the membranes of B. $megaterium$ participates in electron transfer between dehydrogenases and cytochrome (see Scheme 7b, Kröger and Dadak, 1969).

The fact that 80% of the menaquinone is reduced from each substrate is evidence, according to these workers, that a menaquinone pool exists in the bacterial membranes to collect and distribute the reducing equivalents coming from the substrates to the cytochrome system through diffusion in the lipid layer of the membrane. The five- to tenfold excess of menaquinone over cytochromes indicates that the size of the menaquinone pool may vary. These facts suggest the absence of fixed stoichiometric ratios between quinones and cytochromes in the respiratory chain, as has been fairly recently suggested for membranes of the cristae (Hatefi, 1963).

After a 10-year investigation of the respiratory chain of M. $phlei$, it was found that dihydromenaquinone-9 participates in the oxidation chain of malate and $NADH_2$ (see Scheme 2, page 183). Oxidation of succinate includes another photosensitive component; it has not yet been identified (Brodie, 1965, 1969).

The function of the cis- and $trans$-isomers of dihydromenaquinone-9 in M. $phlei$ has been investigated in detail (Dunphy et al., 1967, 1968; Mi Mari and Rapoport, 1968). Comparison of the cis- and $trans$-isomers showed that the former restore only oxidation in the membrane fraction from which the quinones have first been removed, whereas $trans$-menaquinones can also restore phosphorylation. It thus becomes clear that the restoration of oxidation by quinones is a less specific process than the maintenance of oxidative phosphorylation (Phillips et al., 1970). However, the previous hypothesis that an intermediate product of oxidative phosphorylation is present in M. $phlei$ as a quinone methide has not yet been confirmed (Horth et al., 1966).

Two quinones are found in E. $coli:$ UQ-8 and MQ-9. The membranes from E. $coli$ have been divided into two types of particles, which differ in the composition of their quinones and in their oxidative activity. The fraction of large particles oxidized succinate and contained ubiquinone; the small particles contained both quinones, and oxidized succinate and $NADH_2$, and the menaquinone was a component of the $NADH_2$ oxidation system (Brodie, 1965). The membrane of E. $coli$ is evidently mosaic-like in structure. Some of its loci have oxidative enzymes of one composition mounted in them, while chains of another composition are mounted in other areas. Itagaki and Sato showed that the presence of ubiquinone-8 is essential for the oxidation of formate by the membrane fraction of $E.$ $coli$ (see Scheme 1, page 182). Cytochrome b_1 acts as the terminal electron acceptor in the oxidation chain of formate with oxygen or nitrate (Itagaki et al., 1961; Itagaki and Sato, 1962; Itagaki, 1964). The view is held that the ubiquinone which is present as a complex with nonhemin iron participates twice in the respiratory chain of $E.$ $coli$ (Cox et al., 1970).

The hydrogen-oxidizing bacteria $Hydrogenomonas$ H-20 contain ubiquinone-8 (see Scheme 11, page 189) in their respiratory chain (Bongers, 1967). It is considered that the menaquinone-6 in the strict anaerobes $D.$ $vulgaris$ and $D.$ $gigas$ participates in the oxidation of hydrogen with sulfite as the electron acceptor. The ETC of these species contains cytochrome c_3 and ferredoxin (or flavodoxin) as well as MQ-6 (Weber et al., 1970).

One of the factors which strengthens the view that quinones participate in bacterial respiratory chains is the discovery of natural complexes of ubiquinone and menaquinone reductases in membranes. Earlier information on quinone reductases in bacteria is given by Brodie (1965).

A macromolecular dihydroorotate—ubiquinone reductase complex has been isolated from the membranes of $E.$ $coli$ B and purified 100 times. Besides a flavoprotein, the complex contains ubiquinone-6, nonhemin iron, and traces of cytochrome b_{560} (Kerr and Miller, 1968).

Ubiquinone and menaquinone reductases have been located in the particle fraction of $E.$ $coli,$ $B.$ $megaterium,$ $M.$ $lysodeikticus,$ and $Hydrogenomonas$ $eutropha;$ membrane-bound $NADH_2$ menaquinone reductase has been found in $Halobacterium$ $cutirubrum$ (Repaske and Lizotte, 1965; Pandya and King, 1966; Bragg and Hou, 1967a; Lanyi, 1969). Quinone reductases are weakly bound with the membranes and are easily solubilized after injury to the salt bonds, as for example, the flavoprotein menadione reductase in $E.$ $coli$ or malate and $NADH_2$ menaquinone reductases in $M.$ $phlei$ and in several other species (Asano and Brodie, 1963, 1964, 1965; Adelson et al., 1964; Benziman and Perez, 1965; Bragg and Hou, 1967a,b,c; Murthy et al., 1969).

L. Smith (1968) considers that the importance of menadione reductase activity as confirmation of the biological role of quinones must be interpreted with caution because of the low specificity of the flavoproteins toward the hydrogen acceptor. The danger of formation of ETC shunts by the participation of exogenous quinones must also be borne in mind (Brodie, 1965, 1969).

Model experiments have shown that quinones readily oxidize reduced flavins as the result of a favorable difference in their redox potentials (Gibian and Rynd, 1969); this is evidently connected with the stimulant action of exogenous quinones on oxidative activities in intact membranes; inhibitors do not act on these activities (Brodie, 1965; Lanyi, 1969).

Cytochromes

Definition. The Cytochrome Complements of Bacteria

The cytochromes are characteristic components of the respiratory chain. Their role is connected with their ability to perform reversible oxidation, in the course of which the valency of the iron atom contained in the cytochrome molecule changes.

Work on the cytochromes, including the bacterial cytochromes, has been summarized from the chemical, physicochemical, and biochemical aspects in several surveys and in the proceedings of two symposia (Yonetani, 1963; Margoliash and Schejter, 1966; Keilin, 1966; Okunuki, 1966; Bartsch, 1968; L. Smith, 1968; Kamen et al., 1970; Kamen and Horio, 1970; Horio and Kamen, 1970; Chance et al., 1966; Okunuki et al., 1968).

So far as the present book is concerned, cytochromes attract attention as the most comprehensively studied protein components of bacterial membranes. We shall therefore discuss the following problems: the localization of the cytochromes, interactions holding cytochromes in the membrane, agents extracting cytochromes from membranes, the properties of isolated cytochromes and of dehydrogenase–cytochrome complexes, and finally, changes in the membrane cytochromes depending on conditions.

Just as in organisms possessing mitochrondria, bacterial cytochromes are components of the respiratory chains which are functionally connected with the processes of oxidation and energy-linked reactions. In some groups of bacteria, additional functions are associated with the cytochromes: a reflection of the many different forms of metabolism in bacteria, and of the comparative simplicity of the cell's conversion from aerobic to anaerobic conditions of existence and from the heterotrophic to the autotrophic state. Cytochromes are responsible for electron transfer under anaerobic conditions in very many species of bacteria: heterotrophs, chemo- and photoautotrophs. In this case nitrates, sulfates, and other compounds act as the terminal electron acceptor (Peck, 1968; Repaske, 1966; Aleem, 1970). Cytochromes are widely distributed in photoautotrophic bacteria which are obligate anaerobes. In this case, they participate in photosynthesis.

According to the recommendations of the subcommittee on nomenclature of cytochromes of the International Union of Biochemists, four main classes of cytochromes are distinguished: cytochrome c, in which the heme (mesoheme) is linked to the peptide chain by a covalent bond; and three cytochromes whose hemes are less firmly bound with the protein, namely: cytochrome b (prosthetic group, protoheme), cytochrome a (prosthetic group, heme), and cytochrome d, with a ferrochlorin prosthetic group. The variety of the cytochromes, especially those of bacterial origin, arises through the combination of four identical hemes with proteins which differ widely in their properties (Falk, 1964).

Cytochromes are detected by analysis of their absorption spectra (α, β-bands 500-650 nm, γ-band 400-500 nm). To identify the cytochromes, besides their absorption maxima in the oxidized and reduced form, the maxima of the compounds of their hemes with cyanide and pyridine (pyridine ferrohemochromes and cyanide ferrohemochromes) also are studied. The group to which the cytochrome belongs is determined by a series of special tests (Blyumenfel'd and Purmal', 1964; Falk, 1964; Nomenclature of Enzymes, 1966).

Two features distinguishing bacterial cytochrome systems deserve attention: the inconstancy of the qualitative and quantitative composition of the cytochromes and the presence of several terminal oxidases (L. Smith, 1963; White and Smith, 1962, 1964). Besides the typical mitochondrial assortment of components (a, a_3, b, c_1, c) such as the b of *B. subtilis* and *Sarcina lutea,* some bacteria also contain the above-mentioned cytochromes in various combinations with cytochromes a_1, a_2, o, b_1, b_4, c_3, c_4, and c_5. The cytochrome components of bacteria given below are taken from L. Smith (1968, partly) and Davidson and Hartree (1968).

It will be seen that virtually all possible cytochrome combinations are encountered. Similar investigations have been made of cytochrome sets in actinomycete cells (Taptykova, 1968). Details of the bacterial cytochromes in the various groups and some properties of the cell cytochromes are discussed in the literature and will not receive special consideration here (Gel'man et al., 1966; Bartsch, 1968; Kamen and Horio, 1970).

Bacterium	Cytochrome	Source
Bacillus subtilis	a, a_3, b, c, c_1	Smith L., 1955; Chaix and Petit, 1956, 1957; Castor, Chance, 1959
Bacillus megaterium	a, a_3, o, b_1	Broberg, Smith, 1967; Kröger, Dadak, 1969
Bacillus cereus ATCC4342	a, b, c	Lang et al., 1972
Bacillus cereus, cyanide-resistant strain	a, b	McFeters et al., 1970
Bacillus brevis gB	a, a_3, b, c, o	Vasil'eva et al., 1970; Seddon and Fynn, 1971;
Salmonella typhimurium	a_1, a_2, b_1, c	Drabikowska, 1970
Escherichia coli	a_1, a_2, o, b_1	Keilin and Harpley, 1941; Castor and Chance, 1959
Aerobacter aerogenes	a_1, a_2, o, b_1	Tissières, 1951; Castor and Chance, 1959
Proteus vulgaris	a_1, a_2, o, b_1	Moyed and O'Kane, 1956; Castor and Chance, 1959
Proteus mirabilis	a_1, a_2, o, b	Heinen, 1967; de Groot and Stouthamer, 1970
Proteus rettgeri	b_1, a_1, a_2, o	Kröger et al., 1971
Azobacter vinelandii	a_1, a_2, o, b, c_3, c_4	Tissières, 1956; Castor and Chance, 1959
Haemophilus parainfluenzae	a_1, a_2, o, b, c, c_1	White and Smith, 1962
Mycobacterium tuberculosis	a, a_3, b, c	Kearney and Goldman, 1970
Mycobacterium phlei	a, a_3, b, c, c_1, o	Asano and Brodie, 1964; Revsin et al., 1970a,b
Mycobacterium flavum	a, a_3, b, c, o	Erickson, 1971

Bacterium	Cytochrome	Source
Acetobacter suboxydans, including strain ATCC621	o, b, c, c_1	L. Smith, 1963; Castor and Chance, 1959; Daniel, 1970
Acetobacter suboxydans strain YAM 1828	a, b, o	Iwasaki, 1966
Pseudomonas aeroginosa	a_1, a_2, b, c, c_1	Horio, 1958a; Azoulay, 1964
Pseudomonas riboflavina	a, a_2, a_3	Arima and Oka, 1965b
Pseudomonas putida	b, a(?), c, o, P_{450}	Baginsky and Rodwell, 1966; Peterson, 1970
Pseudomonas oleovorans	b, c, o	Peterson, 1970
Pseudomonas Bal-31	c, b_1, a_2, o	Franklin et al., 1971
Halobacterium cutirubrum	a_3, or a_1, b, b_1	
Halobacterium halobium	c, c_1, o	Lanyi, 1969; Cheah, 1970 a,b,c
Halobacterium salinarium	a_1, b, b_1, o	
Leptospira	a_3, b, b_1, c, o	
Brucella abortus	a_1, c, c_1, o	Baseman and Cox, 1969
Brucella melitensis	a_2, b, c	Dranovskaya and Kushnarev, 1968
Brucella suis		
Micrococcus denitrificans	a, a_3, o, b, c, c-d	Sato, 1956; Sapshead and Wimpenny, 1972
Micrococcus lysodeikticus	a, b, b_1, c	Jackson and Lawton, 1958; Mitchell, 1962; Ichikawa and Lehninger,1962; Bishop and King, 1962; Gel'man et al., 1963a; Fujita et al., 1966; Lukoyanova and Taptykova, 1968
Staphylococcus aureus	a, b, o	Taber and Morrison, 1964
Micrococcus luteus (Sarcina lutea)	a, b, b_1, c, c_1	Erickson and Parker, 1969
Treponema reiteri	a_2, b	Kawata, 1967
Microbacterium thermosphactum	a, a_3, b_1	Davidson and Hartree, 1968
Agrobacterium tumefaciens	a, c, c_1, b, o	Hirata and Fukui, 1968
Spirillum itersonii	c, b, o	Clark-Walker et al., 1967
Achromobacter metalcaligenes	b	Doss and Phillipp-Dormston, 1971
Rhizobium not fixing nitrogen	a, a_3, b, b_1, c	Appleby, 1969a, Romanov et al., 1970; Kretovich et al., 1972
Rhizobium fixing nitrogen	b, c, c_1 ; $P_{420} P_{428} P_{450}$	
Thermus aquaticus	a, a_3, b, c	McFeters and Ulrich, 1972
Rhodopseudomonas spheroides (aerobiosis, darkness)	a, c, c_2 (RHP)b_1, b	Kamen, 1962; Newton and Kamen, 1963; Kikuchi et al., 1965; Whale and Jones, 1970
Rhodospirillum rubrum (aerobiosis, darkness, exponential phase)	c_2, b, o	Taniguchi and Kamen, 1965; Kikuchi et al., 1965; Horio and Taylor, 1965; Thore et al., 1969
Thiobacillus denitrificans	a, c, o	Peeters and Aleem, 1970
Thiobacillus ferrooxidans	a_1, c_1, c, b	Vernon et al., 1960; Lyalikova, 1959; Tikhonova et al., 1967; Tikhonova, 1967
Thiobacillus concretivorus	a_1, c, b	Moriarty and Nicholas, 1970
Ferrobacillus ferrooxidans	a, c	Din et al., 1967
Nitrosomonas	a, a_1, c, b, o	Aleem et al., 1963; Aleem and Lees, 1963; Falcone et al., 1963; Anderson, 1964; Rees and Nason, 1965
Nitrosomonas europea	a, c	Lozinov and Ermachenko, 1960, 1962; Lozinov et al., 1966; Ermachenko, 1967; Ermachenko et al., 1966

Bacterium	Cytochrome	Source
Nitrobacter winogradsi	a, a$_1$, c	Lees and Simpson, 1957; Zavarzin, 1958; Butt and Lees, 1958; Aleem and Nason, 1960; Van Gool and Laudelout, 1966
Nitrosocystis oceanus	a$_3$, c	Remsen et al., 1967
Hydrogenomonas eutropha	b$_1$, c, o	Repaske, 1966
Hydrogenomonas H-20	a, b, c	Lees, 1958; Packer, 1958; Bongers, 1967
Desulfovibrio desulfuricans	c$_3$	Ishimoto et al., 1954; Postgate, 1956; Horio and Kamen, 1961
Desulfovibrio gigas	b	Hatchikian and Le Gall, 1972

The Number of Cytochromes in Membranes

By modern spectrophotometric methods, it is possible to measure the quantitative content of cytochromes in turbid preparations of membranes and their fragments (Chance, 1951, 1952; L. Smith, 1954a,b; Borisov and Mokhova, 1964; Lisenkova and Mokhova, 1964; Ermachenko et al., 1966; Lisenkova and Lozinov, 1966; Lisenkova, 1967; Taptykova, 1968). The differential spectra of the membrane cytochromes of *M. lysodeikticus* are given in Fig. 20 as examples.

The quantitative content of cytochromes in several bacterial membranes is known (Table 8). Admittedly, it must be noted that these statistics for bacteria are to some extent conventional in character, for the composition and content of cytochromes can vary on account of several different factors (see page 176). As Table 8 shows, the content of cytochromes in bacterial membranes is perfectly comparable with that in mitochondria, with the exception, of course, of cytochrome oxidases; their content in bacterial membranes is 5–10 times less than in heart muscle mitochondria. In some species of bacteria, this is compensated by the presence of cytochrome o. The membranes of *A. vinelandii* are rich in cytochromes, which agrees well with their high oxidase activity (Jones and Redfearn, 1967a). Besides these membranes which are rich in cytochromes, there are others which are poor in cytochromes such as the membranes of *Mycobacterium tuberculosis*. However, the membranes of the related species *M. phlei* have the normal bacterial content of cytochromes.

The possibility cannot be ruled out that these differences in cytochrome content are connected with certain conditions of cultivation of the cells which have not been taken into consideration. The concentration of cytochromes in different batches of membranes from *M. lysodeikticus* (stationary culture, 18 h) varied for cytochrome a from 0.14 to 0.23, for cytochrome b + b$_1$ from 0.30 to 0.58, and for cytochrome c from 0.30 to 0.65 μmole/g protein (Tikhonova et al., 1970). Unfortunately the cytochrome content in membranes is sometimes given in conventional units, and it was therefore impossible to include all these results in Table 8 (White, 1966; Pandya and

King, 1966; C. Gray et al., 1966; Hendler and Nanninga, 1970; de Groot and Stouthamer, 1970; Sapshead and Wimpenny, 1972).

Information is available on the cytochrome content in intact bacterial cells, especially lithotrophs. All lithotrophs have a high content of cyto-chromes, especially those which exhibit reverse electron transfer: *Nitro-somonas, Nitrobacter, Thiobacillus ferrooxidans* (Aubert et al., 1959; Vernon et al., 1960; Lozinov and Lisenkova, 1966; Lisenkova and Lozinov, 1966; Ermachenko et al., 1966; Van Gool and Laudelout, 1966; Lisenkova, 1967; Ermachenko, 1967). The high cytochrome content is associated with the special metabolic features of the lithotrophs (Newton and Kamen, 1963).

Localization of Cytochromes. Their Binding
with the Membrane and Isolation

Like mitochondrial cytochromes, the bacterial cytochromes are localized in the membranes, and most of these proteins are components of polyen-zyme complexes. The localization of cytochromes within the cell is a par-ticularly interesting problem: are they found in mesosomes only, or can they also be found in mesosomes only, or can they also be found in the cytoplas-mic membrane as well as in mesosomes?

A link with the structure of the respiratory carriers is an essential condi-tion for electron transfer and for energy generation in the respiratory chain. In membranes that have electron-transport functions, it is important to understand the properties of the cytochrome proteins (even though in bac-

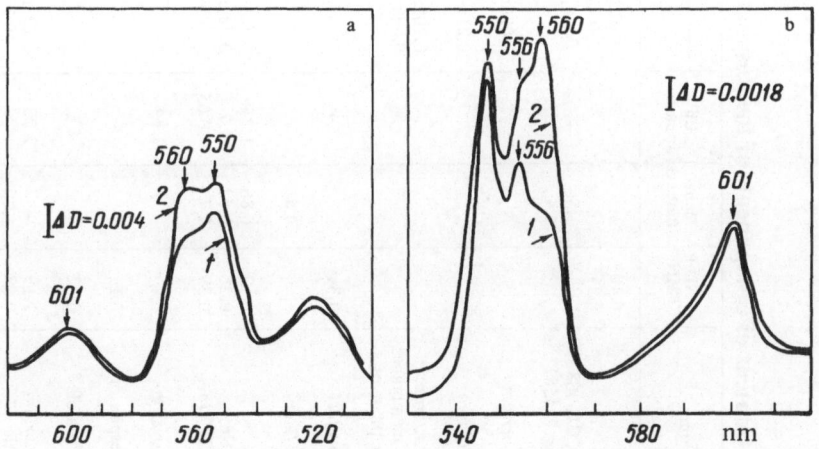

Fig. 20. Differential spectra of membrane cytochromes of *Micrococcus lysodeikticus:* a) at $20°C$; b) at $170°C$; 1) reducing agent, substrate ($NADH_2$); 2) reducing agent, dithionite.

TABLE 8. Content of Components of the Respiratory Chain in Membranes or Their Fragments (in nmoles/mg protein)

Bacterium	FAD + FMN	Mena-quinones	Ubi-quinones	Cytochromes				Nonheme iron	Source
				a	b	c	o (P_{450}, P_{420})		
Bacillus cereus	0.40	—	—	0.19	0.39	0.18	——	—	Doi and Halvorson, 1961
Bacillus subtilis	0.31	1.5		0.21	0.40	0.38	—	—	Arima and Sato, 1970
Bacillus brevis ATCC10068	—	5.4	—	0.1(a_{603})	0.26(b_{559})	0.29(c_{552}) ($c + c_1$)	0.14(o)	—	Seddon and Fynn, 1971
Bacillus megaterium (14581 ATCC)					0.87			—	Kröger and Dadak, 1969
Halobacterium salinarium				0.09(a_3)	0.45(b_{561}) 0.09(b_{564})		0.21	—	Cheach, 1970a,b
Halobacterium halobium				0.21(a_1)	0.54(b_{564}) 0.17(b_{561})		0.27	—	
Halobacterium cutirubrum All 3 species — membranes from logarithmic phase of growth				0.10($a3$) 0.08 *a*	0.47(b_{557})	0.30	0.27	—	Cheach, 1969
Agrobacterium tumefaciens	0.31		2.4	—	—	∿0.30	—	7.4	Kurup et al., 1966
Hemophilus parainfluenzae		0.61	—	—	—	—	—	—	White, 1966
Escherichia coli	0.12	0.86	1.26	0.10	0.62	—	0.07	—	Itagaki, 1964
Escherichia coli AN62	0.25	0.62	4.7	0.03(a_2)	0.19(b_1)		0.04	—	Cox et al., 1970
Escherichia coli AN59 (ubi-mutant)	0.39	2.7	<0.05	0.05	0.19				
Escherichia coli (logarithmic phase)			1.84	—	0.19(b_1)		0.04	2.70	Bragg, 1970, 1971
Eschirichia coli (respiratory particles)	0.11	—	4	—	0.445		0.87		Baillie et al., 1971
Escherichia coli ("soluble" complex)	0.42	—	4.5	1.16	1.16		0.24		Baillie et al., 1971
Salmonella typhimurium	1.19	—	2.20	0.13(a_2)	0.47	0.33	—	—	Drabikowska, 1970
Mycobacterium phlei	0.68	12.0	—	0.27	0.18-0.65	0.62	—	—	Asano and Brodie, 1964; Kashket and Brodie, 1960

Organism								Reference
Mycobacterium tuberculosis $H_{37}Ra$			0.04–0.06 ($a+a_3$)	0.06–0.10	0.07	—	—	Kearney and Goldman, 1970; Goldman et al., 1963
Moraxella lwoffi			0.16–0.49 (a_2)	0.14–0.38 (b_1)	0.13	—	—	Whittaker, 1971
Micrococcus lysodeikticus (stationary phase)	1.33	2.7–3.3	0.18 (a_2)	0.48	0.47			Simakova et al., 1969; Tikhonova et al., 1970
Micrococcus denitrificans (aerobically grown membrane fragments, 105,000 g after treatment of cells by ultrasound)	—		0.47	0.53	0.53			Imai et al., 1967
Pseudomonas putida				0.27	0.13	0.33	11.6	Baginsky and Rodwell, 1966
Acetobacter suboxydans ATCC 621 (respiratory particles)		7.7			1.28			Daniel, 1970
Acetobacter xylinum		15.1	0.35	0.45	0.42	0.08	8.53	Benziman and Goldhamer, 1968
Azotobacter vinelandii Membrane respiratory particles			0.53 (a_2)	1.89 (b_1)	1.95			Jones, Redfearn, 1966, 1967a,b
Red particles	—	23.3	0.33	3.09	2.88	Present	16.4	
Green particles	—	14.9	0.82	1.71	0.71	A little present	7.8	
Nitrosomonas	—	—	0.50	0.34 / 0.24 (b_{564})	3.82	—	—	Falcone et al., 1962, 1963
Rhodospirillum rubrum	—	0.30	— ($a+a_3$)	0.30–0.46 (b)	0.10–0.15 (c_{552} reacts with CO)	—	0.08	Taniguchi and Kamen, 1965
Rhizobium — Not fixing nitrogen	—			0.60	0.48			Appleby, 1969a,b
Fixing nitrogen	—		0.22 ($a+a_3$)	0.64 (b)	0.86 / 0.20	0.04 (P_{450}) / 0.10 (P_{420})		
Electron-transporting particles (ETP$_H$) from bovine heart		1.62	0.85	0.63				Green and Wharton, 1963

teria these account for no more than 5–10% of the total membrane protein). However, to isolate membrane-bound proteins while preserving their enzymic activity is a difficult problem. It is therefore worthwhile discussing in more detail, with special reference to the cytochrome proteins.

The strength of the bond joining the cytochrome components to bacterial membranes varies. There can therefore be no standard formula for their solubilization and subsequent purification. A separate procedure must be developed to extract a given cytochrome from every different object. Cytochromes can differ in the strength of their attachment to the same membrane. For example, cytochrome b is not separated from the membranes of *H. parainfluenzae* by repeated freezing and thawing, whereas partial solubilization of cytochrome c_1 from the same object is produced by this treatment (White and Smith, 1964).

Experimental data on the extraction and purification of bacterial cytochromes show that the cytochromes of type c are least firmly bound in the membranes. Most of the isolated bacterial cytochromes are thus of the c type. Cytochromes of type b, as a rule, are more firmly bound with the membrane although it is possible to isolate some of them. The type a cytochromes are the most firmly bound of all with the membrane; for bacteria, there are as yet virtually no details of the isolation of this group of proteins from membranes. Only partially purified preparations, most probably fragments of membranes, have been obtained. Cytochrome d (a_2) occurs together with cytochrome c in a nitrite reductase which has been purified. Because of differences in the strength of the bond joining the cytochromes to membranes, the literature on their isolation will be examined in groups in order to generalize the data for the last ten years. The earlier work is examined in the surveys of Dolin (1963a), L. Smith (1963), Newton and Kamen (1963) and by others (Margoliash, 1961; Yonetani, 1963; Margoliash and Schejter, 1966).

Many cytochromes of the photosynthesizing bacteria have been isolated, their properties determined, and their interactions in the electron transfer chain studied (Kamen et al., 1963; Orlando, 1967; Meyer et al., 1968; Dus et al., 1968; Sletten et al., 1968; Yamanaka and Okunuki, 1968; Cusanovich and Bartsch, 1969). This field has been covered by a number of surveys (Kondrat'eva, 1963; Bartsch, 1968; Borisov, 1969; Samuilov, 1969; Rubin et al., 1970; Horio and Kamen, 1970; Kamen and Horio, 1970; Kamen et al., 1970; Gest et al., 1963; Shibata et al., 1968).

Experiments on the purification of type c cytochromes have shown that even within this group the proteins differ in the strength of their bond with the membrane; they can be divided conventionally into three categories. The first includes the so-called soluble cytochromes c; these are evidently localized *in situ* outside the membrane in the juxtamembranous layer, or they are so weakly bound with the membrane that they are separated from it during fractionation of the cell. An attempt has been made to classify the soluble cytochromes c; they can be divided conventionally into seven groups: cytochrome c_2, cytochrome c, flavin-cytochrome c, cytochrom c with a split α-band; cytochrome c_3, cytochrome c_{555}, and cytochrome c_{554} (Kamen and Horio, 1970). The second category of the cytochromes c consists of proteins bound with the membrane by electrostatic forces. They are not particularly difficult to isolate. Cytochromes linked by ionic bonds can be extracted by solutions of chelating agents. The ionic bonds are ruptured in a medium of high ionic strength such as is created, for example, by the action of ammonium sulfate (Okada and Okunuki, 1969). The ordinary methods of purification are used subsequently: ion-exchange resins, cellulose, and Sephadex (Okunuki, 1963, Kamen et al., 1963).

The third category includes forms of cytochrome c linked to the membrane by hydrophobic bonds. However, the possibility that the cytochromes are fixed to the membrane in two ways — simultaneously by hydrophobic and electrostatic bonds — cannot be ruled out, judging from the potentiation of the solubilization of cytochromes from membranes by detergents in the presence of salts (Okunuki, 1966; Ohnishi, 1966). The methods used to extract these firmly bound cytochromes frequently do not rest on any theoretical basis; it is therefore not clear how any particular cytochrome, hydrophobically bonded in the membrane, can be selectively removed. The fact is that the membrane is composed of hydrophobically bonded proteins. The problem during extraction of a cytochrome is essentially one of solubilizing the membrane as a whole by some form of powerfully acting agent, and then "catching" the required cytochrome from this mixture of 20 or 30 proteins by salting out and fractionation. Mechanical destruction, lipolysis, sonication, treatment with solvents or detergents, either alone or in conjunction with salts — all these procedures break up the membrane as a whole and at the same time solubilize the hydrophobically bonded cytochromes from the lipoprotein complexes. The methods of extracting hydrophobically bonded membrane cytochromes are summarized below:

Treatment	Cytochrome extracted	Source
Bacillus megaterium and *Bacillus subtilis*		
Lipolysis of particles by lipase for 12 h		Vernon and Mangum, 1960
Bacillus megaterium KM		
Lipolysis of particles by lipase for 7 h at 37°C in 10% sodium deoxycholate solution	o	Broberg and Smith, 1967
Escherichia coli		
Treatment with snake venom for 18 h in presence of 0.1% sodium deoxycholate	b_1	Fujita et al., 1963

Treatment	Cytochrome extracted	Source
Azotobacter vinelandii		
Mechanical and ultra-sonic treatment combined with salt extraction	Incomplete extraction of $c_4 + c_5$ cytochromes	Swank and Burris, 1969
50% butanol followed by salting out at pH 8.0	Complete extraction of $c_4 + c_5$ with other proteins at the same time	
Pseudomonas denitrificans		
Treatment with deoxycholate for 30 min at pH 8.0	Nitrate reductase and cytochrome b extracted to the extent of 70%	Radcliffe and Nicholas, 1970
Mycobacterium phlei		
Lipolysis of 144,000 g particles by lipase for 2 h at 37°C in Tris-buffer medium, pH 8	o	Revsin and Brodie, 1969
Treatment of membrane particles with 2.5% Triton X-100 for 15 h followed by proteolysis with trypsin for 2 h	Complex $(a + a_3)$, o	Revsin et al., 1970a, b
Rhodopseudomonas spheroides		
Sonication of cells in medium of 2% Triton X-100, followed by treatment of 105,000 g supernatant with 2% sodium cholate. Finally salting out with 30-40% ammonium sulfate	Cytochrome oxidase + c and a small amount of b b	Kikuchi and Motokawa, 1966, 1968; Sasaki et al., 1970
Nitrosomonas europea		
Sonication of cells for 15 min and lipolysis with pancreatic lipase for 3 h	c	Levchuk et al., 1967
Chromatium		
Treatment of chromatophores with acetone and extraction of cytochromes with 2% cholate from acetone powder	$c_{552.5}$, c_{555}	Kennel and Kamen, 1971

The situation is complicated by the fact that membrane cytochromes, being strongly hydrophobic proteins, show a strong tendency to polymerize on removal of the detergent (Jones and Redfearn, 1967a; Van't Riet and Planta, 1969). It is therefore necessary to conduct all the successive stages of purification in the presence of detergents, and this is not without its effect on the proteins. In addition, the presence of detergents interferes with the determination of the correct parameters of the proteins (for example, determination of the molecular weight from hydrodynamic data). These technical difficulties are the reason why only a few hydrophobically bonded membrane cytochromes of bacteria have so far been isolated and purified.

Properties of Purified Cytochromes

Mainly soluble and ion-bonded cytochromes (type c) have been obtained in a purified form. Basic information about their properties is given in Table 9; only a few additional details will be given in the text.

The amino acid composition, and sometimes even the amino acid sequence, have been established for individual cytochromes. Ambler has shown that the protein of cytochrome c_{551} from *Pseudomonas fluorescens* consists of a single peptide chain with 82 amino acid residues and has a molecular weight of 9000 (Coval et al., 1961; Ambler, 1963; Needleman and Blair, 1969).

Determination of the amino acid sequence of two cytochromes c_3 — from *Desulfovibrio desulfuricans (vulgaris)* and from *Desulfovibrio gigas* — has shown that these two proteins have a very similar amino acid sequence and can be regarded as homologues (Ambler, 1968; Ambler et al., 1969; Drucker et al., 1970b). Analysis of the amino acid composition of cytochrome c_3 from *D. desulfuricans* has shown that it does not contain tyrosine (Takahashi et al., 1959). The amino acid composition is known for cytochrome c_3 from *D. salexigens* and *D. vulgaris* (Drucker et al., 1970c; Yagi and Maruyama, 1971), and the amino acid composition of the soluble cytochrome c from *Hydrogenomonas eutropha* (Fang and Burris, 1968).

Two cytochromes c from *B. subtilis* (Miki and Okuniki, 1969b), and cytochrome c from *Micrococcus denitrificans* (Scholes et al., 1971) have been subjected to amino acid analysis.

It helped considerably to understand which properties in the molecules of the cytochrome proteins are responsible for their ready solubility and, conversely, for their hydrophobic properties. However, it is impossible to explain the connection between these properties on the basis of their molecular amino acid formula.

Analysis of the prosthetic groups of the bacterial cytochromes of the c type has been carried out for purified preparations from species of the genera *Azotobacter, Pseudomonas,* and *Micrococcus,* and also from *E. coli* (L. Smith, 1963; Fujita et al., 1963). The prosthetic group of cytochromes of the c

TABLE 9. Type c Bacterial Cytochromes

Object from which cytochrome isolated	Cytochrome	Absorption maximum in reduced form, nm	Isoelectric point, pH	Molecular weight	Redox potential (E_0) at pH 7, V	Self-oxidizability	Character of bond between cytochrome and membrane (conventionally)	Number of hemes per molecule	Source
Pseudomonas aeruginosa	c_{551}	551, 521, 416, 316, 290, 280	4.7	8100	+0.286	–	Soluble after treatment of cells with acetone	–	Horio et al., 1958a,b; Horio et al., 1960; Coval et al., 1961
Pseudomonas fluorescens	c_{551}	551, 520, 416, 316	4.7	9000	–	–	Ditto	–	Ambler, 1963
Pseudomonas stutzeri	c_{552}, c_{558}	558 (branch), 552, 526, 421	–	70,000 (37,000×2)	–	Self-oxidizes	Soluble	2	Kodama and Mori, 1969
Pseudomonas denitrificans (aerobic conditions)	c_{553}	553, 524, 419	–	45,000	0–0.090	"	"	2	Iwasaki and Shidara, 1969a
Pseudomonas denitrificans (aerobic and anaerobic conditions)	Cryptocytochrome c	550–560, 426	–	28,000	+0.120	Self-oxidizes slowly	"	2	Iwasaki and Shidara 1969b
Micrococcus denitrificans ATCC 13543	c	550, 520, 418	–	14,000	–	–	Soluble	–	Scholes et al., 1971
Azotobacter vinelandii	Monomer c_4	551, 522, 416	Oxidized 4.4 Reduced 4.7	24,000±2000	–	Self-oxidizes sowly	Hydrophobically bonded	2 hemes on monomer	Tissières, 1956, Tissières and Burris, 1956
	Dimer c_5	555, 526, 420	Oxidized 4.4 Reduced 4.2	24,000±1000 Subunit mol. wt. 13,100	–	Ditto	Ditto	2 hemes on dimer	Swank and Burris, 1969
Bordetella bronchiosep-tica	c_{550}	559, 520, 418	–	–	+0.259	Does not self-oxidize	Ion-bonded	–	Sutherland, 1963
Bordetella pertussis	c_{553}	553, 522, 416	–	–	+0.192	Self-oxidizes slowly	"	–	
Bordetella parapertussis									
Escherichia coli (grown anaerobically with NaNO$_3$)	c_{552}	552, 523, 420	–	11,100-12,300	-0.200	Self-oxidizes	Soluble	–	Fujita and Sato, 1963, 1966; Fujita, 1966a,b
Bacillus megaterium	c_{550}	550, 520, 415	–	–	+0.250	Self-oxidizes	Ion-bonded	–	Vernon and Mangum, 1960
Bacillus subtilis	c_{550}	550, 520, 415	–	12,000	–	–		1	Yamaguchi et al., 1966

Organism	Cytochrome	Absorption maxima (nm)	pI	Molecular weight	E_m	Self-oxidation	Solubility	Heme	References
Bacillus subtilis	c_{550}	550, 520, 414, 316, 279	8.65	12,500–13,000	+0.210	—	—	1	Miki and Okunuki, 1969a,b
	c_{554}	554(550 branch) 581, 417, 316, 280	4.44	14,000	-0.080	—	—	1	
Agrobacterium tumefaciens	c_{552}	552, 592, 415	—	16,000	—	—	Soluble	—	Hirata and Fukui, 1968
Spirillum itersonii	c_{550}	550, 522, 416	9.86	10,400–10,800	+0.297	—	"	1	Clark-Walker and Lascellis, 1970
Hydrogenomonas eutropha	c_{553}	553, 523, 416	—	8500	—	—	"	—	Fang and Burris, 1968
Nitrosomonas europea	c_{552}	552, 524, 418	6.3–7.7	—	+0.245	Self-oxidizes	Hydrophobically bonded	—	Lozinov and Ermachenko, 1962; Lozinov et al., 1966
Nitrobacter agilis	c_{550}	550, 521, 416	6.5	—	+0.250	Does not self-oxidize	Soluble	—	Butt and Lees, 1958; Ketchum et al., 1969
	c_{550}	550, 521, 417	—	—	+0.282	—	—	—	
Desulfovibrio salexigens	c_3	—	10.8	13,400–13,900	-0.205	—	—	1	Drucker et al., 1970c
Desulfovibrio desulfuricans (vulgaris)	c_3	553, 525, 419	10.5	12,000	-0.215	Self-oxidizes	Soluble at pH 6	2(3)? heme	Ishimoto et al., 1954; Postgate, 1956; Takahashi et al., 1959; Horio and Kamen, 1961; Coval et al., 1961; Ambler, 1968; Ambler et al., 1969
Desulfovibrio gigas	c_{553}	553, 525	>8.6	9000	0–0.1	—	Soluble	1	Drucker et al., 1970b; Bruschi et al., 1970
	c_{553}	553, 525, 419	5.2	13,050	-0.216	—	"	2	Le Gall et al., 1965; Bruschi-Heriaud and Le Gall, 1967
Thiobacillus neopolitanus	c_{550}	—	—	—	+0.200	Self-oxidizes slowly	"	—	Trudinger, 1958, 1961, 1964
	c_{553}	553, 522, 403	—	—	+0.210	Self-oxidizes quickly	"	—	
	c_{557}	—	—	—	+0.155	Self-oxidizes very slowly	"	—	
Thiobacillus denitrificans	c_{552}	552, 522, 415, 316	10.2	—	+0.270	Does not self-oxidize	"	—	Aubert et al., 1960
Thiobacillus ferrooxidans	c_{552}	552, 523, 417	—	—	+0.310	—	Soluble after sonication	—	Vernon et al., 1960
Thiobacillus thiooxidans	c	550, 521, 415	10.6	—	+0.253	—	Soluble	—	Tano et al., 1968

type is a substituted mesoheme. Analysis of the protein components of these cytochromes has shown that they differ from mitochondrial cytochromes: the isoelectric points, electrochemical potential, electrophoretic mobility, and enzyme specificity were all different. Meanwhile cytochrome c from *Thiobacillus thiooxidans* resembles the heart muscle cytochrome c in its absorption spectrum, electrophoretic mobility, isoelectric point, and potential (Tano et al., 1968).

A characteristic feature of some bacterial cytochromes c is the presence of two hemes per protein molecule. According to some workers, the cytochrome c_3 from *D. desulfuricans* contains actually three heme groups per molecule (Drucker et al., 1970b); the cytochrome c_3 from *D. vulgaris* has four heme groups (Yagi and Maruyama, 1971). As Table 9 shows, the four bacterial cytochromes of type c have an isoelectric point in the alkaline pH region (pH 9-10), but most of them have isoelectric points in the acid or neutral pH regions.

The redox potential of cytochromes of the c group varies within wide limits, depending on the species of bacterium. This indicates that the protein moieties of the cytochromes differ widely. It is interesting to note that the potential of the cytochrome c from *T. thiooxidans* is +0.310 V, whereas the potential of iron is +0.770 V. To explain the reduction mechanism of cytochrome c, results showing an increase in its potential in an acid medium have been invoked. It is suggested that low pH values may arise in local areas of the membranes of these bacteria (Vernon et al., 1960).

Some bacterial cytochromes of the c type can interact with CO; a similar reaction is found with cytochromes c_{552} and c_{558} from *Pseudomonas stutzeri* (Kodama and Mori, 1969). Cytochrome c_{553} from *Pseudomonas denitrificans* can react with CO and O_2 after purification, possibly on account of denaturation changes in the protein. In general, this protein is more labile to the action of chemical and physical agents than mammalian cytochrome c; keeping at alkaline pH values or dialysis against deionized water led to denaturation (Iwasaki and Shidara, 1969a). Another hemoprotein from the same bacterium is cryptocytochrome c. This is a CO-binding dihemic protein with a hemoglobin-like spectrum. The biological function of cryptocytochrome c is not known; it can be reduced enzymically by $NADH_2$ (Iwasaki and Shidara, 1969b).

Experiments have shown that vertebrate cytochromes c are hardly attacked by proteinases; this is explained by the absence of free lysine and guanidine groups in the molecule (Okunuki, 1966; Margoliash, 1966). The cytochrome c_3 from *D. vulgaris* has been found to be resistant to proteolysis (Ambler, 1968). Proteolysis by trypsin, chymotrypsin, and thermolysin occurred only after removal of the heme and oxidation of the protein by performic acid. Cytochrome b_{562} from *E. coli* underwent proteolysis by α-chymotrypsin or trypsin (3% solution, 24 h) after removal of the heme and denaturation of the protein (Itagaki and Hager, 1968).

As mentioned earlier, most purified bacterial cytochromes of the c type
are soluble proteins. The list of soluble cytochromes in Table 9 shows that
individual members of this group are linked to the membrane by electrostatic bonds. The character of the bonds of the other soluble cytochromes
is unknown, and all that can be said is that they go into solution on destruction of the cell. Only a few type c cytochromes are fixed in the membranes
by strong hydrophobic bonds. It is difficult to determine the reason for the
differences in the manner of binding of the cytochromes with the membrane.
It is equally likely to be connected with differences in the structure of the
cytochromes themselves, or in the areas of the membrane with which they
interact. At the moment there is no evidence to provide a solution to this
problem.

The phenomenon of incomplete solubilization of certain cytochromes,
such as the cytochrome c from $A.$ $vinelandii$ (Jurtshuk et al., 1969) or the
cytochrome c_{552} from $P.$ $denitrificans$ (Miyata and Mori, 1969), is also difficult to explain. The possibility cannot be ruled out that the soluble cytochromes are located on the outer side of the cytoplasmic membrane, as has
been postulated for the cytochrome c_{552} from $E.$ $coli;$ this pigment is easily
removed from spheroplasts (Fujita and Sato, 1966). Physiological observations also help to explain the weakness of the bond linking the cytochromes
c to the membrane. It is surprising that most "soluble" cytochromes function
as components of the nitrate reductase system, and this fact has not yet been
convincingly explained. The "soluble" cytochromes are known to include not
only the c group, but also two special dihemic proteins with the properties
of nitrite reductase and cytochrome oxidase simultaneously; they contain
hemes d (a_2) and c at the same time (Table 10). In facultative anaerobes,
such as $E.$ $coli$ and $Spirillum$ $itersonii$, the synthesis of "soluble" cytochromes
begins if the microorganism is kept under conditions of low aeration with nitrate
as the electron acceptor (Fujita, 1966a,b; Clark-Walker and Lascelles, 1970).

It has also been shown for some denitrifying bacteria that during growth
under anaerobic conditions in the presence of nitrate, soluble cytochromes are
synthesized (Vernon, 1956; Horio et al., 1960; White and Smith, 1962; Scholes
and Smith, 1968b). A similar situation is found also for the photosynthetic
bacteria (Dus and Kamen, 1963; Henderson and Nankiville, 1966; Yamanaka,
1967). It is possible that the membranes have a limited "capacity" for binding cytochromes, and during the surplus biosynthesis required to provide for
certain biochemical functions, some cytochrome remains in the soluble form
(Clark-Walker and Lascelles, 1970).

In contrast with hydrophobic proteins, the soluble weakly bound cytochromes of membranes do not form polymers in aqueous media. An exception is the cytochrome c_{552} from $E.$ $coli$, which forms aggregates with a
molecular weight of 130,000. The molecular weight of the monomer, calculated on the basis of its iron content, is 11,000-12,000 (Fujita and Sato,
1966).

TABLE 10. Bacterial Cytochromes of the b and d Types

Object from which cytochrome isolated	Cytochrome	Absorption maximum in reduced form, nm	Isoelectric point, pH	Molecular weight	Redox potential (E_0) at pH 7, V	Self-oxidizability	Character of bond linking cytochrome to membrane (conventionally)	Number of hemes per molecule	Source
Escherichia coli	b_{557}	557, 527, 425	-	62,000	-0.340	Self-oxidized	Hydrophobically bonded	1 heme	Deeb and Hager, 1964
Escherichia coli	b_{562}	562, 531.5, 427, 324	-	12,000	+0.113	-	Soluble	1 heme	Itagaki and Hager, 1966, 1968; Itagaki et al., 1967
Rhodopseudomonas spheroides	b_{559}	559, 529, 426	-	-	-	-	Soluble	-	Orlando and Horio, 1963
Rhodopseudomonas rubrum	$b_{557.5}$	557.5, 527, 425	4.4	450,000	-0.210	-	-	-	Kamen and Horio, 1970; Bartsh et al., 1971
Rhodospirillum palustris	$b_{558.5}$	558.5, 528, 425	-	-	-	-	-	-	
Pseudomonas aeruginosa	cd Nitrite reductase (1. 9. 3.2)	645, 554, 549, 523, 420	5.8	67,325	-	Self-oxidized quickly	Soluble after treatment of cells with acetone	Two hemes d,c	Horio et al., 1958a,b; Horio et al., 1961a, Yamanaka and Okunuki, 1963a,b; Yamanaka, 1964; Yamanaka et al., 1962, 1963; Nagata et al., 1970
Micrococcus denitrificans	Nitrite reductase	pH 7.5 615–654, 548–553, 521, 417 (460)	3.85	120,000–130,000	-	Ditto	Soluble	Two hemes d,c	Kijimoto, 1968a,b; Newton, 1969; Lam, Nicholas, 1969a,b
Pseudomonas putida P_p g 786	P_{450}	-	4.55	45,000	-	-	-	1 heme	Katagiri et al., 1968; Dus et al., 1970; Vu et al., 1970

Cytochromes of type c, linked to the membrane by ionic bonds, have been obtained in a purified form and their properties determined (Table 9).

It is known that the cytochromes c are not removed from the membranes of bacilli by methods employed to extract mitochondrial cytochrome c. The membrane fraction of *B. megaterium* and *B. subtilis* is known to be readily soluble in weak alkaline solutions (0.02 M NaOH, pH 8.0) after prolonged dialysis, and cytochrome c can be isolated with a high yield from this soluble membrane fraction (Yamaguchi et al., 1966). In some parameters, these proteins resemble mitochondrial cytochrome c; however, both these cytochromes were found to be auto-oxidizable and to react weakly with heart muscle cytochrome oxidase (Yamanaka, 1967).

Solubilization of cytochromes c_{550} and c_{554} has been reported for *B. subtilis* after destruction of the membranes in the presence of 5 mM EDTA and 0.8% KCl (Miki and Okunuki, 1969a,b). Under these circumstances, the cytochrome b remains bound with the membranes. Subsequent purification was carried out on DEAE-cellulose and Amberlite. Degrees of purification of 40 times for c_{550} and 30 times for c_{554} have ben achieved. Attempts to extract these proteins from the membranes with acetone were unsuccessful.

According to Smith, some cytochrome c can be extracted from cytoplasmic membranes or spheroplasts of *M. denitrificans* by treatment with salts (L. Smith, 1968). This fraction of cytochrome c is evidently bound to the membrane by electrostatic bonds.

Besides the cytochromes mentioned above, which are extractable comparatively easily, there are species of bacteria in whose membrane the cytochromes c are just as firmly as the remaining cytochrome proteins. Extraction with trichloroacetic acid, treatment with butanol, and alkaline extraction from a butanol powder have all been used without success in the attempt to extract cytochrome c from the membranes of *M. tuberculosis* (Kearney and Goldman, 1970). Cytochrome c_{550} is extracted from the membranes of *M. lysodeikticus* only by detergents (Nachbar and Salton, 1970; Tikhonova et al., 1970). Many more such examples could be given. Despite great advances in the development of methods of solubilizing hydrophobically bound cytochromes, in most cases they have not been sufficiently purified. Since only a few hydrophobically bound cytochromes c have so far been obtained in a purified form, these will be described in more detail.

Cytochromes c_4 and c_5 from *A. vinelandii* were purified as long as 15 years ago (Tissières, 1956). The procedure used to isolate these proteins has subsequently been improved, so that the degree of purification has been increased to 300 times for c_4 and 150 times for c_5 (Swank and Burris, 1969). Ultrasonic or mechanical treatment of a preparation of particles in salt solution leads to only partial solubilization of these proteins. Treatment with 50% butanol caused their complete solubilization. Cytochrome c_4 is a monomer with two hemes per molecule and a molecular weight of 24,400 ± 2000.

Cytochrome c_5 is a dimer with molecular weight 24,400 ± 1000; under denaturation conditions, it breaks up into monomers with half the molecular
weight: each monomer contains one heme group. Both cytochromes are
weakly auto-oxidizable and virtually do nor react with CO. The amino acid
composition of both proteins has been determined. The homogeneity of both
proteins has been shown by gel electrophoresis and by isoelectric focusing.
Besides these two cytochromes, a third cytochrome protein, described as
minor cytochrome c_4, has also been isolated from *Azotobacter*. Its spectrum
is identical with that of the major cytochrome c_4, but it has only half the
molecular weight.

Solubilization of cytochrome c from the membranes of *Nitrosomonas
europea* was accomplished only after sonication for 15 min, followed by
lipolysis with pancreatic lipase. Even after this vigorous treatment, 22% of
the original quantity of cytochrome c remained in the membranes. The
purified cytochrome had a redox potential of +0.245 V and an isoelectric
point in the region of about 7.3. The cytochrome was self-oxidizable; the
possibility cannot be ruled out that this was due to the vigorous treatment,
especially with ultrasound (Ermachenko, 1967; Levchuk et al., 1967).

To understand the organization of the respiratory chain, data on the conformation of its constituent proteins (especially the cytochromes) are very
valuable. Investigations of this type can be carried out only on purified
proteins, although there is no certainty that the results can then be applied
to the determination of the properties of the same proteins when incorporated
into the membrane. Until recently, conformation has been studied mainly
on cytochromes of the photosynthesizing bacteria. Although it is not within
the scope of this book to examine the cytochromes of photosynthesizing
organisms, they will be discussed briefly since the data on the conformation
of these proteins are of general interest.

The circular dichroism spectra of cytochrome cc′ from *Rhodospirillum
rubrum* have been investigated. This protein consists of a single polypeptide
chain including 112 amino acid residues. The prosthetic group is protoheme
IX; its molecular weight is 12,840. Peptides obtained by proteolysis of the
protein with trypsin and thermolysin have been isolated and purified; their
composition and amino acid sequence have been studied (Sletten et al.,
1968; Dus et al., 1968). According to some observations, cytochrome c′
contains 63% of coiled segments which surround and protect the heme. Contact with 6 *M* urea for 20 h and an extremely alkaline pH largely abolish the
helical structure (Imai et al., 1969; Yong and King, 1970). Other workers
found 36% of helical segments; during oxidation—reduction of this cytochrome, the packing arrangement of the peptide changes (Flatmark and
Robinson, 1968). X-ray structural analysis of the cytochrome c_2 from
R. rubrum shows that this molecule is an elongated ellipsoid measuring 24 ×
31 × 38 Å. The heme group is incorporated into the protein moiety on one

side; the center of the ellipsoid contains hydrophobic chains of amino acids (Kraut et al., 1968). Measurement of the circular dichroism spectra showed that the oxido-reduction of the heme in cytochrome c_{554} from $B.$ $subtilis$ is not accompanied by conformational changes in the polypeptide chain of the molecule (Miki and Okunuki, 1969a). Van Gelder et al. (1968) found 25% of α-helices in the molecule of cytochrome c_4 isolated from $A.$ $vinelandii.$

An atypical hemoprotein, cryptocytochrome c, has been isolated from $P.$ $denitrificans$ grown under anaerobic conditions (Iwasaki and Shidara, 1969b). Conformation investigations of this group of cytochromes showed that the anomaly in their absorption spectra is due to interactions between the heme c and hydrophobic groups of their protein moieties, and is independent of the total structure of their molecules (Imai et al., 1969). Work has begun on investigating the conformation of the cytochrome c_3 from $D.$ $desulfuricans,$ $D.$ $vulgaris,$ and $D.$ $salexigens$ (Drucker et al., 1970a).

Although conformation studies have been carried out on only a few cytochrome proteins, it is nevertheless evident that these membrane proteins have a largely coiled structure. The thickness of the membrane is probably considerable if it includes cytochromes of the same size as the cytochrome c_2 from $R.$ $rubrum.$

The three categories of type c cytochromes mentioned above (soluble, linked by ionic bonds, and linked by hydrophobic bonds) evidently relate both to cytochromes of the remaining types and also to the membrane proteins in general. The type b cytochromes include several bacterial cytochromes which have been sufficiently well purified. These include cytochrome $bb_{1(557)}$, hydrophobically bound to the membrane; soluble cytochrome b_{562} from $E.$ $coli;$ and three cytochromes b from photosynthetic microorganisms. The characteristics of these cytochromes are given in Table 10. Cytochrome b_1 can be isolated from the membranes of $E.$ $coli$ either by lipolysis with snake venom and subsequent treatment with deoxycholate, or by sonication (Fujita et al., 1963; Deeb and Hager, 1964). The fact that a hydrophobically bound cytochrome has been successfully extracted from membranes is evidence that ultrasound ruptures hydrophobic bonds.

The soluble cytochrome b_{562} has been studied in detail (Itagaki and Hager, 1966, 1968). Its prosthetic group is ferroprotoporphyrin IX. The protein moiety had an absorption maximum at 277 nm, and combined with heme to reconstitute the original cytochrome. The protein molecule contains 110 amino acid residues; in some peptides, a homology with myoglobin and hemoglobin was found. The protein of cytochrome b_{562} does not contain cystene, cysteine, or tryptophan.

Three cytochromes b have been purified from the photosynthesizing bacteria $Rhodopseudomonas$ $spheroides,$ $Rhodospirillum$ $rubrum,$ and $Rhodopseudomonas$ $palustris$ (Orlando and Horio, 1963; Kamen and Horio, 1970). A remarkable feature of these cytochromes is that they polymerize with the formation of aggregates whose molecular weight is 500,000. No information is yet available from which to judge how the last two cytochromes are bound with the membrane. The cytochrome b from *R. spheroides* is evidently soluble, for it is extracted from lyophilized cells in $0.1 M$ neutral phosphate buffer. Cytochrome b_{450} with protoporphyrin IX as its prosthetic group has been obtained in a crystalline form from *Pseudomonas putida*. Asparagine and valine are its terminal amino acids (Katagiri et al., 1968; Dus et al., 1970; Vu and Gunsalus, 1970).

Besides these carefully purified cytochromes, partially purified preparations of type b have been described. Hori (1961) partially purified two type b cytochromes from a halotolerant species of micrococcus. It was later shown that cytochrome c is present in these specimens as an impurity (Falk et al., 1961). Further investigations of the cytochromes of the halotolerant micrococcus led to the discovery of an unexpected phenomenon: reversible interconversion of cytochrome b (maxima in the reduced form at 574, 537, and 418 nm; $E_m = +0.045$ V) into cytochrome $c_{552,548}$ (double maximum; $E_m = +0.223$ V) (Mori and Hirai, 1968). Cytochrome b_1 from *Corynebacterium diphtheriae* has been incompletely purified by Pappenheimer (1955), and cytochrome b from *M. lysodeikticus* by Jackson and Lawton (1959). The purified bacterial cytochromes b, like the mitochondrial pigment, contained protoheme IX as their prosthetic group (Vernon, 1956; Deeb and Hager, 1964; Jacobs and Wolin, 1963).

The so-called cytochromes o have been found in bacteria. These are hemoproteins which also contain protoheme as their prosthetic group, but unlike the true cytochromes b, they react with carbon monoxide and, because of their auto-oxidizability in some species, they perform the role of a terminal oxidase. Cytochromes o in bacteria are being intensively investigated (Horio and Kamen, 1970) but they have not yet been isolated from membranes, with the exception of the cytochrome o from acetic acid bacteria, partially purified by Y. Iwasaki (1960, 1966) and the cytochrome o from the membranes of *Vitreoscilla* (Webster and Hackett, 1966). No type a cytochrome of bacterial origin has yet been purified. The bonds fixing the cytochromes of this group to the membrane are evidently entirely hydrophobic, and this interferes with their isolation. Terminal oxidases of some bacterial species have been incompletely purified. By combined treatment with Triton X-100 and proteolysis with trypsin, a preparation containing cytochrome o as well as cytochromes a + a_3 has been isolated from the membranes of *M. phlei*. Details of the isolation and the characteristics of the complex are described by Revsin et al. (1970a,b).

The terminal cytochrome oxidase from *P. aeruginosa,* cytochrome $c:O_2$ oxidoreductase, 1.9.3.1, has been partially purified (Azoulay and Couchoud-Beamont, 1965). The cytochrome spectrum had maxima at 420, 522, 552, and 595 nm. The enzyme contained heme of type a as the prosthetic group. Cyanide, azide, and carbon monoxide inhibited the activity of the enzyme.

It was synthesized only under aerobic conditions. Under anaerobic conditions and in the presence of nitrate, *P. aeruginosa* synthesizes another terminal oxidase — the soluble enzyme nitrite reductase (cytochrome oxidase), which has been thoroughly studied (Table 10). This enzyme has been shown to contain two hemes (c and d). It does not contain copper, although the addition of copper sulfate to the culture medium stimulated its biosynthesis. One of the hemes (d) is a chlorin derivative, dihydroporphyrin (Barret, 1956; Yamanaka, 1964). The mechanism of the reaction between the isolated enzyme, and oxygen and nitrite has been investigated, with the result that heme d has been shown to be the component that reacts with these two substances. Both acceptors compete for the same place on the enzyme (Kijimoto, 1968a). This enzyme has been split with 1% sodium dodecylsulfate; the heme c and heme d have been shown to be bound with different protein subunits, which are linked together hydrophobically and form the active enzyme. The combination of components c and d showed cytochrome oxidase activity, whereas nitrite reductase activity was detected even in the absence of component c. These results can be interpreted from an evolutionary point of view as follows: primitive nitrite reductase contained only heme d in its molecule. In the course of evolution, this hemoprotein was then combined with cytochrome c and it acquired additional cytochrome oxidase activity (Kijimoto, 1968b). The amino acid composition of the crystalline enzyme preparation has been determined (Nagata et al., 1970).

A cytochrome c with two heme groups, very similar to this hemoprotein and possessing nitrite reductase activity, has been isolated from *Micrococcus denitrificans* and purified 140 times (N. Newton, 1967, 1969; Lam and Nicholas, 1969a,b). The protein contains hemes c and d; its molecular weight is 120,000–130,000. Just as for the nitrite reductase of *P. aeruginosa,* most of the protein (70%) is found in the supernatant after fractionation; 30% remained bound with the membrane. For this reason, it can be conventionally regarded as a soluble protein. The spectral characteristics of this cytochrome were found to be dependent on pH: they varied slightly with the acidity or alkalinity of the medium. The purified cytochrome nevertheless contained 10-20% of lipid material.

Purification of nitrite reductase from *Achromobacter fischeri* showed that the enzyme is not analogous to the nitrite reductase from *P. aeruginosa,* but it contained only one heme group of type c, and in the reduced form it gave absorption maxima at 420, 522, and 552 nm (Prakash et al., 1966). A nitrate reductase could be solubilized to the extent of 70% by deoxycholate from particles of *P. denitrificans.* This enzyme, purified 60 times, contained a flavin, sulfhydryl groups, an unidentified metal, and cytochrome b (Radcliffe and Nicholas, 1970).

It can be concluded from the results of the investigation of these four groups of cytochromes described above that cytochromes of group c have been more thoroughly investigated than the rest, because many of them

belong to the soluble or ion-bound group. Meanwhile, among the cytochromes a of the bacterial species so far studied, none are soluble or ion-
bound. This makes the study of this group of bacterial hemoproteins more
difficult.

In the course of isolation and purification of the cytochromes, their
properties often change, for most hemoproteins are allotopic in nature.
These changes affect several parameters: reactivity, potential, auto-oxidizability, interaction with ligands, and reducibility with ascorbate.

After isolation from the membrane, the cytochromes c of *B. subtilis* lose
their property of substrate reduction (Miki and Okunuki, 1969a,b). Inhibition of the reactivity of cytochromes after their extraction from membranes
has also been observed for other species (Repaske and Josten, 1958; Trudinger, 1961). Electron-transporting particles isolated from the membranes of
A. vinelandii in the course of purification lost their ability to react rapidly
with cytochromes c_4 and c_5 isolated from the same microorganism, but by
contrast, and despite all the data on specificity, they rapidly oxidized heart
muscle cytochrome c (Jurtshuk and Old, 1968).

The redox potential of the purified cytochrome b from *E. coli* was −0.340
V, but in less purified preparations the potential was much higher: 0.010 to
0.016 V. The addition of a special "potential-modifying" protein isolated
from membrane extract lowered the potential of crystalline cytochrome b
(Deeb and Hager, 1964).

After isolation, many cytochromes of types b and c acquired the property of auto-oxidizability, which they had not shown previously when the
hemoprotein was incorporated in the membrane. Examples of this behavior
are given by the cytochrome c_{553} from *P. denitrificans* (Iwasaki and Shidara,
1969a,b) and several cytochromes b (Vernon, 1956; Jackson and Lawton,
1959; Orlando and Horio, 1963).

Ability to react with CO, which some cytochromes c acquire after isolation, is also regarded as an indication of loss of the native state of the cytochrome (Ben-Gurshom, 1961; Iwasaki and Shidara, 1969a,b; Kodama and
Mori, 1969).

The view is held that ascorbate oxidase activity of cytochromes (of mito-
chrondrial cytochrome c, for example) indicates the degree of denaturation
of this protein. After proteolysis with pepsin, for instance, when its enzymic
activity in the succinate oxidase system was virtually zero, maximal ascorbate oxidase activity was observed, whereas in native cytochrome c it was
minimal (Margoliash et al., 1959; Skov and Williams, 1968).

All these observations indicate a change in the properties of the cytochromes during their extraction, due to the allotopic character of these typical-
ly membranous components and also, in some cases, to the denaturing action
of the extraction procedure. It is difficult at present to choose between these
two possibilities.

Dehydrogenase—Cytochrome Complexes

During fractionation of the respiratory chains, the bacterial dehydrogenases are often found to be firmly bound with the cytochrome components. Data on these dehydrogenase—cytochrome complexes are given below:

Object	Complex	Source
Corynebacterium diphteriae	Succinate dehydrogenase and cytochrome b	Pappenheimer and Hendee, 1949; Pappenheimer et al., 1962
Propionibacterium pentosaceum	Ditto	Lara, 1959
Escherichia coli	Formate dehydrogenase and cytochrome b_1	Linnane and Wrigley, 1963
Escherichia coli	Formate dehydrogenase, nitrate reductase, and cytochrome b_1	Itagaki et al., 1961; Fujita et al., 1963
Micrococcus lysodeikticus	Malate dehydrogenase, $NADH_2$ dehydrogenase, and cytochrome b_{556}	Ostrovskii et al., 1968a; Lukoyanova and Taptykova, 1968; Tsfasman et al., 1972a,b
Micrococcus lysodeikticus	Succinate dehydrogenase and cytochrome b	Pollock et al., 1971
Bacterium anitratum	Glucose dehydrogenase and cytochrome b	Hauge, 1960; Hauge and Hallberg, 1964
Acetobacter sp.	Alcohol dehydrogenase and cytochrome c_{553}	Nakayama, 1960; Takeyoshi, 1961
Acetobacter suboxydans	Lactate dehydrogenase and cytochrome c_{554}	Iwasaki, 1960
Xanthomonas phaseoli	Succinate dehydrogenase and cytochrome b	Madsen, 1960
Rhodospirillum rubrum	Succinate dehydrogenase and cytochrome b	Taniguchi and Kamen, 1965
Chromatium	Flavin and cytochrome c_{552}	Bartsch, 1963, 1968
Chlorobium	Flavin and cytochrome c_{553}	Radcliffe and Nicholas, 1970
Pseudomonas denitrificans	Nitrate reductase and cytochrome b	
Escherichia coli	Dihydroorotate ubiquinone reductase and cytochrome b_{560}	Kerr and Miller, 1968

The cytochromes in these complexes are reduced by dehydrogenases, just as they are in the intact membrane. Many workers have described the very close structural and functional connection between bacterial cytochrome b and succinate dehydrogenase (Pappenheimer et al., 1962). This strong bond between the dehydrogenases and cytochromes of the b group indicates that this member of the series of cytochromes occupies a part of the respiratory chain next to the dehydrogenase, which correlates with the low redox potential of cytochrome b.

In individual bacteria, dehydrogenases form complexes with type c cytochrome. According to some observations, no cytochromes of the b group are present in the respiratory chains of these bacteria, and cytochrome c lies next to the dehydrogenase.

A hemoprotein isolated from strains of *Acetobacter suboxidans* (Nakayama, 1961) and *Acetobacter peroxidans* (De Ley and Schel, 1965), performing the function of cytochrome c, is linked with alcohol dehydrogenase.

Alcohol dehydrogenase exhibits wide specificity relative to saturated and unsaturated straight-chain alcohols. The enzyme has not been investigated for its flavin content, and the participation of flavin is assumed on the basis of the fact that firmly bound dehydrogenase and cytochrome coexist in this complex.

A complex of formate dehydrogenase and cytochrome b_1 has been found in *E. coli*. The complex is free from other cytochromes; it is extracted from membrane fractions by the action of deoxycholate and ammonium sulfate solution in the presence of formate or after treatment of the particles with snake venom. Formate added to the purified preparation reduced the cytochrome (Itagaki et al., 1962; Linnane and Wrigley, 1963). The dehydrogenase is evidently linked to the cytochrome by means of sulfhydryl groups, and the bond is disturbed under aerobic conditions and in the absence of the substrate (formate). Nitrate respiration takes place by means of this complex.

Nitrate reductase and cytochrome oxidase are solubilized from particles of halotolerant micrococci by the combined action of lipase, snake venom, and deoxycholate (Hori, 1963). Characteristic complexes of cytochrome and flavin are found in photosynthesizing bacteria: *Chromatium* and *Chlorobium thiosulfatophilum* (Horio and Kamen, 1970; Bartsch, 1963, 1968). Cytochrome c_{552} from *Chromatium* contains one flavin (probably bound FMN) and two heme groups in a molecular weight of 72,000; cytochrome c_{553} from *Chlorobium* contains one flavin and one heme group in a molecular weight of 50,000. It is considered that the bond between the protein moiety of the hemoprotein and the flavins, which are not removed from the protein by extraction with TCA or with acid ammonium sulfate (Yong and King, 1970), is covalent in character. The flavins are removed after treatment for 12-24 h with saturated urea or 6 M guanidine chloride, or in the presence of a 20-fold molar excess of mercury derivatives, such as p-mercuriphenylsulfonic acid, at neutral pH, followed by chromatographic fractionation of the heme protein from the flavin. The function of these complexes is unknown.

An active complex containing malate dehydrogenase, $NADH_2$ dehydrogenase, and cytochrome b_{556} has been isolated by means of EDTA and Tween-80 from the membrane fraction of *M. lysodeikticus*. Cytochrome was reduced in the presence of malate or $NADH_2$, and the process did not require the addition of quinones. The content of cytochrome b_{556} was 0.60 $\mu mole/g$ protein (Ostrovskii et al., 1968a; Lukoyanova and Taptykova, 1968; Tsfasman et al., 1972a,b). It is interesting to note that on treatment of membranes with EDTA only, a lipoprotein complex containing dehydrogenase, but without cytochrome b_{556}, is extracted. A flavin has been found in the complex (Gel'man et al., 1962). Tween-80 extracts a very small quantity of dehydrogenases,

also free from cytochromes b. Hence it follows that the complex of dehydrogenases and cytochrome b is solubilized from the membrane by detergent
only after rupture of the salt bridges binding it to the membrane (Tsfasman
et al., 1972a,b).

Lipids are found in the complexes of dehydrogenases with cytochromes:
3% of lipid has been found in the complex of formate dehydrogenase and
cytochrome b_1 from *E. coli* (Linnane and Wrigley, 1963). Lipids are found
in a complex with cytochromes of electron-transporting particles from *Bacillus cereus* (Doi and Halvorson, 1961).

Cytochrome-Concentrating Membrane Fragments

As has been mentioned above, the attempt to isolate pure cytochromes
has been successful in only a few cases. Usually, membrane fragments with
a somewhat higher concentration of the components of the respiratory chain
are all that can be obtained. I shall refer to only three papers which give a
more or less complete description of the characteristics of the fragments.

Fourfold purification of the cytochrome oxidase fragment of the ETC
of *R. spheroides* grown aerobically in darkness was carried out by successive
treatment of the membrane fragments with Triton X-100 and sodium cholate.
The preparation had an increased content of cytochromes of types a and c,
and a small quantity of type b cytochrome. No cytochrome o was found
(Motokawa and Kikuchi, 1966; Kukuchi and Motokawa, 1968).

Cytochrome oxidase of the same type has been isolated from membrane fragments of *M. denitrificans*. The cytochrome oxidase was solubilized
by 0.05 M EDTA solution at pH 10.5; for further purification, ammonium
sulfate was used. The specific activity of the cytochrome oxidase was increased tenfold. The preparation nevertheless still contained cytochromes
b_{560} and c_{550}, but it was free from nitrate reductase activity, this enzyme
having been solubilized already during fractionation of the cell extract (Lam
and Nicholas, 1969a,b).

Two fragments of the membrane, green and red in color, were isolated
from the membrane fraction of *A. vinelandii* by sodium deoxycholate treatment. The difference in color of the particles reflects the presence of different cytochromes. The red particles contain little cytochrome a_1 + a_2, but
are rich in cytochromes c_4 + c_5, b_1, and o. The green particles are rich in
cytochromes a_1 and a_2, but poor in cytochromes o and c, although they
contain cytochrome b_1. Besides cytochromes, the red particles were also
found to contain ubiquinone-8 and nonhemin iron. Enzyme analysis of
the red particles showed that they originate from components lying next
to the substrate end of the ETC; the green particles are from the terminal end.
The red particles, when examined in the electron microscope, still had the shape
of reticular disks measuring 1000-2000 Å in diameter; the green particles

appeared as flat disks only 90-150 Å in diameter (Jones and Redfearn, 1967a). A "soluble" complex containing $NADH_2$ oxidase and succinate oxidase has been isolated from small fragments of the membranes of *E. coli.* The molecular weight of the complex is about 2×10^6. It contains 30% of phospholipids (Baillie et al., 1971).

A membrane fragment containing lactate dehydrogenase, cytochrome b, and cytochrome o has been isolated from *Azotobacter* cells (Iwasaki, 1966).

A complex isolated from *E. coli* includes formate dehydrogenase, two cytochromes b_{555} possessing identical spectral characteristics but differing in potential, and nitrate reductase. These components were found to be strongly linked together in the membrane and difficult to separate. The complex has been isolated as a catalytic unit (Ruiz-Herrera and De Moss, 1969).

A very similar nitrate—reductase complex bound with the respiratory chain has been isolated from membrane fragments of *P. denitrificans* by treatment with deoxycholate and purified 60 times. The complex contained cytochrome b, flavin, and an unidentified metallic component (Radcliffe and Nicholas, 1970).

An enzyme complex of hydroxylamine—cytochrome c reductase, containing flavin, cytochromes b and c, and possibly iron, has been isolated from two lithotrophic bacteria: *Nitrosomonas europea* and *Nitrocystis oceanus* (Hooper and Nason, 1965). An enzyme complex containing cytochromes c, b, o, and a_1 has been isolated from *Nitrobacter agilis* (Straat and Nason, 1965).

It is considered that the purification of bacterial respiratory systems is more difficult than the purification of such systems from mammalian mitochondria (L. Smith, 1968), although electron-transport particles can be isolated from individual species by the use of methods developed for mitochondria (Jones and Redfearn, 1967a,b). These results are important to the investigation of the molecular organization of membranes, for presumably lipoprotein fragments of the membrane reflect its state.

Evaluation of the Action of Detergents on Cytochromes

It will be clear from the foregoing facts that the cytochromes of most importance to us (i.e., those firmly bound with the membrane) are difficult to isolate. The use of detergents and solvents is unavoidable. We know, although the facts are few, that these agents can induce denaturation of proteins, and this fact must be remembered when they are used to isolate cytochromes.

During the action of any detergent on any enzyme, three possible situations may arise theoretically: no effect, activation, and inhibition, depending on the concentration, the duration of action, and various other conditions (see page 97). There is evidence that activity of the cytochromes increases after detergent treatment of membranes. Mitochondrial and bacterial cytochrome oxidases are activated by 2-10 times in the presence of some detergents. These and other similar facts relating to activation have been interpreted as solubilization of the membrane enzyme, making it more accessible to the substrate (Wainio and Aronoff, 1955; Yonetani, 1959; L. Smith et al., 1966; L. Smith and Newton, 1968).

The action of deoxycholate of cytochrome c oxidase activity of membrane particles from *M. denitrificans* is very demonstrative in this respect. Many writers have previously observed the low reactivity of bacterial oxidases with mammalian cytochrome c. On this basis, a scheme of evolution of cytochromes c from various objects was proposed (Yamanaka, 1967; Yamanaka and Okunuki, 1964, 1968). However, the example of *M. denitrificans* showed that this specificity in the reaction between oxidase and cytochrome c can often be explained by the absence of contact between the oxidase and the soluble cytochrome c. Detergents solubilizing the oxidase activate this reaction (L. Smith et al., 1966). These results show that the structural arrangement of the cytochrome c oxidase in the bacterial membrane differs from the arrangement of the analogous enzyme in mitochondria. Sometimes there was a similar increase in cytochrome c oxidase activity during natural "aging" of the preparation, when solubilization of the membrane probably takes place as the result of autolysis. This phenomenon is linked with the activation of $NADH_2$ oxidase of *Azotobacter* (Jurtshuk and Old, 1968).

Investigations of mitochondrial cytochrome oxidase have revealed the possibility of a different activation mechanism of detergent action and its multiplicity (Okunuki et al., 1968; Morrison, 1968). Cytochrome oxidase responded to the activating action of detergents even if isolated from the membrane. The original enzyme is a tetramer with molecular weight 530,000-580,000. During the action of sodium dodecylsulfate (SDS), it splits up into two dimers with molecular weights of 290,000. The dimer form is two or three times more active than the original form. A further increase in the SDS concentration, however, leads to the formation of inactive monomers.

Yet another cause of the activation of cytochrome oxidase by detergents can be mentioned. Mitochondrial cytochrome oxidase exhibits its activity only in the presence of lipids, and behaves as a typical allotopic enzyme (Awasthi et al., 1970). On separation of the lipids from cytochrome oxidase, its activity falls and it can be increased by 3-5 times by the addition of nonionic detergents such as Tween-80 or Asolectin, which evidently simulate lipid encirclement for the enzyme (Morrison, 1968).

Detailed information of this type is not yet available for the multiplicity of detergent effects on bacterial cytochromes. In some cases, detergents have been used to analyze the quaternary structure of bacterial cytochromes. As reported above, nitrite reductase (cytochrome oxidase) from *P. aeruginosa* contains hemes c and d in complex form. The hydrophobic bond between them is ruptured by the action of SDS. Cytochrome c could be separated from cytochrome d by treatment with 1% SDS, followed by centrifugation in a sucrose gradient. The less dense fraction contained cytochrome c, and the denser fraction contained cytochrome d with some cytochrome c as an impurity. If these two fractions were mixed, the nitrite reductase activity was almost completely restored (Kijimoto, 1968b).

Sometimes, changes in the absorption spectra of the cytochromes are observed in the presence of detergents. Substantial changes took place in the spectrum of isolated cytochrome $c_{552, 558}$ from *P. stutzeri* under the influence of SDS in concentrations as low as 0.001%. This fact has not been explained. However, neither deoxycholate nor Emasol changed the spectrum (Kodama and Mori, 1969). Spectral changes have been observed for the isolated cryptocytochrome c from *P. denitrificans* through the action of 50% propanol or 0.01% SDS (Iwasaki and Shidara, 1969b).

In contrast with the action of detergents on isolated cytochromes, the analysis of their action on the respiratory chain in the membrane is complicated by the multiplicity of the effects. In some cases, the activity of the respiratory chain is impaired (Scholes and King, 1965a; Jones and Redfearn, 1967a; Bragg and Hou, 1967a,b; Scholes and Smith, 1968b; Tikhonova et al., 1970; Boll, 1970; Eisenberg et al., 1970a,b). SDS and deoxycholate inhibit the reduction of cytochromes a + a_3 partially purified from the membranes of *M phlei* (Revsin et al., 1970a,b). The inhibitory action of Triton X-100 on $NADH_2$ oxidase and succinate oxidase has been demonstrated in the membrane fraction of *M. denitrificans* (Scholes and Smith, 1968b). The addition of increasing quantities of detergent up to 0.1 mg/ml was found to inhibit $NADH_2$ oxidase. Succinate oxidase was inactivated more slowly, and in low concentrations of Triton, activity was even slightly stimulated. Activation of succinate oxidation is explained by increased accessibility of the substrate to succinate dehydrogenase after disruption of the membranes by the detergent (Scholes and Smith, 1968a,b). $NADH_2$ oxidase and $NADH_2$ cytochrome c reductase in the membrane fraction of *R. rubrum* are inhibited by 50% by Triton X-100 in a concentration of 0.01%.

Quantitative relations become important when the action of detergents on the membrane is evaluated. The activity of succinate cytochrome c reductase in the membranes of *R. rubrum,* for instance, was stimulated by low concentrations and depressed by high concentrations of Triton X-100 (Boll, 1970).

The following factors have thus been shown to be responsible for the multiplicity of detergent action on the respiratory chain: 1) spatial separation of the enzyme complexes composing the ETC (Bragg and Hou, 1967b,c; Tikhonova et al., 1970); 2) removal of the lipids which, in some cases, are essential for the activity of certain enzymes (Greenlees and Wainio, 1959; Machinist and Singer, 1965; Okunuki, 1966); 3) extraction of quinones (these matters are discussed on page 142); 4) inhibition of individual enzymes such

as dehydrogenases in *A. vinelandii* (Jones and Redfearn, 1967a), $NADH_2$
dehydrogenase in *M. denitrificans*, or succinate dehydrogenase and lactate
dehydrogenase from *E. coli* (Kidwai and Murti, 1965; Nasir and Murti, 1965;
Scholes and Smith, 1968b); 5) activation of enzymes (L. Smith et al., 1966;
Jurtshuk and Old, 1968; Scholes and Smith, 1968a,b; Tikhonova et al., 1970;
Lang et al., 1972).

Additional Components of the Respiratory Chain

Besides cytochromes, nonheme metalloproteins and nonheme iron are
found in bacterial respiratory chains (San Pietro, 1965; Rajagopalan and
Handler, 1968). By analogy with mitochondrial electron-transport systems,
a role of nonheme iron in bacterial respiratory chains is also accepted. Nonheme iron is concentrated in membranes containing other components of the
ETC. Treatment of the membranes with chelating agents forming complexes
with iron, which inhibit electron transfer, inhibits the respiratory chain. The
"red particles" obtained from membranes of *A. vinelandii* concentrate nonheme iron up to 16.4 nmoles/mg protein (Jones and Redfearn, 1967b). Membrane particles of *M. phlei*, which concentrate the ETC, at the same time
concentrate the nonheme iron of the cell up to 90% (Kurup and Brodie,
1967a,b). The particle fraction of $NADH_2$ oxidase from *M. tuberculosis*
(strain H_{37}) contained 3.3-5.2 nmoles/mg protein of nonheme iron (Goldman
et al., 1963). From 5 to 25 μmoles/g dry weight of nonheme iron has been
found in preparations containing electron-transport enzymes from *B. subtilis*
(Downey, 1964).

The respiratory chain of *E. coli* contained nonheme iron, 20% of which
would be reduced by $NADH_2$; this reaction was inhibited by quinoline-N-
oxide. On the basis of these observations, it was suggested that nonheme iron
participates in the $NADH_2$ oxidase system of *E. coli*. It is interesting that
this 20% of the total nonheme iron reacted with *o*-phenanthroline. The inaccessibility of *o*-phenanthroline to the remaining 80% of the iron is explained
by assuming that it is packed in the depth of the membrane (Bragg
and Hou, 1967b; Bragg, 1970; Gutman et al., 1968). Nonheme iron in
M. phlei is a component of the succinate and male oxidase chains, but
it is not found in the $NADH_2$ oxidation chain. Its location in the
ETC is between flavoprotein and cytochrome b (Kurup and Brodie,
1967a,b). There is evidence that a protein containing nonheme iron
and sulfur, which has been isolated from the membranes of *A. vinelandii*,
incorporates a nitrogen-fixing nitrate reductase system (Naik and Nicholas,
1966a,b; Shetna et al., 1966, 1968; Shetna, 1970; Ivleva et al., 1969).

Cuproproteins are found in electron-transporting preparations from certain bacteria. Copper-containing proteins are found as components of the
respiratory chain in two members of the genus *Pseudomonas* (Horio, 1958b;
Horio et al., 1961a,b; Coval et al., 1961; Ambler, 1963; Miyata and Mori,

1969). One of these is the "blue protein," containing one copper atom per molecule. It has been purified and crystallized, and its molecular weight (16,000-17,000) and amino acid sequence have been determined. The protein contains 130 amino acids. In its oxidized form, the blue protein has a characteristic maximum at 630 nm. In the reduced form this maximum is absent. The bond between copper and protein can be broken by dialysis of the blue protein against cyanide solution. Its redox potential E_0 = +0.328 V, and its isoelectric point lies at pH 5.4. When reduced by lactate dehydrogenase or cytochrome c_{551}, the blue protein is readily oxidized by cytochrome oxidase from *Pseudomonas aeruginosa* but not by mitochondrial cytochrome oxidase (Horio et al., 1961a,b). Investigation of the conformation of the blue protein molecule from *Pseudomonas* showed that it contains 40% of α-helices, 37% of β-structure, and 23% of random coil (Tang et al., 1968). A similar blue protein, known as azurin, has been isolated from *Bordetella pertussis, Alcaligenes faecalis,* and *Achromobacter denitrificans.* Its molecular weight is 14,000 and it also contains one copper atom per molecule. The blue protein is linked with cytochrome c and is reversibly oxidized by oxygen (Sutherland and Wilkinson, 1963). Azurin obtained from three species of the genus *Bordetella* is a water-soluble protein with a high redox potential (+0.395 V). The absorption maximum in the oxidized form lies at 625 nm. The protein contained 0.45% copper and its molecular weight was 14,600. Azurin was reduced by succinate in the presence of cell-free extract from *Bordetella bronchioseptica* and was oxidized during aeration (Sutherland, 1963; Sutherland and Wilkinson, 1963). Information on the cuproproteins is given by Frieden (1964). Two new classes of electron carriers, the ferredoxins and flavodoxins, have recently been discovered in bacteria. They are the most powerful reducing agents, their redox potential is comparable with that of molecular hydrogen, and during bacterial photosynthesis they have a role in nitrogen fixation and in the metabolism of anaerobic bacteria. Information on the flavodoxins and ferredoxins is summarized by Benemann and Valentine (1971).

The Composition of Bacterial Respiratory Chains

Variation of Respiratory Chains

One of the characteristic features of bacterial respiratory chains, including cytochromes, is their great variation with the composition of the medium, the degree of aeration, the phase of development of the culture, and other factors. This variation probably reflects a specialized evolutionary mechanism giving adequate adaptation to the very diverse conditions of bacterial life.

Rapid changes in the composition of respiratory chains must be reflected in the molecular organization of the membrane, and the existing models of biological membranes must evidently give a satisfactory explanation of these rapid transformations.

The qualitative and quantitative set of components of the ETC can vary depending on the conditions of aeration (Wimpenny, 1969), the composition of the growth medium and the source of carbon (White, 1967), the electron acceptor (White, 1962; Azoulay and Couchoud-Beamont, 1965; Gray et al., 1966; Sinclair and White, 1970), and the source of nitrogen (Knowles and Redfearn, 1968).

Addition of a source of nitrogen to the medium affects cytochrome biosynthesis in *A. vinelandii:* during growth on urea or ammonia the synthesis of cytochromes is inhibited, but during growth on a nitrogen-free medium the concentration of cytochrome a_2 is increased fourfold (Lisenkova and Khmel', 1967; Knowles and Redfearn, 1968, 1969; Swank and Burris, 1969). Most workers now consider that this increase in the cytochrome concentration during nitrogen fixation simply reflects the need for an increase in the utilization of ATP generated during oxidative phosphorylation (Gvozdev et al., 1968; Swank and Burris, 1969).

The complement of ETC's for root-nodule bacteria (*Rhizobium*) of various species has been investigated in detail (Appleby, 1969a,b). A classical electron-transport chain with cytochromes,a, a_3, b, c, and o (see Scheme 9) was present in cells not fixing nitrogen. In nitrogen-fixing cells, cytochromes a, a_3, and o are absent, and the role of oxidases is played by cytochrome c_{552} and pigment P_{428}, which react with CO.

Interesting results have been obtained with respect to the synthesis of terminal oxidases, depending on the conditions of cultivation and aeration. These results explain why several oxidases are present simultaneously in bacteria and why the pathways of biosynthesis of each oxidase are so rigidly maintained. A deficiency of oxygen in the medium stimulates the formation of cytochrome a_2, as has been shown for *E. coli, A. aerogenes,* and *H. parainfluenzae* (Moss, 1956; White and Smith, 1962). In four species of bacteria — *E. coli, P. vulgaris, A. aerogenes,* and *Pseudomonas riboflavina* — cytochrome a_2 is absent in the logarithmic phase and appears in the stationary phase. This fact is linked with a decrease in the oxygen concentration in the medium during the stationary phase of growth (Castor and Chance, 1959; Arima and Oka, 1965b). The relative quantitative proportions of the terminal oxidases may vary in the same bacterium, depending on the conditions of growth, as White and Smith have shown for the facultative anaerobe *H. parainfluenzae.* In a strongly aerated culture, cytochrome o was predominant; if aeration was restricted, a_2 was predominant. During anaerobiosis in the presence of nitrate, there was an increase in the concentration of cytochrome a_1, linked with nitrate reduction (White and Smith, 1962; L. Smith et al., 1970). In a low partial pressure of oxygen, cytochrome c peroxidase, which is evidently localized close to cytochrome oxydase in the membrane (Soininen et al., 1970), takes over the role of terminal oxidase in *P. fluorescens.* One of the terminal oxidase of nitrate reductase from *P. aeroginosa* arises during cultivation on nitrate under anaerobic conditions (*Pseudomonas* cytochrome c_{551}: nitrite O_2 oxidoreductase). In the presence of oxygen, the biosynthesis of another terminal oxidase, cytochrome c: O_2 oxidoreductase 1.9.2 1, is induced (Azoulay, 1964; Azoulay and Couchoud-Beamont, 1965). Similar adaptive biosynthesis of nitrite reductase is found in *M. denitrificans* and *H. parainfluenzae* (N. Newton, 1969; Sinclair and White, 1970).

Generally speaking, the ability to switch from oxygen to nitrate respiration with a change in the complement of respiratory carriers is possessed by very many bacteria. Replacement of the respiratory pathways connected with the switch from oxygen to nitrate respiration has been repeatedly observed in many species of the genus *Pseudomonas*, in *E. coli, A. vinelandii, M. denitrificans*, and other bacteria (Taniguchi and Ohmachi, 1960; De Ley, 1964; Yamanaka, 1964; Lam and Nicholas, 1969a,b; Ruiz-Herrera and De Moss, 1969; Sinclair and White, 1970; Sapshead and Wimpenny, 1972).

In some species, synthesis of an excess of cytochromes is also linked with nitrate respiration. Cytochromes of type c are linked with nitrate reduction in *M. denitrificans, Achromobacter fischeri, P. aeruginosa, E. coli,* and *Spirillum itersonii* (Sato, 1965; Verhoeven and Takeda, 1956; Wimpenny et al., 1963; Cole and Wimpenny, 1968; Scholes and Smith, 1968b; Lam and Nicholas, 1969a,b). Moreover, the content of cytochrome c in these objects increased when the cells were placed under conditions most favorable for the action of nitrate-reducing systems (Verhoeven and Takeda, 1956; Cole, 1968; Gray and O'Hara, 1968; Gauthier et al., 1970). The concentration of cytochrome c in particles from *Vibrio succinogenes* was doubled during growth of the cells on nitrate (Jacobs and Wolin, 1963). The biosynthesis of cryptocytochrome c in *P. denitrificans* is doubled in intensity during anaerobic growth, and under these circumstances slight changes take place in its molecular weight and redox potential (Iwasaki and Shidara, 1969b).

The link between the partial oxygen pressure in the culture medium, the cytochrome content, and respiratory activity has frequently been investigated.

TABLE 11. Effect of Oxygen Pressure on Cytochromes and Oxidase in *Bacillus subtilis* (Downey, 1964)

Oxygen pressure during growth, mm Hg	Cytochromes per 100 mg protein		Oxygen absorption, μl O_2/mg protein/h
	c, $E_{550-574}$	a_3, $E_{600-610}$	
160	20.0	1.7	135.0
80	14.6	1.1	46.3
40	3.1	0.1	2.2
0	0.1	0.1	0

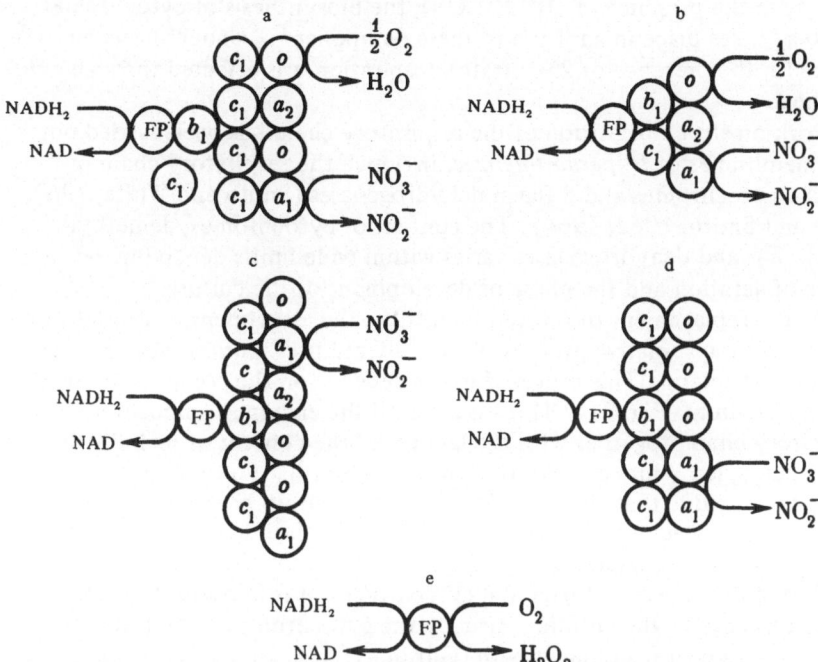

Fig. 21. Composition of the respiratory chain of hemin-dependent species of *Haemophilus* during growth under different conditions (White, 1963). Conditions of growth: a) Loewenthal's medium (aerobiosis); b) proteose, peptone, 0.5 μg hemin/ml (aerobiosis); c) proteose, peptone, 0.5 μg hemin/ml (CO); d) proteose, peptone, 0.5 μg hemin/ml (anaerobiosis); e) proteose, peptone, 0.002 μg hemin/ml, 0.5% glucose (anaerobiosis). In medium containing proteose, peptone, 0.002 μg hemin/ml (aerobiosis); 0.002 μg hemin/ml, 0.5% glucose (anaerobiosis); 0.5 μg hemin/ml, 0.005 M KCN (anaerobiosis), no respiratory chain enzymes are formed.

No general principle applicable to all species has been found among these factors. Results for *B. subtilis* show a direct relationship between the partial pressure of oxygen in the cultivation medium, the cytochrome concentration, and respiratory activity (Table 11).

Similar observations have been made on *Bacillus macerans* and *Streptococcus epidermis*. Under anaerobic conditions of growth only traces of cytochromes were found, while respiration on glucose was reduced by ten times (Jacobs and Conti, 1965; Conti et al., 1968; Jacobs et al., 1967a,b).

The partial pressure of oxygen has been shown to determine both the quantity of cytochromes formed and the nature of the terminal oxidase in *Achromobacter* (Arima and Oka, 1965a). If aeration is strong, respiration is effected by cytochromes b_1 and o; if aeration is weak (in the stationary phase or in the presence of $10^{-3} M$ KCN), the biosynthesis of cytochromes a_1 and a_2 takes place in addition to these components. Under anaerobic conditions (in the presence of 2% nitrate), respiration was effected through cytochrome a.

Work on the composition of the respiratory chain has been carried out with membranes of *H. parainfluenzae*, in which the respiratory chain includes 6 cytochromes and 5 flavin dehydrogenases (Smith and White, 1962; White and Smith, 1962, 1964). The content of cytochromes, demethylvitamin K_2, and dehydrogenases varies within wide limits depending on the degree of aeration and the phase of development of the culture.

On the replacement of oxygen by nitrate, the assortment of respiratory carriers and their relative proportions are altered in *H. parainfluenzae*, just as in other bacteria. This process further depends on the composition of the nutrient medium (Fig. 21). This variation of the enzyme composition of the respiratory chain, together with the lability of the content of its individual components, led White and Smith (White and Smith, 1964; L. Smith et al., 1970) to conclude that the stoichiometric relationship between the enzymes postulated by Green and Hechter (1965) for mitochondria is absent in the respiratory chain of bacteria.

During the analysis of bacterial cytochromes, it is important to take into account the age of the culture. Usually the cytochrome content increases after the period of maximal growth (Gibson, 1961). In the case of *A. aerogenes,* the cytochrome concentration after growth for 48 h was twice that observed after 20 h (Smith, 1954c). The content of oxidases and cytochrome c in the stationary phase of growth of *H. parainfluenzae* is increased (White, 1963). Investigation of the cytochrome system of *B. megaterium* KM under different conditions of cultivation (6, 12, 24, and 48 h) showed that cytochrome a is predominant in the early periods and cytochrome o is predominant in the later periods (Broberg and Smith, 1967). The opposite picture is seen in *Pseudomonas syringae:* cytochrome o was predominant in young cultures and cytochrome a in older cultures (Sands et al., 1967). The maximal content of cytochromes c and a in cells of *Nitrobacter* is found in the middle of the logarithmic phase of growth (Tsien and Laudelout, 1971).

For *A. vinelandii* and *M. lysodeikticus,* on the other hand, it has been shown that the age of the culture is unimportant, for the content of components of the ETC remains throughout the stationary phase at the same level as in the early logarithmic phase (Knowles and Redfearn, 1969; Lukoyanova et al., 1972).

In many cases a decrease in the cytochrome content in *C. diphtheriae, Aerobacter indolegenes,* and *A. aerogenes* has been observed on account of an iron deficiency in the medium (Waring and Werkman, 1944; Tissières, 1951; Mitsuhashi et al., 1956; Lisenkova and Khmel', 1967). Conversely, the addition of hemin to the medium stimulates cytochrome biosynthesis in

Leuconostoc mesenteroides (Superstein, 1970). Synchronized synthesis of the heme and protein moieties of the cytochromes was observed in *S. itersonii* (Clark-Walker et al., 1967).

Different strains of the same bacterium may have different assortments of cytochrome components. This has been shown for *C. diphtheriae* (Pappenheimer et al., 1962; Kleczkowska et al., 1965). Some differences in the cytochrome complement have been found between the original strain and a mutant of *H. parainfluenzae* (White and Smith, 1964).

During spore or cyst formation, the membranes lose their cytochromes and quinones, as has been shown for *B. megaterium* and *Myxococcus xanthus* (Doi and Halvorson, 1961; Dworkin and Niederpruem, 1964).

In some, but not all, of the photosynthesizing bacteria interchange of terminal oxidases takes place depending on the autotrophic or heterotrophic character of feeding. In the membrane fraction of photosynthesizing cells of *R. spheroides* cytochrome o was formed, but under heterotrophic conditions cytochrome a was formed (Sasaki et al., 1970). These workers consider that cytochrome a is the terminal oxidase for bacteria adapted to aerobic conditions, whereas cytochrome o is found in bacteria poorly adapted to aerobic conditions, e.g., *R. rubrum*.

These facts, together with data on variation in the lipids, indicate that the chemical and enzymic composition of the respiratory chains varies considerably with the conditions even within the same species of bacteria. Since respiratory chains are localized in the membrane, it must be recognized that bacterial membranes cannot be regarded as static structures; on the contrary, by assessing them as dynamic and changing structures, ways and means must be sought to studying membranes under dynamic conditions.

Schemes of Respiratory Chains

Investigation of the molecular organization of biological membranes is facilitated by the fact that they contain catalytic proteins which can act as markers for particular areas of the membranes. Accordingly, the components of the respiratory chain are widely used in the study of bacterial membranes. Much information has been gathered on the composition, function, and distinguishing features of bacterial respiratory chains. This information is summarized below, although some of the schemes of the ETC are provisional in character. The sites of action of inhibitors, alternative pathways of electron transport, reverse electron transport in autotrophs, etc., are shown on the corresponding schemes.

To illustrate the complexity of bacterial respiratory chains, let us examine the respiratory chain of *E. coli,* one of those which has received the most study (Scheme 1) (Kashket and Brodie, 1963; Itagaki, 1964; Birdsell and Cota-Robles, 1970; Cox et al., 1970). This scheme, based largely on the work of Itagaki, Kashket, and Brodie, reflects the multiplicity of electron-transport

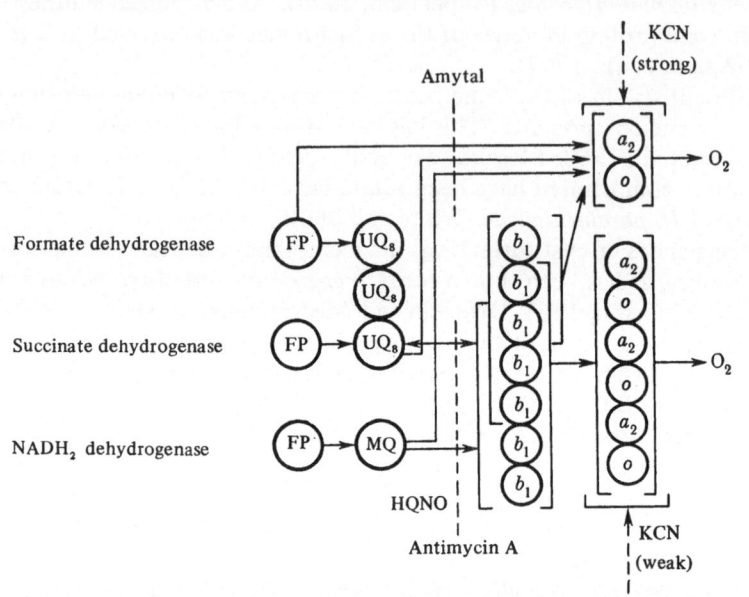

Scheme 1. Respiratory chain of *Escherichia coli,* aerobic conditions (Birdsell and Cota-Robles, 1970); FP = flavoprotein.

pathways and the existence of shunts and "crossroads." It is surprising that,
given the identical composition of the terminal cytochromes, the oxidases
differ in their sensitivity to inhibitors. The process of oxidation of formate
is ten times more sensitive to cyanide and azide than the oxidation of succinate and $NADH_2$ (Birdsell and Cota-Robles, 1970). It has also been shown
by the EPR method and by inhibitor analysis that ubiquinone can participate
twice in the respiratory chain of *E. coli:* before and after cytochrome b_1
during the oxidation of $NADH_2$, lactate, and malate (Cox et al., 1970). However, on the basis of the sensitivity of cytochrome a_2 to irradiation, Bragg
(1971) considers that UQ_8 participates in the respiratory chain of *E. coli*
only in the region before cytochrome b_1.

The respiratory chain of *M. phlei* (Scheme 2) bears the greatest resem-
blance to the mitochondrial pattern.

The respiratory chain of *C. diphtheriae* is basically similar to that of
M. phlei but, depending on the strain, the composition of the enzymes varies
somewhat (Scheme 3). The fast-growing strain $PW8_5P$ has the complete complement of cytochromes in the composition of its respiratory chain (b_{564},
c_{552}, a, a_3), and an intermediate factor between b and c. This factor was absent in the slowly growing strain. The toxic strain had a single cytochrome

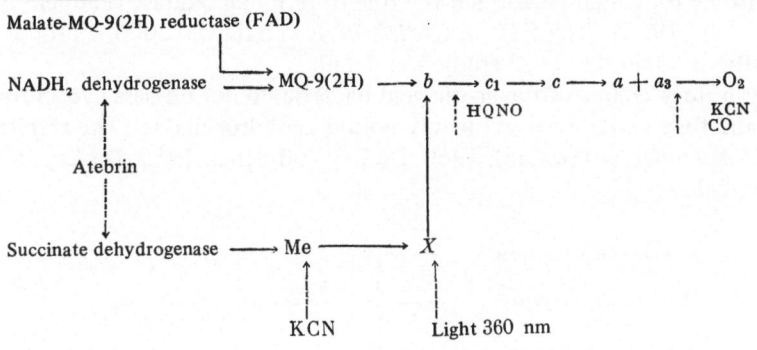

Scheme 2. Respiratory chain of *Mycobacterium phlei* (Asano and Brodie, 1964, 1965; Brodie, 1965). Here and subsequently, HQNO refers to 2-*n*-heptyl-4-hydroxyquinoline-*n*-oxide; Me to methyl, and X to an unidentified component.

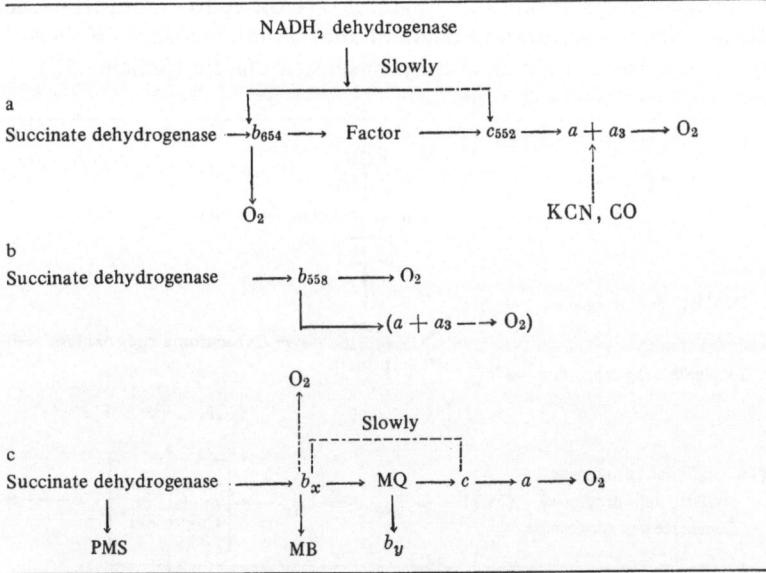

Scheme 3. Respiratory cahins of *Corynebacterium diphtheriae:* a) *C.diphtheriae* (fast-growing wild strain) (Pappenheimer et al., 1962); b) *C. diphtheriae* PW8 (Pd) (slow-growing strain) (Pappenheimer et al., 1962; Kleczkowska et al., 1965; c) succinate oxidase of *C. diphtheriae* (Scholes and King, 1965); PMS = phenazine methosulfate; MB = methylene blue.

(cytochrome b), which performed the role of terminal oxidase (Pappen-heimer et al., 1962). The ETC in *C. diphtheriae* strain CN 2000, a producer of diphtheria toxin, has been studied in detail.

Respiratory chains of the acetic acid bacteria do not possess cytochromes $a + a_3$, and they contain several firmly bound dehydrogenases. The respiratory chain of *Acetobacter* (Iwasaki, 1960; De Ley and Schel, 1959; De Ley, 1960) is shown below:

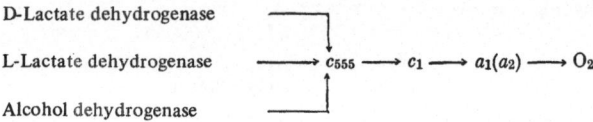

A distinguishing feature of the ETC of some species of *Pseudomonas* (Scheme 4) is that they contain a cuproprotein (blue protein). In *P. aeruginosa*, there are two terminal oxidases whose biosynthesis determines the terminal electron acceptor. The character of the terminal oxidase in the respiratory chain of *P. putida* is uncertain.

The respiratory chain of *M. denitrificans* is remarkable for the fact that, like the mitochondrial respiratory chain, it is sensitive to inhibitors (Scheme 5a). Meanwhile, the respiratory chain of another micrococcus (*M. lyso-deikticus*) does not contain an antimycin-sensitive factor (Scheme 5b).

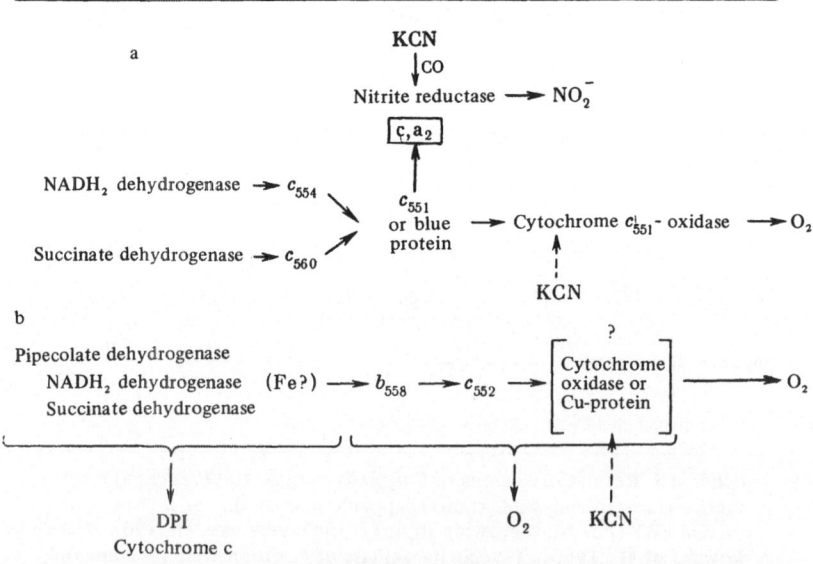

Scheme 4. Respiratory chains of *Pseudomonas aeruginosa* (a) (Yamanaka, 1964; Horio et al., 1961a; Azoulay, 1964, 1965) and *Pseudomonas putida* (b); DPI refers to 2,6-dichlorophenolindophenol (Baginsky and Rodwell, 1966).

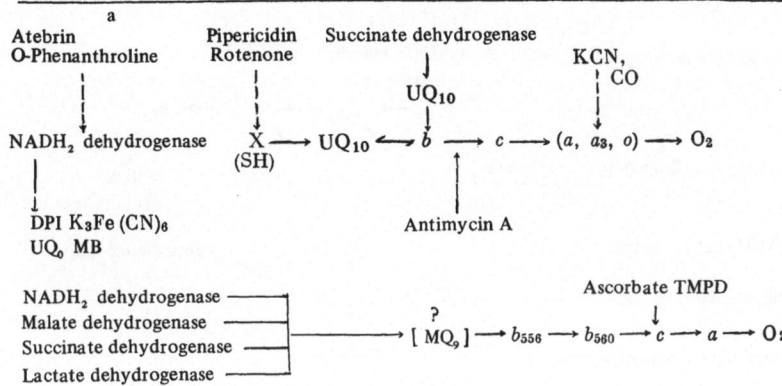

Scheme 5. Respiratory chains of *Micrococcus denitrificans* and *Micrococcus lyso-deikticus:* a) *M. denitrificans* (cells grown aerobically) (Asano et al., 1967a,b; Imai et al., 1967, 1968a,b; Scholes and Smith, 1968b; Lam and Nicholas, 1969b); b) *M lysodeikticus* (Mitchell, 1962; Ishikawa and Lehninger, 1962; Bishop and King, 1962; Gel'man et al., 1963a; Fujita et al.,1966; Lukoyanova and Taptykova,1968; Tikhonova et al., 1970).

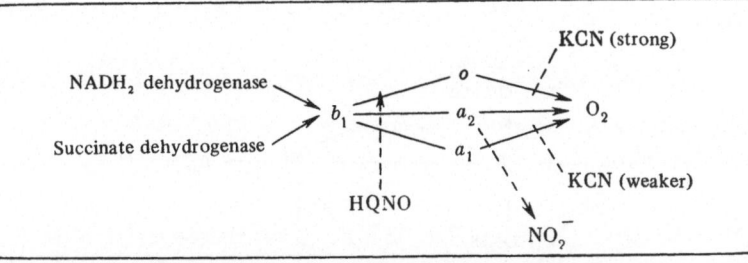

Scheme 6. Respiratory chain of cyanide-resistant cells of *Achromo-bacter* (Arima and Oka, 1965a,b).

Compared with the chains of *M. lysodeikticus, M. denitrificans, M. phlei,* and *P. putida,* with their many components, the respiratory chain of *Agrobacterium tumefaciens* is very modest in its structure (Kurup et al., 1966):

$$NADH_2 \text{ dehydrogenase} \rightarrow UQ_{10} \rightarrow \text{Cytochrome c} \rightarrow O_2.$$

The respiratory chain of cyanide-resistant cells of *Achromobacter* contains three terminal oxidases: cytochromes o, a_1, and a_2. Cytochrome a_2 is a cyanide-resistant oxidase (Scheme 6).

Numerous experiments have shown that individual strains of the respiratory chain of *B. subtilis* (Scheme 7a) contain an antimycin-sensitive factor; however, an antimycin-insensitive shunt is also present (Arima and Sato, 1970). A similar respiratory chain is present in *B. megaterium* (Scheme 7b).

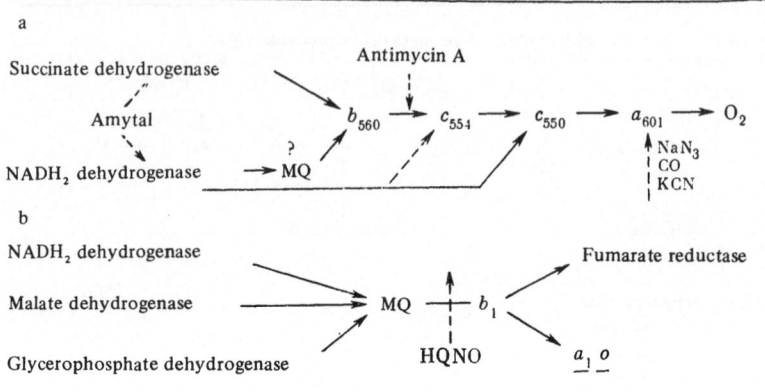

Scheme 7. Respiratory chains of *Bacillus subtilis* (a) (Miki and Okunuki, 1969b) and *Bacillus megaterium* (b) (Kröger and Dadak, 1969).

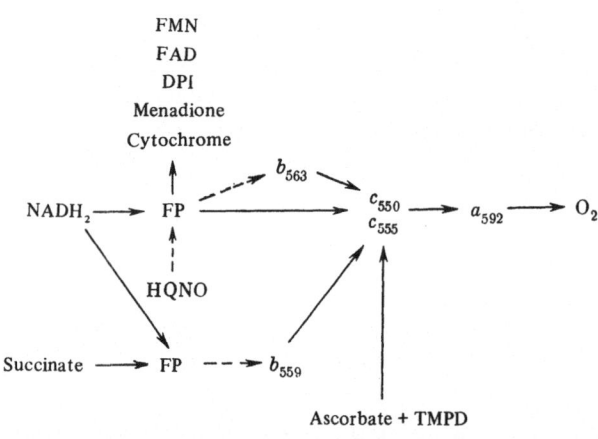

Scheme 8. Hypothetical scheme of respiratory chain of *Halobacterium cutirubrum* (Lanyi, 1969). Broken arrows show alternative slow pathways of oxidation.

   The respiratory chains of the halophile bacteria, adapted to life in a high NaCl concentration, show some special features (Scheme 8). This adaptation extends also to the activity of the respiratory enzymes, which increases in a salt medium (Cheah, 1970b; Lanyi, 1969; Holmes and Halvorson, 1963, 1965; Hochstein and Dalton, 1968). The assortment of respiratory carriers in *Halobacterium cutirubrum* varies with the pH of the medium (Lanyi, 1968).

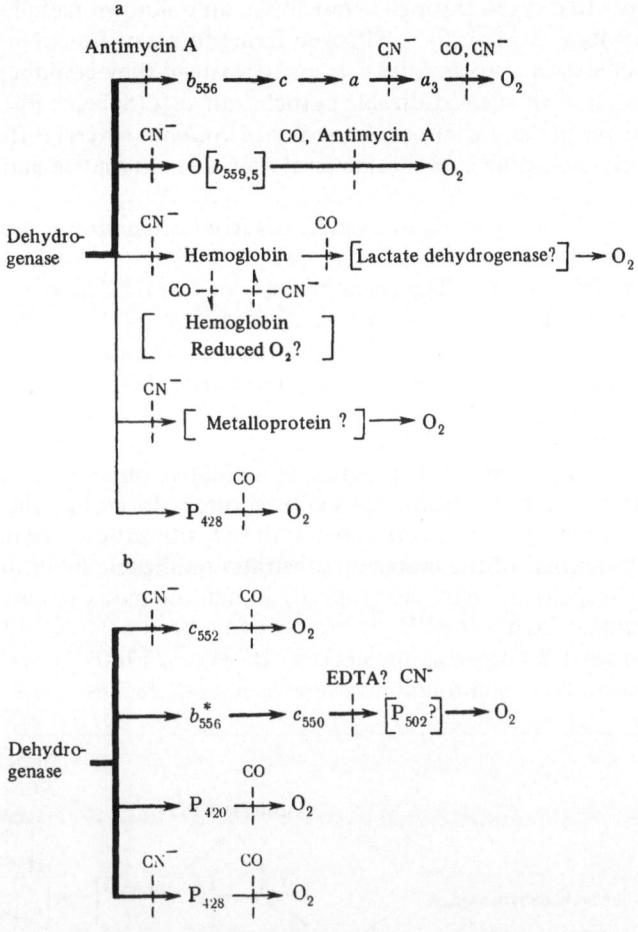

Scheme 9. Respiratory chains of *Rhizobium japonicum:* a) free-living cells (Appleby, 1969b); b) nitrogen-fixing cells (Appleby, 1969a); thick lines show principal pathways of electron transport; thin lines show possible alternative pathways; cytochrome b_{556} is probably identical with the cytochrome b_{556} of the free-living cells.

Respiratory chains of free-living and nitrogen-fixing cells of *Rhizobium* have been investigated in detail (Scheme 9). Electron transfer in the former takes place through a classical assortment of carriers; however, besides this main pathway there are several shortcuts and slower pathways of electron transport to oxygen through hemoglobin, an unknown metalloprotein, and pigment P_{428} (Scheme 9a). Nitrogen-fixing forms of *Rhizobium* do not contain oxidases of cytochromes a-a_3 and o; instead they contain pigments P_{420} and P_{428}, and an auto-oxidizable cytochrome c_{552} (Scheme 9b).

The respiratory chain of *A. vinelandii* contains several different cytochromes, and at the cytochrome level there are alternative pathways (Scheme 10).

The respiratory chain in membranes from *H. parainfluenzae* cells grown under aerobic conditions contains six cytochromes, five dehydrogenases, and demethyl-vitamin K_2. The characteristics of this ETC have been given above (see page 145).

The respiratory chains of the lithotrophs (Schemes 11-15) are extremely varied (Zavarzin, 1964; Doman and Tikhonova, 1965; Tikhonova, 1967; Repaske, 1966; Kiesow, 1967; Peck, 1968; Wallace and Nicholas, 1969; Aleem, 1970).

Besides ATP, which they obtain by oxidative phosphorylation in the respiratory chain, the lithotrophs also require reduced pyridine nucleotides for various biosyntheses connected with CO_2 utilization. Meanwhile, the redox potentials of the inorganic substrates oxidizable by lithotrophs (except hydrogen) are more electropositive than the redox potential of pyridine nucleotide, which is 0.32 V. It has therefore been postulated that lithotrophs contain what is known as an "energy lift" (Lees, 1960; Zavarzin, 1964). Such a lift has in fact been found in several species (Schemes 12-15).

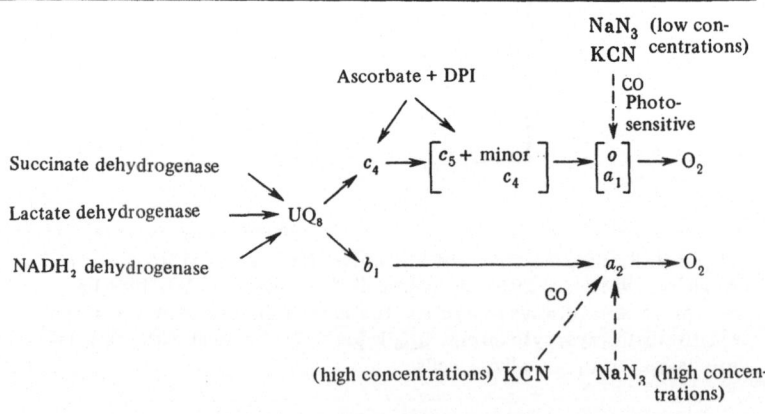

Scheme 10. Respiratory chain of *Azotobacter vinelandii* (Jones and Redfearn, 1967a,b; Jurtshuk and Harper, 1968; Swank and Burris, 1969).

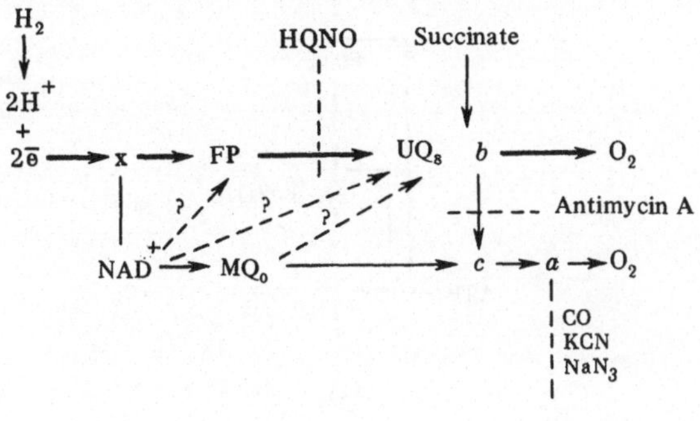

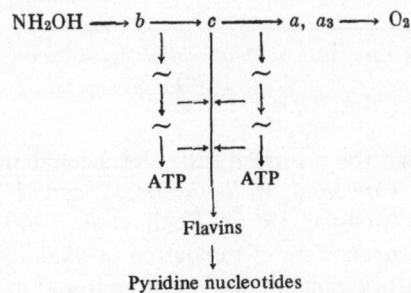

Scheme 11. Hypothetical respiratory chain of *Hydrogenomonas* H-20 (Bongers, 1967): X) unidentified primary electron acceptor.

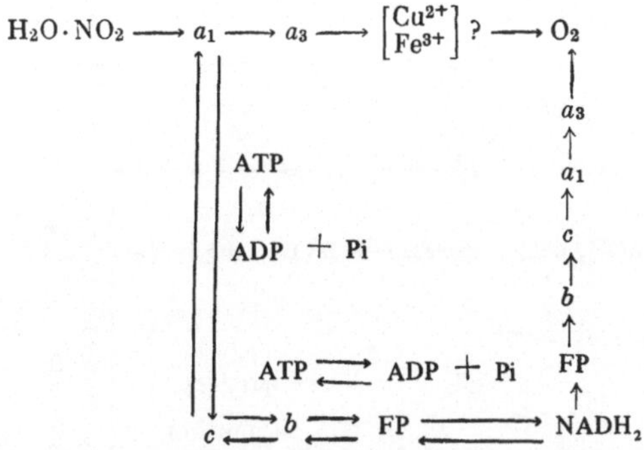

Scheme 12. Respiratory chain of *Nitrobacter agilis* (Aleem, 1968).

Scheme 13. Respiratory chain and pathways of energy transformation in *Nitrosomonas europea* (Aleem, 1960a).

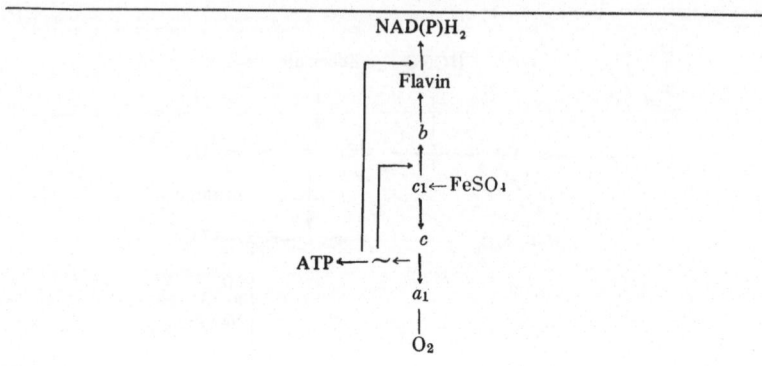

Scheme 14. Respiratory chain of *Thiobacillus ferrooxidans* (Aleem et al., 1963; Tikhonova et al., 1967).

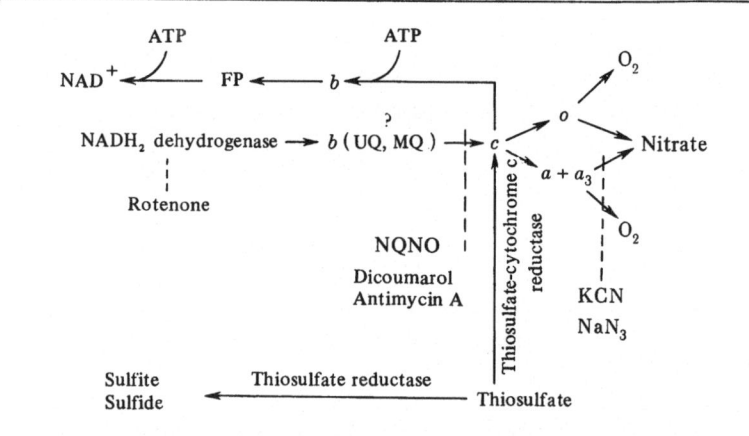

Scheme 15. Respiratory chain of *Thiobacillus denitrificans* (Peeters and Aleem, 1970).

It is remarkable that the reverse transfer has been demonstrated for some heterotrophs also: *A. vinelandii, M. denitrificans,* and *M. phlei* (Asano et al., 1967a,b; Naik and Nicholas, 1966b; Bogin et al., 1969). For a long time this vitally important mechanism of regulation of oxidative metabolism as a whole, namely respiratory control, could not be found in bacterial cells (Asano and Brodie, 1964; Kufe and Howland, 1968). Low respiratory control was found in *M. denitrificans* and *Azotobacter chroococcum* (John and Hamilton, 1970). Respiratory activity was stimulated in *T. ferrooxidans, Brevibac-terium ammoniagenes,* and five more species of Gram-negative bacteria by inorganic phosphate (Oishi et al., 1970; Beck and Shafia, 1964; Moses and Prevost, 1966). It has recently been shown that the detection of a respiratory control in bacterial preparations requires special conditions (Jones et al., 1971a,b). On the whole, bacterial respiratory chains are much more complex than the corresponding mitochondrial chains because of the branched nature of the pathways and the great variety of dehydrogenases and cytochromes (White and Sinclair, 1971). Variation also exists in sensitivity to inhibitors. The terminal oxidases differ in their sensitivity to cyanide and azide; anti-mycin either does not inhibit the respiratory chain of bacteria, or does so only in high concentrations (Sazykin, 1968; Rieske, 1969).

Role of Membrane Structures in the Action of the Respiratory Chain

There is no doubt that the polyenzyme system of the electron transport chain can act effectively only if the membrane retains its native structure. Any disturbance of this structure will impair its activity.

The importance of structure for the action of the respiratory chain is due to many factors, the most important of which is the creation of the necessary conformation of the proteins in the membrane through lipid–protein and protein–protein interactions. Extraction of lipids or individual proteins from the membrane, proteolysis and lipolysis of the membrane, and removal of cations by chelating agents therefore usually cause loss of activity of the respiratory chain. The system of enzymes can also be inactivated by defor-mation of the membrane through osmotic shock during cell lysis, by ultrasonic waves, or by high pressures.

There have been many investigations in which lipids have been removed from membranes by solvents or solubilized by detergents. As a rule, activity of the respiratory chain has been found to depend on the presence of lipids in the membrane. These results have been partially discussed above; at this point, merely a few examples will be given. Destruction of membranes of mycobacteria by detergents led to inactivation of the respiratory chain (Gold-man et al., 1963); Emazol and Tween-80 inhibited lactate oxidase of *A. sub-oxidans* (Iwasaki, 1960). Triton X-100 inhibited succinate oxidase of mem-branes from *C. diphtheriae* by 70-80% (Scholes and King, 1965a).

Disruption of the membranes of *M. lysodeikticus* was accompanied by inhibition of succinate oxidase, malate oxidase, and $NADH_2$ oxidase (Gel'man et al., 1960b; Lukoyanova et al., 1961; Tikhonova et al., 1970).

On extraction of lipids from various preparations of bacterial membranes containing a respiratory chain, electron transfer to oxygen is inhibited. In some cases it is difficult to understand whether the whole lipid fraction is necessary, or only ubiquinones and menaquinones. Treatment of particles of *E. coli* with isooctane abolished the oxidation of succinate and malate.

Reconstruction took place after the addition of menaquinone and phospholipids (Kashket and Brodie, 1963). Similar results have been obtained with
Bacillus stearothermophilus (Downey et al., 1962). Extraction with solvents
inhibited succinate oxidase. Activity was restored by the addition of menaquinone or ubiquinone and lipids of the extract. In experiments in which
lipids were extracted, the process could be reconstructed only by the action
of quinones (Van Demark and Smith, 1964; Itagaki, 1964; Segel and Goldman,
1963).

More precise information on the role of lipids in the respiratory chain
can evidently be obtained by the use of lipolytic enzymes, which hydrolyze
lipids without denaturing proteins or damaging quinones. To investigate the
function of lipids, membrane lipids of *M. lysodeikticus* were treated with
pancreatic lipase and phospholipase A (Ostrovskii et al., 1964; Ostrovskii,
1964; Lukoyanova, 1964; Oparin et al., 1965b). As a result of treatment
with these enzymes, electron transport in the membrane was inhibited by
80-90%, in agreement with the results of their action directly on the protoplasts
of *M. lysodeikticus* (Ostrovskii et al., 1964). Neither solubilization of the
cytochromes nor any marked degree of destruction of the membranes was
observed under these circumstances (Lukoyanova, 1964; Lukoyanova et al.,
1967). However, in some bacterial membranes, lipolysis was accompanied
by extraction of the cytochromes. For instance, treatment of membranes
from *B. megaterium* with lipase leads to extraction of cytochrome o, although
the bond linking cytochrome a_3 to the membrane is not thereby affected
(Broberg and Smith, 1967). Cytochrome o could be extracted from membrane fragments of *M. phlei* also by means of lipase, whereas cytochromes
b, c, and a + a_3 remained in the membrane (Revsin and Brodie, 1969).
These results are evidence that the bonds linking cytochrome proteins to
the membrane differ in their nature. It is interesting to note that the cytochrome oxidase of a halotolerant micrococcus could be solubilized only
by the combined action of lipase, phospholipase A, and deoxycholate (Hori,
1963).

The role of lipids in the function of the respiratory chain is closely linked
with the problem of conformational changes in the proteins as the result of
their incorporation in the membrane. This phenomenon was described by
Racker and Schatz as "allotopy" (from the Greek *allos* = another, and *topos* =
place) (Racker, 1969a; Pullman and Schatz, 1967). Allotopy of enzymes is
presently being closely investigated in model experiments (Poltorak and
Chukhrai, 1969; Deborin, 1967).

The dependence of some isolated membrane fragments on particular
lipid components for their activity is a clear example of allotopy. Succinate
dehydrogenase, β-hydroxybutyrate dehydrogenase, and cytochrome oxidase
of mitochondria all require phospholipids (Cerletti et al., 1965, 1967; Jurt-
chuk et al., 1961, 1963; Cohen and Wainio, 1963; Brierley and Merola, 1962).

Similar results have been obtained for other bacterial membrane enzymes which have been isolated.

Malate dehydrogenase from *Pseudomonas* required phospholipids to exhibit its activity (Francis and Phizackerley, 1965). Isolated malate dehydrogenase from *Mycobacterium avium* and from acetic acid bacteria behaves in the same way (Tobari, 1964; Benziman and Galanter, 1964; Asano and Brodie, 1964; Asano et al., 1965). Malate dehydrogenase in a preparation obtained from the membranes of *M. lysodeikticus* is inactivated by phospholipase A (Oparin et al., 1965a).

In the course of purification of bacterial cytochromes they were found to be bound with lipids (Kamen et al., 1963). The need has been demonstrated for a lipid fraction extracted from the cells for the reduction of cytochrome b_1 under the influence of formate dehydrogenase from *E. coli* (Itagaki et al., 1961). It is postulated that the formation of a complex between the enzyme protein and phospholipid determines the conformation required for activity to be exhibited. Allosteric inhibition of the ATPase from *E. coli* membranes by Na ions depended on the lipid composition of the membranes (Farias et al., 1972).

Evidently not only the change in enzyme activity is related to allotopy, but so also are changes in many other parameters of the proteins during their incorporation in the membrane, compared with the properties of isolated proteins (latency of activity, thermolability, sensitivity to inhibitors, resistance to proteolysis, etc.). A clear example from this point of view is mitochondrial phosphorylating ATPase; its properties are strikingly different when it is in the membrane and when it is isolated from it (Racker, 1967, 1969b). Similar changes have been described for the bacterial ATPase from *B. megaterium* (Ishida and Muzushima, 1969). Membrane-bound ATPase is stable to cold and is activated by Mg^{++} and Ca^{++}. Solubilized ATPase is rapidly inactivated in the cold, and it can be activated only by Ca^{++}. Some substances — for example, pentachlorophenol, gramicidin, ADP, and ethanol — differed sharply in their action on the membrane-bound and isolated fragment.

The membrane ATPase of *E. coli* becomes cold-labile after solubilization (Evans, 1970).

ATPase isolated from the membranes of *M. lysodeikticus* was digested by trypsin. After reincorporation in the membrane, the same enzyme was resistant to proteolysis. In the membrane it is evidently shielded from digestion by certain other components of the membrane. The sensitivity of this enzyme to gramicidin changed similarly: the isolated enzyme was resistant to it. After reincorporation in the membrane, the ATPase became sensitive to gramicidin (Ishikawa, 1970).

ATPase from the membranes of *S. faecalis* in a soluble state is insensitive to *n,n* -dicyclohexylcarbodiimide (DCCD), as well-known inhibitor of mitochondrial ATPase; however, the same enzyme in the membrane is almost

completely inhibited (Harold et al., 1969b). Since the ATPase can be extracted from the membrane in an active state after treatment with snake
venom, these workers consider that DCCD interacts in the membrane with
some other protein, producing conformational changes in it which lead to
inactivation of the adjacent ATPase.

The properties of such typical membrane proteins as cytochrome are
also modified in some cases by changes in their state. Such changes are known,
for example, for the redox potentials of the cytochromes. For instance, cytochrome c_{550} with a redox potential of $+0.34$ V has been isolated from $R.$
$spheroides;$ the same cytochrome, when bound in particles of the membrane,
had a potential of $+0.25$ V (Motokawa and Kikuchi, 1966). A crystalline
preparation of cytochrome b_1 from $E.$ $coli$ has a potential of 0.34 V, whereas
the same cytochrome, bound in the membrane, has a potential of 0.10 V. A
specific potential-modifying protein, which on complex formation with cytochrome b_1 gives rise to a corresponding change in the potential, has been
isolated from the membranes of $E.$ $coli$ (Fujita et al., 1963; Deeb and Hager,
1964).

During purification of cytochromes c_{550} and c_{554} isolated from $B.$ $subtilis,$
the rate of their reaction with substrates fell (Miki and Okunuki, 1969b).
Membrane-bound cytochromes c do not react with CO because of the strong
configuration of their protein moiety; isolation facilitates modification of the
protein and enables it to react with CO. This fact has been observed in several bacterial cytochromes c, for example, in the cytochrome c_{553} from $P.$
$denitrificans$ (Iwasaki and Shidara, 1969a).

The importance of the allotopy phenomenon as a means of understanding
the structure of the respiratory chain is clearly seen in the case of mitochondrial cytochrome. This cytochrome is damaged by treatment of the membrane,
even with weak agents (Ernsterc et al., 1969; Racker, 1969b). It ceases to take
part in the ETC, whereas in the intact mitochondrial membrane the whole of
the cytochrome b is in a functioning state (Wainio et al., 1968; Urban and
Klingenberg, 1969). Since lyophilization of the membranes, treatment with
ultrasound, mechanical disintegration, and treatment of the membrane with
inhibitors are factors which change the activity of cytochrome b, it is obvious
that any modification of the membranes must have harmful consequences
for the proteins. Thus it is difficult, on the basis of these observations, to
construct a scheme for the respiratory chain in $vivo.$

The arrangement of the components of the respiratory chain in the membrane is interesting. Facts relating to the orientation of the ETC in the membrane are few in number and contradictory in nature. In $M.$ $denitrificans,$
$B.$ $megaterium,$ and $B.$ $subtilis,$ the orientation of the respiratory chain is
considered to be across the membrane (Scholes and Smith, 1968b; Broberg and
Smith, 1967, 1968), the dehydrogenases being found on the inner surface of
the membrane vesicle and the oxidases on the outer surface. This conclusion
is confirmed by the fact that dehydrogenase activity is increased after the
action of agents disrupting the membrane (ultrasound, freezing and thawing),

whereas these treatments do not affect oxidase activity. $NADH_2$-menadione reductase is located on the inner surface of the cytoplasmic membrane of *H. Halobium* (Lanyi, 1972), while succinate dehydrogenase is located on the inner surface of the membrane of *M. phlei* (Hirata and Brodie, 1972).

Some very interesting results regarding the localization of the respiratory chain of *M. lysodeikticus* are given by Shah and King (1965, 1966). Succinate oxidase is inhibited in the course of lysis of the cell by lysozyme, but its activity is unchanged if the cell is broken up by ultrasound. If, however, the sonicated fragments are then treated with lysozyme, succinate oxidase is inhibited. Remembering that mucopeptide, the substrate on which the lysozyme acts, is found only in the cell wall, these results suggest that the respiratory chain is located on the outer side of the cytoplasmic membrane. However, neither malate oxidase nor $NADH_2$ oxidase is inhibited during lysis of the cell (Gel'man et al., 1963b), so that perhaps only succinate dehydrogenase is located on the outer side of the membrane. By contrast, the observations of Mitchell showed that succinate dehydrogenase is located on the inner surface of the cytoplasmic membrane of *M. lysodeikticus* (Mitchell, 1962).

Opinions thus differ, even as regards the localization of dehydrogenases in the bacterial membrane. Clearly this problem of the arrangements of the components of the respiratory chain in the bacterial membrane requires further investigation, stimulated by Mitchell's hypothesis of the loop-like arrangement of the chain in the membrane (Mitchell, 1966).

Oxidative Phosphorylation

General Remarks

Transport of electrons from substances undergoing oxidation along the chain of respiratory enzymes is accompanied, in bacteria just as in the cells of higher organisms, by the formation of high-energy adenosine triphosphate (ATP).

Three fundamentally different hypotheses have been put forward in the attempt to explain the coupling of electron transport with ATP synthesis:

1. The hypothesis of "chemical" coupling (Slater, 1953);
2. The electrochemical (chemo-osmotic) hypothesis (Mitchell, 1961, 1970);
3. The conformation (mechanochemical) hypothesis (Boyer, 1964).

According to the first hypothesis, a high-energy bond is formed on some electron carriers during respiration, and as a result of the successive reactions of this carrier with inorganic phosphorus and ADP (oxidative phosphorylation), this energy is ultimately concentrated in the ATP molecule.

The conformation hypothesis postulates that electron transport leads to contraction of special proteins; structural changes in these proteins cause the formation of a high-energy bond which is subsequently handed on from the protein to ADP. Conformation changes during oxidation in mitochondrial membranes have been described by Green and Baum (1970).

Finally, the chemo-osmotic hypothesis is based on the assumption that the biological membrane carrying the respiratory enzymes has special properties of permeability. It is assumed that respiration creates a potential difference on the membrane surface as the result of the unidirectional transport of protons, and this potential difference leads to ATP synthesis in specific areas of the membrane with limited accessibility to water and other reagents.

The development of bioenergetics during the last 5 years has confirmed our predictions regarding the successful outlook for the concept of vectorial metabolism (Gel'man et al., 1966). Before the mid-1960's most workers accepted the chemical hypothesis of coupling, but more recently interest in the views of Boyer and, in particular, of Mitchell has increased sharply. This is due, to some extent, to the development of new approaches to the study of conformation changes and of the structure of biological membranes in general.

The fundamental trends in bioenergetics and, in particular, oxidative phosphorylation in bacteria have been fully discussed in Soviet and Western surveys (Kotel'nikova, 1969; Skulachev, 1969; Gel'man et al., 1966; Pullman and Schatz, 1967; Green and Baum, 1970). We can therefore concentrate our attention on only those aspects of the problem which are directly concerned with the membrane as a whole.

Localization of Oxidative Phosphorylation

The membranes in bacterial cells form a single continuous system represented by the cytoplasmic membrane and its various invaginations (see Chapter I). Mitochondria, the centers of oxidative phosphorylation in higher organisms, are absent from bacteria, and the problem of localization of oxidative phosphorylation is reduced to the choice between the mesosomal and cell membrane. Since aerobic ATP synthesis and respiration are evidently spatially coupled processes, the whole controversy regarding the role of bacterial mesosomes and the cytoplasmic membrane in respiration (see Chapters I and IV), for the resolution of which so many cytochemical and biochemical data have been invoked, applies even more to the localization of oxidative phosphorylation (Ostrovskii and Gel'man, 1963, 1965; Ostrovskii, 1964).

Without discussing the more cytochemical data here, we shall mention other approaches which have been made to the solution of this problem. Feo et al. (1962) consider that the sharp decrease in the efficiency of oxidative phosphorylation when normal cells of *Proteus morganii* are converted into L-forms or penicillin-induced spheroplasts is due to disturbances in the structure of the cytoplasmic membrane under the influence of osmotic pressure. These workers thus regard the results of their own experiments as evidence that respiratory enzymes are located on the surface.

The discovery of a system of internal membranes in anaerobically grown microorganisms, e.g., *Listeria monocytogenes* and *Actinomyces bovis*, or in obligate anaerobes, even if without cytochromes (Edwards and Stevens, 1963;

Fitz-James, 1962) and, on the other hand, the absence of mesosomes in some
aerobic organisms (Hines et al., 1964) are also powerful arguments against
the participation of mesosomes in electron transport and in oxidative phosphorylation.

Biryuzova et al., (1964) found fungus-like outgrowths in a preparation of
membranes of *M. lysodeikticus* similar to those on the internal membranes
of mitochondria (Fernández-Morán et al., 1964). Bladen et al. (1964) found
them on the membranes of *Eubacterium* sp., and Abram (1965) found them
on eight other representatives of the aerobic bacteria. It is interesting to note
that in the first case the outgrowths occurred on the internal membranes and
were evidently derivatives of the mesosomes, whereas in the second case they
were found also on the cytoplasmic membrane. These fungus-like outgrowths
are also found on the cytoplasmic membrane of *B. megaterium* and *Sporo-
bacillus inulinus* (Fitz-James, 1968). Whatever the function of the fungus-
like outgrowths, their existence on membrane structures of bacteria is evidence
that these structures perform mitochondrial functions. This may provide indirect proof that oxidative phosphorylation is localized both in the cytoplasmic membrane and in the mesosomes. Much thus remains to be explained
in the problem of the localization of the enzymes of respiration and oxidative
phosphorylation in bacteria. It can be assumed, as did Stanier (1964),
that the enzymes of electron transport and oxidative phosphorylation are
located in the cytoplasmic membrane itself, and also in its derivatives consisting of different types of invaginations. However, the problem of distribution of these enzymes will clearly not be solved until preparative separation
of discrete membrane structures has been accomplished.

Factors of Oxidative Phosphorylation

The efficiency of oxidative phosphorylation in preparations of bacterial
membranes is significantly increased by the addition of certain substances
which remain in solution when bacterial homogenates are centrifuged. These
substances are called phosphorylation factors, although other components
of the system such as Mg^{++} ions, neutral buffer, etc., are also essential for
this process. It is considered that phosphorylation factors play a dual role
in bacterial systems (Bogin et al., 1970). On one hand, they possibly catalyze
individual reactions of oxidative phosphorylation, while on the other hand, they
may be specific stabilizers of the membrane structure which may undergo disruption during its extraction from the cell (Ishikawa, 1970).

Parsons (1966) showed by the use of specific antibodies that factor F_1
(ATPase), the principal enzyme of mitochondrial oxidative phosphorylation,
is located in the fungus-like outgrowths. This enzyme has been purified to
the individual protein state. It is a decamer with molecular weight 280,000
(Kagawa, 1969).

Very little is known regarding the chemical nature of the bacterial coupling
factors. The only compound to have been completely identified is a tetra-
nucleotide from *Alcaligenes faecalis* (Shibko and Pinchot, 1961), which con-

sists of uridylic, adenylic and guanylic acids in the ratio of $1:2:1$. The molecular weight (5600) and amino acid composition of a thermostable factor from *M. lysodeikticus* have been determined (Ishikawa, 1970). The remaining factors have not been sufficiently purified, but from what is already known (for example, regarding the sensitivity of some factors to treatment with proteinases), they are proteins. Watanabe et al.(1965) have reported the existence of a nonprotein phosphorylation factor in *M. phlei,* whose only additional characteristics are absorption maxima at 260 and 340 nm.

Ota (1965) isolated three protein coupling factors from *E. coli.* Two of them (F_1 and F_2) had, in addition to the usual absorption maximum at $\lambda = 280$ nm, another maximum at $\lambda = 410$ nm. The third (F_4) showed three absorption maxima at $\lambda = 352, 371,$ and 393 nm.

Recently Higachi et al. (1969) showed that the phosphorylation factor from *M. phlei* contains at least three components, equally essential for different coupling points. Two protein factors (one thermostable, the other thermolabile) have been found in *M. lysodeikticus,* and the thermolabile factor lost its coupling property on keeping, although it still remained an activator of the ADP-ATP exchange reaction, suggesting the internal complexity of the actual labile factor itself (Ishikawa and Lehninger, 1962; Yamashita and Ishikawa, 1965; Ishikawa, 1970).

Brodie (1959) found and investigated the complete interchangeability of particle fractions and factors from *A. vinelandii, M. phlei* and *Corynebacterium creatinovorans.* Also, it has recently been shown (Bogin et al., 1970) that protein factors from bacteria and from animal mitochondria can replace each other with an efficiency of 30-50%. This universality of the bacterial factors for different species of bacteria is of great interest, but the universality itself must be treated with caution. For instance, a polynucleotide from *A. faecalis* is required as a factor coupling oxidation with phosphorylation (together with a protein factor) only by *A. faecalis* (Pinchot, 1957a,b), while a protein factor from *M. lysodeikticus* does not stimulate respiration, unlike the factor from *M. phlei,* although it strongly stimulates the reaction of ADP-ATP exchange, unlike the factor from *A. faecalis* (Ishikawa and Lehninger, 1962).

An interesting problem arises in connection with the observation by Ishikawa and Lehninger (1962) that a preparation of particles from *M. lysodeikticus* requires a high concentration of Mg^{++} ions, and that Mg^{++} can be replaced by spermine and spermidine in its role as stabilizer of oxidative phosphorylation (Ishikawa, 1970). Evidently Mg^{++} and these polyamines in some way facilitate interaction between the factor and the membrane carrying the respiratory chain.

Spermine and spermidine are widely distributed components of bacterial cells. In very low concentrations (10^{-3} M) they can prevent osmotic shock in the protoplasts of *M. lysodeikticus* and the spheroplasts of *E. coli,* a property not shared by Mg^{++} ions. They also prevent lysis of the membranes of

Pseudomonas striata through the action of EDTA (Tabor, 1962; Adams and Newberry, 1961). The stabilizers perhaps form a bridge linking the two structures, or neutralize the negative charge on the membrane or on the factor, with or without producing any change in the configuration of the structures. An energy-dependent transhydrogenase is present in *E. coli* (Bragg and Hou, 1968; Fisher and Sanadi, 1971).

Whatever the facts, the question arises: what is there in common between Mg^{++} and, for example, spermidine $[H_2N-(CH_2)_3-NH(CH_2)_4-NH_2]$ so that they have the same effect on oxidative phosphorylation in bacteria, whereas many other compounds, also carrying positive charges, are ineffective?

On the basis of these facts and our own investigations, we consider that phosphorylation factors are hydrogen-bonded with the membrane, whereas Mg^{++} or spermidine simply neutralizes negatively charged groups on the membrane and on the factor, thereby enabling them to come together and facilitating their interaction. Urea solutions, even in the presence of high concentrations of Mg^{++} ions in Tris-HCl buffer, cause complete (but reversible) separation of the phosphorylation factors from the membranes. Treatment of phosphorylating preparations from membranes of *M. lysodeikticus* with bacterial proteinase (Ostrovskii and Gel'man, 1965) leads to the uncoupling of oxidative phosphorylation, suggesting the absence of binding of the protein factors with lipids which could protect them against hydrolysis. This is confirmed by results obtained by Yamashita and Ishikawa (1965), who found inactivation of a thermostable factor after treatment with proteolytic enzymes, and with our own observations showing the absence of inactivation of the respiratory chain enzymes on treatment of protoplasts with proteinases (Ostrovskii et al., 1964).

Because of the uncertainty regarding the coupling points of oxidation with phosphorylation in bacteria, it is difficult to assess the role of the factors in cell energy metabolism. Since the need for coupling factors arises as the result of vigorous action on bacterial membranes, it is perfectly possible that most factors are not directly concerned with phosphorylation reactions but simply reflect the need for preserving the structure of the biological membrane. This feature is particularly emphasized in the electrochemical hypothesis of the mechanism of oxidative phosphorylation (Mitchell, 1961) which, like the other hypotheses mentioned above, claims to be applicable to all organisms, not only bacteria.

Some Special Features Distinguishing Bacterial
Oxidative Phosphorylation

Many bacteria, especially *A. faecalis*, *E. coli*, *M. phlei*, and *A. vinelandii*, have been used as objects for the study of oxidative phosphorylation. Whereas the mitochondria of different species of animals or from different organs generally possess identical or very similar properties, this is by no means so in the case of preparations from bacteria. Oxidative phosphorylation in bacteria not only differs from that in higher organisms but also varies considerably in many respects (for example, requirement of cofactors and sensitivity to inhibitors) within the same group of bacteria. It must be pointed out that, with respect to some characteristics, this process in individual mitochondrial microorganisms (fungi) occupies an intermediate position between oxidative phosphorylation in higher organisms and in bacteria (Kotel'nikova, 1969). In the writer's view, this fact indicates that the special features of their oxidative phosphorylation are natural properties of bacterial membranes and are not produced as the result of using imperfect techniques, as many workers consider.

The efficiency of coupling between respiration and synthesis of high-energy ATP in preparations of bacterial membranes is low. Consequently, in most cases the $P:O$ ratio does not exceed 1, even when under the same conditions the mitochondria of higher organisms give a $P:O$ ratio of 3. This is partly due to the smaller number of coupling points in bacteria. Bragg and Hou (1968), for instance, suggest that there are two points of ATP synthesis in the membranes of $E.\ coli,$ one linked with $NADH_2$ oxidase and the other with an energy-dependent transhydrogenase reaction. Two points of phosphorylation have also been found in the particles of $A.\ vinelandii:$ one between $NADH_2$ and flavoprotein, and the other appearing during oxidation of succinic acid, i.e., located somewhere between flavoprotein and oxygen (Pandit-Hovenkamp and Eilerman, 1968). Further evidence of two coupling areas in the cells of $A.\ vinelandii$ is given by the $P:O$ ratio of approximately 2 during oxidation of β-hydroxybutyric acid, calculated from the increase in ATP concentration in the whole cells when transferred from anaerobic to aerobic conditions of growth (Knowles and Smith, 1970). Similar observations, indicating the possible coupling of phosphorylation with electron transport in two areas of the respiratory chain, have been made on $M.\ denitrificans$ (Imai et al., 1967, 1968a,b). The existence of the first site (between $NADH_2$ and ubiquinone) in these investigations is confirmed by the discovery of phosphorylation during $NADH_2$ oxidation in the presence of ubiquinone as acceptor. The second site is postulated on the basis of the results of recording phosphorylation during oxidation of succinate ($P:O = 0.4$). Sensitivity of the preparation from $M.\ denitrificans$ to typical uncouplers and inhibitors, as well as the system of reverse electron transport, make it very similar to mitochondria.

Some investigators have described finding three coupling points in bacterial membranes. This has been demonstrated in preparations from the chemoautotroph $Nitrobacter\ agilis$ (Aleem, 1968). The existence of three independent segments of respiratory chains capable of phosphorylating electron transport proves the existence of three energy coupling points, differing in their sensitivity to inhibitors, to cofactors, and to coupling proteins. The same has been found in *M. phlei* also (Asano and Brodie, 1965; Bogin et al., 1968). Meanwhile, in bacterial such as in *C. diphtheriae,* not more than one coupling point (between cytochromes b and c) has yet been found (Kufe and Howland, 1968).

The so-called respiratory control (dependence of the rate of substrate oxidation on the presence of ADP as phosphoryl acceptor) and reverse electron transport, (the transport of electrons on account of the energy of ATP from components of the chain with a positive electrochemical potential toward the oxidation substrates) are indicators of rigid coupling between ATP synthesis and respiration in mitochondria. In bacterial preparations, only very weak stimulation of respiration by ADP, or no stimulation whatever, has been found (Kotel'nikova and Ivanova, 1964; Cole and Aleem, 1970; Eilerman et al., 1970; Scocca and Pinchot, 1968; Adolfson and Moudrianakis, 1971a). In other words, the control of respiration by the phosphorylation system is weak, although stronger coupling has been described in the membranes of *M. denitrificans* (John and Whatley, 1970) and *A. vinelandii* (Jones et al., 1971a,b).

Reversal of electron transport in the respiratory chain of *Nitrobacter agilis, Nitrosomonas europea, Thiobacillus novellus,* and other species of chemoautotrophs has been demonstrated (see Schemes 11-15). It is interesting to note that reduction of NAD on account of electrons of thiosulfate and energy of ATP takes place very efficiently: the ratio of ATP used to NAD reduced reaches 1. It might be expected that during direct electron transport the coefficient P:2e would be 1 only at this site, but this has not been shown. On the contrary, as Hempfling and Vishniac (1965) showed, oxidation of $NADH_2$, sulfide, sulfite, and thiosulfate by cell-free extracts of *Thiobacillus* eggs is completely unaccompanied by phosphorylation. A P:O ratio of only 0.44 is given by the oxidation of 2-mercaptoethanol.

Reverse electron transport has been reliably demonstrated in preparations from *M. denitrificans* in the succinic acid–NAD site (Asano et al., 1967a,b) and in preparations of particles from *E. coli* (Sweetman and Griffiths, 1971a). Reduction of NAD by reverse electron transport has also been observed by Tikhonova (1967) in cells of *T. ferrooxidans.* The source of energy for this process was ATP, and electrons were taken up at the cytochrome c level from the reaction $Fe^{++} \rightarrow Fe^{+++}$. Analogous electron transport from the cytochrome a level has been shown in preparations from *N. agilis* (Aleem, 1968).

Reverse electron transport in the short segment $NADH_2 \rightarrow NADP$ (the energy-dependent transdehydrogenase reaction) has also been described in some bacterial preparations; in the membranes of *M. denitrificans* (Asano et al., 1967b) and *E. coli* (Bragg and Hou, 1968; Fisher et al., 1970; Sweetman and Griffiths, 1971b), and in the chromatophores of *R. rubrum* (Fisher and Guillory, 1969) and *A. vinelandii* (Chung, 1970).

One of the most obvious features which distinguishes bacterial energy metabolism is the ability of some bacteria to utilize mineral substances as respiration substrates or as terminal electron acceptors. For instance, *E. coli*, *P. denitrificans, P. aeruginosa,* and *M. denitrificans* (Hori, 1961; Fewson and Nicholas, 1961a,b) can transfer electrons to nitrate, synthesizing high-energy phosphates in the process. Ohnishi and co-workers (Ohnishi and Mori, 1960, 1962; Ohnishi, 1963; and Yamanaka et al., 1962, 1964) showed that this process takes place on sonicated fragments of the membranes of *P. denitrificans* and *P. aeruginosa* in the presence of soluble factors during the oxidation of $NADH_2$, lactate, and succinate. The $P : NO_3^-$ ratio is 0.2–0.3, although E_0 for the system $NO_2^- : NO_3^-$ is +0.421 V, i.e., the ratio $P : NO_3^-$ would be expected to be 2. The preparation from *E. coli* under analogous conditions phosphorylates more energetically, with $P : NO_3^- = 0.55$ (Yamanaka et al., 1964), while the preparation from *M. denitrificans* Has $P : NO_3^- = 0.5$ (Whatley, 1962). The sensitivity of the system to DNP is very low, and uncoupling takes place only at concentrations of about $1 \times 10^{-3} M$.

However, despite the abundance of evidence of the low efficiency of oxidative phosphorylation in nitrate-reducing bacteria, an abundance which gives the appearance of a universal rule, some workers have nevertheless obtained bacterial preparations with highly efficient phosphorylation. Kiesow (1964), for instance, showed that in *N. winogradskii* there is an $NADH_2-NO_3^-$ oxidoreductase system of enzymes producing 1.2–2 molecules of ATP during the oxidation of one molecule of $NADH_2$.

On the basis of indirect evidence, Hadjipetrou and Stouthamer (1965) consider that whole cells of *A. aerogenes* synthesize three molecules of ATP during the reduction of a single molecule of nitrate into nitrite. Oxidation of thiosulfate by preparations from *T. novellus* also proceeds with high efficiency ($P : O = 0.96$) (Cole and Aleem, 1970).

Cells of *N. europea* oxidize hydroxylamine into nitrite with $P : 2e = 0.2$. (Ramaiah and Nicholas, 1964), while *N. agilis* utilizes nitrite as its oxidative substrate (Aleem and Nason, 1960; Malavolta et al., 1960), transferring electrons along the chain of carriers (see Scheme 12). The synthesis of high-energy phosphate taking place under these conditions ($P : O = 0.2$) can be obtained in the fraction of membrane fragments without the addition of soluble factors. Just as in many other bacterial systems, phosphorylation is insensitive to $5 \times 10^{-5} M$ DNP, thyroxine, and dicoumarol; oxidation is inhibited only by very high (for this poison) concentrations of antimycin A ($20 \mu g/ml$).

Kiesow (1964) considers that nitrite is oxidized in two stages, one of which is accompanied by expenditure of energy and leads to the reduction of NAD, while the other is the oxidation of $NADH_2$ with oxygen, through the participation of the usual electron transport chain. The energy metabolism of the lithotrophs, organisms utilizing mineral oxidation substrates, has been discussed in some aspects above (page 188).

TABLE 12. Efficiency of Oxidative Phosphorylation (P:O) in Preparations from Some Bacteria

Organism	Preparation	Succinic acid	Malic acid	NADH$_2$	Source
Escherichia coli	E	0.24	0.4	0	Kashket and Brodie, 1963
	P_L	0.15	0.25	0.1	
	P_L + S	0.5	1.1	0.3	
	P_H + S	0	0	–	
	Cells	–	–	2.5-3	Hempfling, 1970
Azotobacter vinelandii	E	0.55	0.59	0.43	Kotel'nikova and Ivanova,
	P	0.87	0	0.41	1964; Brodie, 1963
	P + S	0.87	0.41	0.32	
Azotobacter vinelandii	P	–	0.53	1.10	Ackrell and Jones, 1971
Mycobacterium phlei	E	1.78	1.8	1.5	Brodie, 1959
	P	0.56	0	0	
	P + S	0.98	1.6	1.1	
Aerobacter aerogenes	E	–	–	0.33	Nossal et al., 1956
	P			0.16	
Corynebacterium creatinovorum	E	1.4	–	–	Brodie and Gray 1956a,b
	P	0.56			
Proteus vulgaris	E			0.62	Nossal et al., 1956
	P	–	–	0	
	P + S			0.52	
Micrococcus denitrificans	P	0.57	–	1.60	John and Whately, 1970
Micrococcus denitrificans (autotrophic strain)	P	1.0	–	1.2	Knoblock et al., 1971
Alcaligenes faecalis	E			0.36	Pinchot, 1957a,b,c,d
	P	–	–	0.03	
	P + S			0.33-0.78	
Micrococcus lyso-deikticus	E		0.3	0.43	Ishikawa and Lehninger,
	P		0.1-0.65	0.13-0.73	1962; Ostrovskii and Gel'man, 1965;
Proteus morganii	E	1	–	–	Feo et al., 1962
L-form	E	0.5			
Rhodospirillum rubrum	P + S	0.3	0.35	0.22	Geller, 1962; Tanigu-
	P + S	0.1-0.4	—	0.1-0.4	chi and Kamen, 1965
Gluconobacter lique-faciens	E	0.14	–	0.09	Stouthamer, 1962
	P	–	–	0.13	
Bacillus subtilis	P + S	–	1.85	–	Downey, 1964
Nitrobacter winog-radskii	P	–	–	2.2-2.8	Kiesow, 1964
Streptococcus faecalis	E			0.2	Gallin and Van Demark, 1964
	Cells	–	–	0.4-0.7	Smalley et al., 1968;
	P	–	–	0.19	Faust and Vandemark, 1970
Thiobacillus novellus	S	1.91	–	0.25	Cole and Aleem, 1970
Corynebacterium diphtheriae C$_7$	P	0.42	–	0.51	Kufe and Howland, 1968

Note: E) cell-free extract; P) subcellular particles: L) light, H) heavy; S) solution of cytoplasm.

Oxidative phosphorylation in bacterial membrane preparations has low efficiency and low sensitivity to uncouplers of the 2,4-DNP type. To this must be added the almost complete absence of respiratory control, comparatively easy fractionation, and the ability to utilize mineral substrates and electron acceptors (Table 12). In this connection the results obtained with whole cells of *A. vinelandii* (Zaitseva et al., 1961, 1963; Knowles and Smith, 1970), *B. megaterium* (Mal'tseva, 1963), *A. suboxidans* (Klungsoyr et al., 1957), and *P. denitrificans* (Ohnishi and Mori, 1960) are particularly interesting. The P:O ratio of this "intact" oxidative phosphorylation, as a rule, does not exceed unity.

The study of the efficiency of oxidative phosphorylation in the intact cell is a very difficult problem, and its solution is presently being sought in three ways.

The first method is by analysis of the distribution of radioactive phosphorus incorporated by bacteria during very short time intervals (seconds). If P^{32} is found entirely or predominantly in the ATP molecule, the radioactivity of ATP can serve as an indicator of its synthesis, provided, of course, that no fast metabolic reactions have occurred.

The second approach is based on comparison of the rate of increase of the bacterial mass with the energy capacity of the substrate utilized. By calculation of the energy expenditure of a culture of *Gluconobacter liquefaciens* (Stouthamer, 1962) when grown on a medium with glucose, the P:O ratio was found to be 0.22. The efficiency of oxidative phosphorylation of cells of *S. faecalis* 10C1, also on medium with glucose or mannitol, was 0.4-0.7 (Smalley et al., 1968), but oxidation of substrates in a culture of *A. aerogenes* is accompanied by the more rapid accumulation of ATP, so that the value of P:O reaches 3 (Hadijipetrou et al., 1964, 1965). By using a technique of rapid incorporation of radioactive phosphorus into the cell, followed by analysis of the primary products of P^{32} acceptance, Zaitseva et al. (1961, 1963) showed that the efficiency of oxidative phosphorylation reaches a maximum in the period of preparation for division in a synchronous culture of *A. vinelandii* (P:O up to 1.5). In the phase of more intensive division, the efficiency of the process was much lower (P:O < 1). During cooling of the culture, the P:O ratio falls to values of 0.5 or below.

A new method of determining oxidative phosphorylation has recently been developed for cells of *E. coli* on the basis of changes in the oxidation-reduction state of the respiratory enzymes and the content of phosphorylation groups in the nucleotide fraction during very short time intervals (Hempfling, 1970). It was shown in this way that the P:O ratio for the work of the complete respiratory chain is 2.5-3. The value of P:O for cells of *A. vinelandii* measured by this method is 3.0 (Baak and Postma, 1971).

Oxidative phosphorylation in whole cells is thus a distinctly inconstant process whose efficiency depends on the species of bacteria and the conditions of cultivation. The great importance of the conditions of cultivation, the phase of growth, and the rate and method of treating the bacteria during the study of oxidative phosphorylation has been demonstrated by the work of Zaitseva (1963, 1965).

If the imperfect organization of bacterial phosphorylation is accepted as the true state of affairs, the need arises to explain it in chemical and anatomical terms. The bacterial respiratory chain is represented by most investigators as something very similar to that of higher organisms, but in bacteria there are all manner of variations from the classical chain, especially in the region of the cytochromes, as was pointed out in the previous chapter.

We consider that bacteria are able, to a far greater degree than higher organisms, to utilize pathways of what is known as free oxidation (Skulachev, 1962), the energy of which is either dissipated as heat or utilized in a different form than by ATP, such as in the incorporation of C^{14}-amino acids into mitochondrial protein (Bronk, 1963; Kroon, 1964), in the activation of fatty acids (Wojtczak et al., 1965), or in ion transport (Judah and Ahmed, 1964; Rossi and Lehninger, 1964). It is possible that the succinate oxidase of *E. coli* and the lactate oxidase of *M. phlei* (Brodie, 1959), which do not give phosphorylation under any conditions, are also representatives of free oxidation. The presence of phosphorylating and nonphosphorylating respiratory chains has also been observed in membranes of *A. vinelandii* (Ackrell and Jones, 1971). In this connection, it is interesting to note that the efficiency of oxidation of different substrates, when assessed from the value of P:O, varies within very wide limits. For instance, the soluble system of oxidative phosphorylation from *T. novellus* (Cole and Aleem, 1970) oxidizes $NADH_2$ with P:O=0.25, but oxidizes succinic acid with P:O=1.91. The rate of free oxidation, calculated by Skulachev's (1962) equation, in bacterial preparations oxidizing NAD-bound substrates with P:O of about 1, is twice the rate of phosphorylating oxidation, i.e., two-thirds of the substrates are oxidized by a pathway unconnected with ATP synthesis.

Mitchell's hypothesis has been widely accepted by the Soviet biochemists, and it has been further developed by the work of Skulachev and Liberman (see Skulachev, 1969, 1972; Liberman, 1970).

The basic propositions of this hypothesis are as follows:

1. Electron transport along the respiratory enzyme chain takes place across the membrane and is accompanied by liberation of $6H^+$ on the outside of the membrane (Fig. 22).
2. The membrane is assumed to be impermeable to H^+ and OH^-, so that during respiration the membrane becomes polarized on account of the separation of these ions.
3. Unknown substances, which are weak acids in the dissociated form (YO^- and X^-), are protonated on the outside of the membrane and then combine with the loss of a molecule of water to form the substance $X-Y$.
4. The compound $X-Y$ is transported by a special translocase to the inside of the membrane, where the $X-Y$ bond becomes of the high-energy type because of the low concentration of hydrolysis products (XH and YOH).
5. Finally, the high-energy compound $X{\sim}Y$ reacts with ADP and P_{inorg} with the formation of ATP, while the reaction products XH and YOH, after interacting with oxygen (O^-), revert to the form of ions X^- and YO^-.

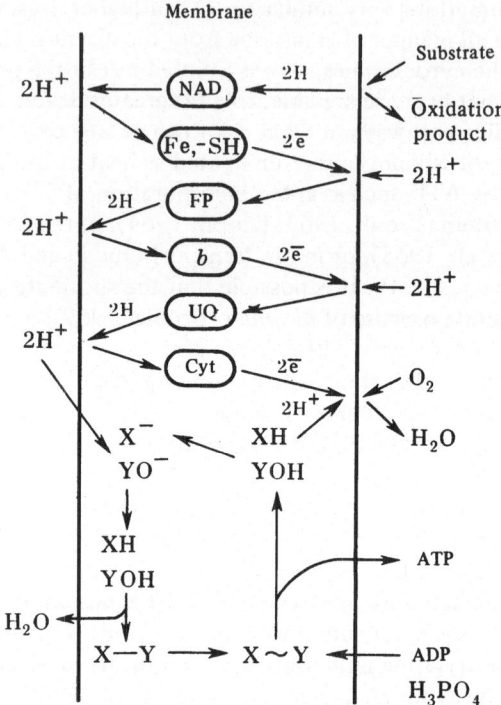

Fig. 22. Diagram showing how electron transfer in the respiratory chain is coupled with ATP synthesis according to Mitchell's hypothesis. Explanation in text.

The most important conditions for the normal working of this mechanism as a whole are that the structure of the membrane is intact and the internal space is completely enclosed. Breakdown of the membranes as the result of mechanical action, or treatment with uncouplers of oxidative phosphorylation, which is considered to facilitate the nonspecific transfer of protons, leads to depolarization of the membranes and arrests ATP synthesis.

Can this mechanism of oxidative phosphorylation function in the bacterial cell? The external space relative to the bacterial membrane is the cultivation medium with all the fluctuations in its chemical composition and, in particular, in its hydrogen ion concentration. The pH of the cytoplasm of a bacterial cell is directly dependent on the pH of the medium, and in the case of $E.$ $coli$ cells, for example, it can vary by at least 2 units, from pH 6 to pH 8 (Kashket and Wong, 1969); the external pH may be either lower or higher than the internal pH (Harold et al., 1970).

This state of affairs perhaps reflects the primitive character of the oxidative phosphorylation process in bacteria, its instability, and its dependence on external environmental conditions outside the control of the organism, although this is pure speculation. So far as we know, nobody has yet measured the efficiency of oxidative phosphorylation as a function of the proton concentration in the culture medium. However, the possibility cannot be ruled out that the external surface potential of the bacterial membrane is not so completely passively dependent on the composition of the medium. The cell wall of Gram-positive bacteria is thick enough, and the membrane of Gram-negative bacteria is complex enough, for the juxtamembranous space to be to some extent independent of the culture medium (Glauert and Thornley, 1969). This independence evidently increases with the formation of invaginations of the membrane or of the mesosomes, and in this way the development of an internal membrane apparatus in bacteria must be regarded not only as the intensification of a particular function, but as a means of acquiring control over the vital processes of metabolism.

While demonstrating that his hypothesis could be applied to bacterial energy metabolism, Mitchell showed that during respiration of a suspension of *M. lysodeikticus* cells, the medium becomes acid (Fig. 22). It was later shown that the ratio $H^+:O$ for cells of *M. denitrificans* is 4.0 (Scholes and Mitchell, 1970). A study of oxidative phosphorylation in bacterial preparations, however, raises doubts about the validity of the chemo-osmotic hypothesis. Aleem, for instance, observed oxidative phosphorylation and reverse electron transport in the supernatant obtained after centrifuging a homogenate of *T. novellus* cells at 144,000 g for 5 h (Aleem, 1960a; Cole and Aleem, 1970). It is unlikely, as Skulachev (1969) points out, that membrane fragments large enough to form enclosed vesicles, which would satisfy the demands of the chemo-osmotic hypothesis, should remain in such a system.

Commenting on the chemo-osmotic hypothesis, Skulachev (1969) observes that the role of the compounds XH and YOH (Fig. 22) must be played by coenzymes of low molecular weight which can undergo acid dissociation and form anhydrides which are soluble in lipids and possess high mobility in the lipoprotein membrane, and that the enzyme molecules would be too large to perform the functions ascribed to them. This condition, in our opinion, is not absolutely essential. On the contrary, we regard as extremely likely the existence of a protein with two reaction groups capable of forming high-energy bonds, a protein capable of reorienting itself and of changing its conformation in the membrane in response to a change in membrane potential (Fig. 23). Such a protein, able to contract and to form a high-energy thioester bond, is in fact postulated by Boyer's (1964) hypothesis; we are only "hybridizing" his idea.

The existence not only of local, but also of general, conformation changes in bacterial membranes can now be put to the experimental test (Chapter III).

The spectral characteristics of the carotenoids in membranes of the spheroplasts of *R. spheroides* vary with the ionic gradient, i.e., with the potential

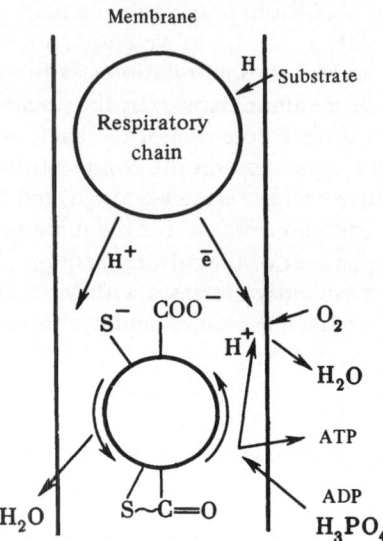

Fig. 23. Hypothetical mechanism of the coupling between oxidation and phosphorylation allowing for possible conformation changes of the ATP-synthesizing enzyme.

gradient on the membrane (Jackson and Crofts, 1969). In this case, the carotenoid possibly behaves as an indicator of the conformation changes in the membrane arising in response to the change of potential.

One reason for the special features of bacterial oxidative phosphorylation could be the highly distinctive chemical composition of bacterial membranes (see Chapter I). For example, the membranes of *M. lysodeikticus* contain neither nitrogenous phospholipids nor sterols — those very compounds which, according to observations made in Green's laboratory, play a very important role in oxidative phosphorylation in mitochondria (Green, 1962).

The unusual and variable lipid composition may give bacterial membranes their special electrical properties. The degree of orderliness of the lipid aggregations, and the size and distribution of the charged groups undoubtedly exert a tremendous influence on the molecular organization of bacterial membranes as a whole (see Chapter III). The sensitivity of individual membrane enzyme systems to the action of inhibitors is also linked with the specific compounds which they contain. The nature of the metabolism of certain species of *Mycoplasma* is such that they can change the concentration of steroid compounds in their membranes. This has revealed a very interesting phenomenon: membranes with a high steroid content have been shown to be sensitive to polyene antibiotics (Kinsky, 1969), whereas cells not containing steroids were resistant to the action of these substances.

The essential fact is that oxidative phosphorylation in bacteria takes place in a unitary lipoprotein membrane. This is a perfectly obvious feature of bacteria such as *A. suboxidans,* in which the mesosomes are simply small invaginations of the cytoplasmic membrane (Claus and Roth, 1964) and no double membranes whatever are found.

Doubts may arise with regard to bacteria with developed mesosomes, such as *M. lysodeikticus,* but the instantaneous unrolling of the mesosomes during lysis of the cells, as well as numerous observations on serial sections, provide convincing arguments that there is only one continuous membrane in the mesosomes (Chapter I). In the mitochondria of higher organisms, on the other hand, this process takes place in the membranes of the cristae, which are protected from the cytoplasm by an additional external mitochondrial membrane.

The intermembranous space possibly allows the electrical potential on the internal membrane to be regulated more precisely. The role of this space can be compared with the role of the intermediate decompression chambers in caisson work or with the role of a forevacuum pump.

To sum up, we can say that the mechanism of oxidative phosphorylation in bacteria is unknown, but it will undoubtedly be discovered when the molecular organization of the bacterial membrane as a whole is elucidated. In this connection, we consider that investigations of conformation changes in the membrane, linked with the work of the respiratory and other enzyme chains, are extremely promising (see Chapter III).

Conclusion

The study of biological membranes is engaging the attention of an ever-increasing number of investigators, but theories explaining the structure of these membranes have reached a point of crisis.

A number of surveys on the molecular organization of biological membranes have recently been published. These surveys can be subdivided into three groups. The first group draws attention to the limitations of the traditional Danielli–Davson model of the membrane (Danielli and Davson, 1935; Davson and Danielli, 1952). The authors of these surveys summarize the factual evidence against this model and suggest that most biological membranes are systems composed of lipoprotein mosaics. The actual models vary somewhat (Green and Goldberger, 1968; Prezbindowski et al., 1969; Arntzen et al., 1969; Sjöstrand and Barajas, 1970; Crane et al., 1968; Lenaz, 1968; Benson, 1966; Lenard, 1969; Glauert, 1968; Glauert and Lucy, 1969). A second group of authors, on the other hand, insists that the Danielli–Davson model, with a lipid bilayer as the skeleton of the membrane and with protein molecules on both sides, is applicable to most biological membranes. Electron-microscopic data, the results of low-angle x-ray diffraction experiments, investigations of thermal phase transitions, and most important of all, analogies in certain parameters of the model and biological membranes all support the existence of a lipid bilayer covered by a protein monolayer (Tien and Diana, 1968; Stoeckenius and Engelman, 1969; Robertson, 1959, 1967; Finean, 1969).

The position occupied by the third group of authors of these surveys is somewhere between the prior two (Rothfield and Finkelstein, 1968; Korn, 1969; Borovyagin, 1969). After examining the two hypotheses mentioned above, so apparently self-contradictory at first glance, these authors conclude that membrane structures probably incorporate both models. This conclusion is not unexpected if chemical and functional differences between membranes are considered and if the wide variations in the relative proportion of lipids and proteins are remembered. [Side by side with myelin, in which lipids are predominant, there are other natural membranes constructed entirely of proteins (Stoeckenius and Kunau, 1968).] Changes in the structure

of membranes dependent on various external factors are clearly illustrated
by the plasma membranes of animal cells (Benedetti and Emmelot, 1968).

There is evidently little point at the present time in trying to build any
form of general model claiming to represent the organization and the function
of all biological membranes without exception. Such a universal interpretation would make sense only when the investigation of these structures was
in its infancy and very little information on biological membranes was available. At the present time, a more promising course of action is the detailed
investigation of certain concrete membranes, with no attempt to extend these
data to the whole vast family of widely different biological membranes.
Such a generalization must inevitably have a somewhat schematic appearance.
It is a striking fact that the survey in the Annual Review of Biochemistry for
1969 on cell membranes was actually written in this form (Korn, 1969).
Without claiming to make broad and premature generalizations, Korn gives
data on every type of membrane which has been investigated: plasma, mitochondrial, and chloroplast membranes, membranes of erythrocytes and bacteria, and myelin sheaths. Such an analysis brings out both the common features and the profound differences between these structures. Korn considers
that most of the data on metabolically active membranes indicate that they
are built of lipoprotein and protein subunits, whereas metabolically inert
membranes such as myelin are most probably built on the principle of the
Danielli—Davson model. The properties of membranes of this class are well
reproduced by model membranes. O'Brien (1967) also describes differences
in principle between the structure of metabolically active and passive membranes.

We would also like to emphasize the following fact: in the study of any special
function of a membrane, such as the function of electron transport in the mitochondrial or bacterial membrane, a model which satisfactorily explains the packing
arrangement of concrete functional proteins can often be built, such as the model
of Crane and Prezbindowski (Prezbindoski et al., 1969). With this type of argument there is a tendency to forget the important fact that any membrane performs several functions and is an integral system (Salton, 1967a,b; Ryter, 1969).
In this case, not only the enzymes of the electron-transport chain must be
incorporated into the membrane, but also the carrier proteins responsible
for the transport of materials. In other words, the function of permeability
and many other important membrane functions must be explained in physical
terms. It is remarkable that models explaining this aspect of membrane function
are quite different from that described above and are unconnected with it in
every way; models explaining permeability will serve as examples (Pardee,
1969; Stein, 1969). The necessity of explaining the immunological properties of membranes had led to the appearance of yet a third type of molecular
arrangement of the membrane. There is thus no escape from the conclusion

that no single model can embrace one membrane as a whole and also other biological membranes. Must we therefore conclude that every concrete membrane has a mosaic structure and that each site on the membrane performs its own particular function? This could well be true, but apart from a few figures we know extremely little about this problem. For example, the content of electron transport chain proteins in the membranes of, for example, *Micrococcus lysodeikticus* does not exceed 10-20% of the total protein content of the membranes; for this reason, data on the arrangement of these proteins clearly cannot be extended to all proteins composing the membrane (Tikhonova et al., 1970; Gel'man et al., 1970; Simakova, 1970).

The differences in structure of individual types of biological membranes and the specificity of structure of sites in the same membrane with different functions have compelled us, when writing this book, to confine our attention to a limited range of problems connected with bacterial membranes and to pay particular attention to the organization of sites carrying enzymes of the respiratory chain.

To complete the examination of the chapter on molecular organization of bacterial membranes, the following general conclusions can be drawn:

1. The structural basis of the bacterial membrane consists of proteins, about 30 in number, and ranging in molecular weight from 10,000 to 200,000; no one component is quantitatively predominant.
2. The polypeptide chains of protein molecules in the membrane form random coils, interspersed with segments formed of α-helices.
3. The protein molecules are linked with each other and with lipids mainly by hydrophobic interactions, although ionic and hydrogen bonds are also present.
4. The lipid located close to the hydrophobic sites of the protein may form large aggregates.
5. Components of membranes are mobile, and this movement may lead to a change in the charge and lyophilicity of the individual sites of the membrane as they perform their functions.

In the principle of their organization, bacterial membranes are thus similar to other metabolically active membranes, especially to mitochondrial membranes, and they are evidently built in the from of a complex system of lipoproteins. From a consideration of all the data on the organization of bacterial membranes and the mechanism of their biogenesis, it is not yet possible to conclude that the lipoproteins are arranged in the membrane so as to form discrete, geometrically identical globular subunits. At the same time, there is some evidence that local aggregates of lipids do exist, and the formation of bimolecular lipid layers in certain sites of the membrane cannot therefore be ruled out. The assumption is equally justified that regions consisting only of protein molecules may exist in the membrane, and these are held together in aggregates by powerful protein—protein interactions of hydrophobic character.

The organization of sites of the membrane carrying the respiratory chain is presumably determined to some extent by the quantity and properties of the proteins composing the chain.

One possible approach to the more detailed study of molecular topography in the bacterial membrane is to choose those areas of the membrane that contain marker proteins, e.g., cytochromes and dehydrogenases forming the complex polyenzymic electron transport system known in aerobic heterotrophic bacteria as the respiratory chain. It can be concluded from the results obtained by treating membranes of *M. lysodiekticus* with agents rupturing electrostatic and hydrophobic bonds that the protein components of the respiratory chain are fixed to the membrane by bonds of different types. Cytochrome proteins are held entirely by hydrophobic interactions. Some of the malate and $NADH_2$ dehydrogenases are fixed to the membrane by hydrophobic bonds also, while the rest is fixed by salt bridges. ATPase is fixed mainly by salt bridges. Even after a superficial examination of the forces stabilizing the membrane, the heterogeneity of its structure is thus apparent. If some molecules of an enzyme are fixed by electrostatic bonds and others by hydrophobic bonds, it can naturally be concluded that these components differ in their arrangement in the membrane and also, presumably, in their molecular environment.

Experiments of fragmentation of the membrane of *M. lysodeikticus* have shown that the respiratory chain is incorporated in the membrane so that on relaxation of hydrophobic interactions produced by detergents, the membrane breaks up into two fragments, one carrying malate and $NADH_2$ dehydrogenases and cytochrome b_{556}, and the other carrying cytochromes b_{560}, c_{550}, and a_{601}. Both fragments can still function actively, i.e., spatial separation with rupture of the chain into fragments takes place only between cytochromes b_{556} and b_{560}.

On the basis of a functional criterion (interaction between enzymes), it thus seems that the respiratory chain of *M. lysodeikticus* is built of two units (substrate and terminal) which are separated on relaxation of hydrophobic bonds.

These units of the respiratory chain differ not only in the composition of their components, but also in the strength of the bond joining the unit as a whole to the membrane. To judge from its solubilizability by detergents, the substrate unit incorporting dehydrogenases and cytochrome b_{556} is less securely fixed to the membrane than the terminal unit containing cytochromes b_{560}, c_{550}, and a_{601}. The difference in the strength of attachment of the individual segments of the respiratory chain can be explained both by the properties of its constituent proteins and by differences in the molecular environment in which the two parts of the chain are situated.

Even if the respiratory chain in the bacterial membrane is composed of lipoprotein units which differ in their enzymic composition and in the strength of their attachment to the membrane, this still does not mean that the membrane as a whole is constructed entirely of units. Nevertheless, the possibility cannot be ruled out that in membranes carrying polyenzymic systems, such as the respiratory chain, the structural organization of the membrane conforms to the pattern of incorporation of the polyenzymic system. This view is supported by the results of detergent fragmentation of functionally active membranes, the membranes of mitochondria and chloroplasts, and the membranes of *Azotobacter vinelandii* (Crane et al., 1968; Shibuya et al., 1968; Jones and Redfearn, 1967).

Bibliography

Abaturov, L. V., 1970. Isotope Exchange in Peptide NH Groups of Globular Proteins. Author's Abstract of Candidate's Dissertation. Moscow.

Abel, K., de Schmertzing, H., and Peterson, J., 1963. J. Bacteriol., 85:1939.

Abram, D., 1965, J. Bacteriol., 89:855.

Abram, D., Koffler, H., and Vatter, A., 1965. J. Bacteriol., 90:1355.

Abram, D., Vatter, A., and Koffler, H., 1966. J. Bacteriol., 91:2045.

Abrams, A., 1965. J. Biol. Chem., 240:3675.

Abrams, A., and Baron, C., 1968a. Biochemistry, 7:501.

Abrams, A., and Baron, C., 1968b. In: Guze, L. (Editor). Microbial Protoplasts, Spheroplasts, and L-Forms. Baltimore, p. 163.

Abrams, A., McNamara, P., and Johnson, F., 1960. J. Biol. Chem., 235:3659.

Abrams, A., Nielsen, L., and Thaemert, J., 1964. Biochim. Biophys. Acta, 80:325.

Ackrell, B., and Jones, C., 1971. Europ. J. Biochem., 20:22.

Ackrell, B., Erickson, S., and Jones, C., 1972. Europ. J. Biochem., 26:387.

Adair, F., 1968. J. Bacteriol., 95:147.

Adams, E., and Newberry, S., 1961. Biochem. Biophys. Res. Commun., 6:1.

Adelson, J., Asano, A., and Brodie, A., 1964. Proc. Nat. Acad. Sci. (Washington), 51:402.

Adolfson, R., and Moudrianakis, E., 1971a. Biochemistry, 10:434.

Adolfson, R., and Moudrianakis, E., 1971b. Biochemistry, 10:440.

Akamatsu, Y., and Nojima, S., 1965. J. Biochem., 57:430.

Akamatsu, Y., Ono, Y., and Nojima, S., 1966. J. Biochem., 59:176.

Aleem, M., 1965. Biochim. Biophys. Acta, 107:14.

Aleem, M., 1966a. Biochim. Biophys. Acta, 128:1.

Aleem, M., 1966b. J. Bacteriol., 91:729.

Aleem, M., 1968. Biochim. Biophys. Acta, 162:338.

Aleem, M., 1970. Anuual Rev. Plant Physiol., 21:67.

Aleem, M., and Lees, H., 1963. Canad. J. Biochem. Physiol., 41:763.

Aleem, M., and Nason, A., 1960. Proc. Nat. Acad. Sci. (Washington), 46:763.

Aleem, M., Lees, H., and Nicholas, D., 1963. Nature, 200:759.

Altenburg, B., and Suit, J., 1970. J. Bacteriol., 103:227.

Ambler, R., 1963. Biochem. J., 89:341, 349.

Ambler, R., 1968, Biochem. J., 109:47p.

Ambler, R., Bruschi, M., and Le Gall, J., 1969. FEBS Letters, 5:115.

Anderson, J., 1964. Biochem. J., 91:8.

Annaev, B. A., Kol'tover, V. K., Raikhman, L. M., and Suskina, V. I., 1971. Dokl. Akad. Nauk SSSR, 196:969.

Ansell, G., and Hawthorne, J., 1964. Phospholipids. Amsterdam, Elsevier.

Antonov, V. K., 1970. Izvest. Akad. Nauk SSSR, Seriya Biol., 6:823.

Appleby, C., 1969a. Biochim. Biophys. Acta, 172:71.

Appleby, C., 1969b. Biochim. Biophys. Acta, 172:88.

Argaman, M., and Razin, S., 1965. J. Gen. Microbiol., 38:153.

Argaman, M., and Razin, S., 1969. J. Gen. Microbiol., 55:45.

Arima, K., and Oka, T., 1965a. J. Bacteriol., 90:734, 744.

Arima, K., and Oka, T., 1965b. J. Biochem., 58:320.

Arima, K., and Sato, E., 1970. Agric. Biol. Chem., 34:739.

Arntzen, C., Dilly, R., and Crane, F., 1969. J. Cell. Biol., 43:16.

Aronson, A., 1966. J. Mol. Biol., 15:505.

Asano, A., and Brodie, A., 1963. Biochem. Biophys. Res. Commun., 13:416.

Asano, A., and Brodie, A., 1964. J. Biol. Chem., 239:4280.

Asano, A., and Brodie, A., 1965. Biochem. Biophys. Res. Commun., 19:121.

Asano, A., Kaneshiro, T., and Brodie, A., 1965. J. Biol. Chem., 240:895.

Asano, A., Imai, K., and Sato, R., 1967a. J. Biochem., 62:210.

Asano, A., Imai, K., and Sato, R., 1967b. Biochim. Biophys. Acta, 143:477.

Ashe, G., and Steim, J., 1971. Biochim. Biophys. Acta, 233:810.

Askogenskaya, N. A., and Petinov, N. S., 1972. Uspekhi Sovr. Biol., 73:288.

Asselineau, J., 1962. Les Lipides Bactériennes. Paris, Hermann.

Aubert, L., Milhaud, G., Moncel, Ch., and Millot, J., 1957. C. R. Acad. Sci., 276:559.

Aubert, L., Millet, J., and Milhaud, G., 1959. Ann. Inst. Pasteur, 96:559.

Avakyan, A. A., and Karavaiko, G. I., 1970. Mikrobiologiya, 29:855.

Avakyan, A. A., Pavlova, I. B., Kats, L. N., and Levina, E. N., 1967a. Zh. Gig., Epidemiol.,
    Mikrobiol., Immunol. (Prague), 11:140.

Avakyan, A., Pavlova, I., Kats, L., and Vysotsky, V., 1967b. Z. ges. Hyg. Grenzgeb.,
    4:272.

Avakyan, A. A., Tordzhyan, I. Kh., and Oparin, A. I., 1970. Dokl. Akad. Nauk SSSR,
    193:1379.

Avakyan, A. A., Tordzhyan, I. Kh., and Kats, L. N., 1971. Zh. Mikrobiol., Epidemiol.,
    Immunobiol., 3:10.

Awasthi, Y., Chuang, T., Keenan, T., and Grane, F., 1970. Biochem. Biophys. Res.
    Commun., 39:822.

Azerad, R., Cyrot, M., and Lederer, E., 1967. Biochem. Biophys. Rès. Commun., 27:249.

Azoulay, E., 1964. Biochim. Biophys. Acta, 92:458.

Azoulay, E., and Couchoud-Beamont, P., 1965. Biochim. Biophys. Acta, 110:301.

Azoulay, E., and Puig, J., 1968. Biochem. Res. Commun., 33:1019.

Azoulay, E., Puig, J., Pichinoty, F., 1967. Biochem. Biophys. Res. Commun., 27:270.

Azzi, A., Chance, B., Radda, G., and Lee, C. P., 1969a. Proc. Nat. Acad. Sci. (Washington),
    62:612.

Azzi, A., Fleischer, S., and Chance, B., 1969b. Biochem. Biophys. Res. Commun., 36:323.

Baak, J., and Postma, P., 1971. FEBS Letters, 19:189.

Bachrach, M., 1970. Annual Rev. Microbiol., 24:109.

Backmuire, F., and MacLeod, R., 1965. Canad. J. Microbiol., 11:677.

Bagdasarian, M., Matheson, N., Synge, R., and Youngson, M., 1964. Biochem. J., 91:91.

Baginsky, M., and Rodwell, V., 1966. J. Bacteriol., 92:424.

Baillie, R., Hou, C., and Bragg, P., 1971. Biochim. Biophys. Acta, 234:46.

Bakerman, S., and Wasemiller, G., 1967. Biochemistry, 6:1100.

Barker, D., and Thorne, K., 1970. J. Cell Sci., 7:755.

Barnes, E., and Kaback, H., 1970. Proc. Nat. Acad. Sci. (Washington), 66:1190.

Barnes, E., and Kaback, H., 1971. J. Biol. Chem., 276:5518.

Baron, C., and Abrams A., 1971. J. Biol. Chem., 246:1542.
Barret, J., 1956. Biochem. J., 64:626.
Barridge, J., and Shively, J., 1968. J. Bacteriol., 95:2182.
Bartsch, R., 1963. In: Gest, H., San Pietro, A., and Vernon, B. (Editors). Bacterial Photosynthesis. Yellow Springs, Antioch Press, p. 475.
Bartsch, R., 1968. Annual Rev. Microbiol., 22:181.
Bartsch, R., Kakuno, T., Horio, T., and Kamen, M., 1971. J. Biol. Chem., 246:4489.
Baseman, J., and Cox, C., 1969. J. Bacteriol., 97:1001.
Baum, R., and Dolin, M., 1963. J. Biol. Chem., 238:PC 4109.
Baum, R., and Dolin, M., 1965. J. Biol. Chem., 240:3425.
Bauman, A., and Simmonds, P., 1969. J. Bacteriol., 98:528.
Bauman, N., Hagen, P., and Goldfine, H., 1965. J. Biol. Chem., 240:1559.
Bayer, M., and Remsen, C., 1970. J. Bacteriol., 101:304.
Beaton, C., 1968. Gen. Microbiol., 50:37.
Beck, I., and Shafia, F., 1964. J. Bacteriol., 88:850.
Bell, R., Mavis, R., Osborn, M., and Vagelos, R., 1971. Biochim. Biophys. Acta, 249:628.
Benedetti, E., and Emmelot, P., 1965. J. Cell. Biol., 26:299.
Benedetti, E., and Emmelot, P., 1968. In: Dalton, A., and Hagenau, F. (Editors). The Membranes. New York, Academic Press, p. 33.
Benemann, J., and Valentine, R., 1971. Adv. Microb. Physiol., 5:135.
Ben-Gurshom, E., 1961. Biochem. J., 78:218.
Bennett, K., Taylor, D., and Hurst, A., 1966. Biochim. Biophys. Acta, 118:512.
Benson, A., 1966. J. Amer. Oil Chem. Soc., 43:265.
Bentley, R., and Schlechter, L., 1960. J. Bacteriol., 79:346.
Benziman, M., and Galanter, Y., 1964. J. Bacteriol., 88:1011.
Benziman, M., and Goldhamer, D., 1968. Biochem. J., 108:311.
Benziman, M., and Karmely, Y., 1968. Europ. J. Biochem., 5:45.
Benziman, M., and Perez, L., 1965. Biochem. Biophys. Res. Commun., 19:127.
Beppa, T., and Arima, K., 1970. Biochim. Biophys. Acta, 219:512.
Bergel'son, L. D., 1967. Uspekhi Sovr. Biol., 64:355.
Berger, K., Barratt, M., and Kamat, V., 1970. Biochem. Biophys. Res. Commun., 40:1273.
Bertsch, L., Bonsen, P., and Kornberg, A., 1969. J. Bacteriol., 98:75.
Beychok, S., 1968. Annual Rev. Biochem., 37:674.
Bezborodov, A. M., and Chermenskaya, G. S., 1970. Mikrobiologiya, 39:316.
Bhattachariya, P., Wendt, L., Whitney, E., and Silver, S., 1970. Science, 168:998.
Biedermann, M., 1970. Arch. Mikrobiol., 75:171.
Biedermann, M., 1971. Hoppe-Seylers Z. Physiol. Chem., 352:567.
Biedermann, M., and Drews, G., 1968. Arch. Mikrobiol., 61:48.
Biggins, D., and Postgate, J., 1971. Europ. J. Biochem., 19:408.
Binyukov, V. I., Zhukova, I. G., and Ostrovskii, D. N., 1971a. Dokl. Akad. Nauk SSSR, 198:1457.
Binyukov, V. I., Borunova, S. F., Gol'dfel'd, M. G., Zhukova, I. G., Kudlai, D. G., Kuznetsov, A. N., Shapiro, M. B., and Ostrovskii, D. N., 1971b. Biokhimiya, 36:1149.
Binyukov, V. I., Kaprenyants, A. S., Kuznetsov, A. N., and Ostrovskii, D. N., 1972. Abstracts of Proceedings of the 4th International Biophysical Congress (in Russian), Vol. 3. Moscow, p. 16.
Birdsell, D., and Cota-Robles, E., 1967. J. Bacteriol., 93:427.
Birdsell, D., and Cota-Robles, E., 1970. Biochim. Biophys. Acta, 216:250.

Biryuzova, V. I., and Meisel', M. N., 1964. In: Molecular Biology. Problems and Perspectives (in Russian), Moscow, Nauka, p. 316.

Biryuzova, V. I., Lukoyanova, M. A., Gel'man, N. S., and Oparin, A. I., 1964. Dokl. Akad. Nauk SSSR, 156:198.

Bishop, D., and King, H., 1962. Biochem. J., 85:550.

Bishop, D., Pandya, K., and King, H., 1962. Biochem. J., 83:606.

Bishop, D., Rutberg, L., and Samuelsson, B., 1967a. Europ. J. Biochem., 2:448.

Bishop, D., Rutberg, L., and Samuelsson, B., 1967b. Europ. J. Biochem., 2:454.

Bladen, H., Nylen, M., and Fitzgerald, R., 1964. J. Bacteriol., 88:763.

Blair, P., Oda, T., Green, D., and Fernandez-Moran, H., 1963. Biochemistry, 2:756.

Bloch, K., 1965. In: Evolving Genes and Proteins. New York, Academic Press, p. 53.

Blout, E., Ioze, C., and Asadourian, A., 1961. J. Amer. Chem. Soc., 83:1895.

Blumenfeld, O., 1968. Biochem. Biophys. Res. Commun., 30:200.

Blyumenfel'd, L. A., and Purmal', A. P., 1964. In: A. E. Braunshtein (Editor). Enzymes (in Russian). Moscow, Nauka, p. 215.

Boatman, F., 1964. J. Cell. Biol., 20:297.

Bodman, M., and Welker, N., 1969. J. Bacteriol., 97:924.

Bogin, E., Higashi, T., and Brodie, A., 1968. Federat. Proc., 27(2):297.

Bogin, E., Higashi, T., and Brodie, A., 1969. Arch. Biochem. Biophys., 129:211.

Bogin, E., Higashi, T., and Brodie, A., 1970. Biochem. Biophys. Res. Commun., 38:478.

Boll, M., 1970. Arch. Mikrobiol., 71:1.

Bongers, L., 1967. J. Bacteriol., 93:1615.

Boon, T., 1972. Proc. Nat. Acad. Sci. (Washington), 69:549.

Borisov, A. Yu., 1969. Uspekhi Sovr. Biol., 68:210.

Borisov, A. Yu., and Mokhova, E. N., 1964. Pribory i Tekhn. Eksperimenta, 2:145.

Borovyagin, V. L., 1969. In: Biological Ultrastructures (in Russian). Moscow, Nauka, p. 46.

Bowman, M., Ottolenghi, A., and Mengel, C., 1971. J. Membr. Biol., 4:156.

Boyer, P. D., 1964. In: Molecular Biology. Problems and Perspectives (in Russian). Moscow, Nauka, p. 227.

Bragg, P., 1965. Biochim. Biophys. Acta, 96:263.

Bragg, P., 1970. Canad. J. Biochem., 48:777.

Bragg, P., 1971. Canad. J. Biochem., 49:492.

Bragg, P., and Hou, C., 1967a. Canad. J. Biochem., 45:1107.

Bragg, P., and Hou, C., 1967b. Arch. Biochem. Biophys., 119:194.

Bragg, P., and Hou, C., 1967c. Arch. Biochem. Biophys., 119:202.

Bragg, P., and Hou, C., 1968. Canad. J. Biochem., 46:631.

Bragg, P., and Hou, C., 1972. Biochim. Biophys. Acta, 274:478.

Branton, D., 1966. Proc. Nat. Acad. Sci. (Washington), 55:1048.

Branton, D., 1969. Annual Rev. Plant Physiol., 20:209.

Bresler, S. E., 1966. Introduction to Molecular Biology (in Russian). Moscow, Nauka.

Brian, B., and Gardner, E., 1969. J. Bacteriol., 96:2181.

Brierley, G., and Merola, A., 1962. Biochim. Biophys. Acta, 64:205.

Bril, C., 1960. Biochim. Biophys. Acta, 39:296.

Broberg, P., and Smith, L., 1967. Biochim. Biophys. Acta, 131:479.

Broberg, P., and Smith, L., 1968. In: Okunuki, K., et al. (Editors). Structure and Function of Cytochromes. Tokyo Univ. Press, p. 183.

Brock, T., and Edwards, M., 1970. J. Bacteriol., 104:509.

Brock, T., Brock, M., Bott, T., and Edwards, M., 1971. J. Bacteriol., 107:303.

Broda, E., 1970. Progr. Biophys., 21:145.

Brodie, A., 1959. J. Biol. Chem., 234:398.
Brodie, A., 1961. Federat. Proc., 20:995.
Brodie, A., 1963. In: Colowick, S., and Kaplan, N. (Editors). Methods in Enzymology, Vol. 6. New York, Academic Press, p. 284.
Brodie, A., 1965. In: Morton, R. (Editor). Biochemistry of Quinones. New York, Academic Press, p. 356.
Brodie, A., 1969. Biochem. J., 113:25.
Brodie, A., and Gray, C., 1956a. Biochim. Biophys. Acta, 19:384.
Brodie, A., and Gray, C., 1956b. J. Biol. Chem., 219:853.
Brodie, A., Revsin, B., Kalra, V., Phy
Brodie, A., Revsin, B., Kalra, V., Phillips, P., Bogin, E., Higashi, T., Krishna Murti, C., Cavari, B., and Marquez, E., 1970. In: Natural Substances Formed Biologically from Mevalonic Acid. London, p. 119.
Bronk, I., 1963. Proc. Nat. Acad. Sci. (Washington), 50:524.
Brown, A., 1964. Biochim. Biophys. Acta, 93:136.
Brown, A., 1965. J. Mol. Biol., 12:491.
Brown, A., and Pearce, R., 1969. Canad. J. Biochem., 47:833.
Brown, A., Shorey, C., and Turner, H., 1965. J. Gen. Microbiol., 41:225.
Brown, J., 1961. Biochim. Biophys. Acta, 52:368.
Brown, J., Edwards, M., and Van Demark, P., 1968. Canad. J. Microbiol., 14:823.
Brown, M., and Winsley, B., 1969. J. Gen. Microbiol., 56:99.
Bruckdorfer, K., Oemel, R., De Gier, J., and Van Deenen, L., 1969. Biochim. Biophys. Acta, 183:334.
Brundish, D., Shaw, N., and Baddiley, J., 1966. Biochem. J., 99:546.
Brundish, D., Shaw, N., and Baddiley, J., 1967. Biochem. J., 105:885.
Bruschi, M., Le Gall, J., and Dus, K., 1970. Biochem. Biophys. Res. Commun., 38:607.
Brushi-Heriaud, M., and Le Gall, J., 1967. Bull. Soc. Chim. Biol., 49:753.
Bulen, W., and Le Comte, J., 1966. Proc. Nat. Acad. Sci. (Washington), 56:979.
Bulen, W., Burns R., Le Comte, J. 1964. Biochem. Biophys. Res. Commun. 17:265.
Burd. G. I., 1967. Zh. Mikrobiol., Epidemiol., Immunobiol., 11:117.
Burdett, J., and Rogers, H., 1970. Ultrastr. Res., 31:354.
Burge, R., and Draper, J., 1967. J. Mol. Biol., 28:173.
Butler, K., Dugas, H., Smith, I., and Schneider, H., 1970. Biochem. Biophys. Res. Commun., 40:770.
Butler, T., Smith, G., and Grula, E., 1967. Canad. J. Microbiol., 13:1471.
Butt, W., and Lees, H., 1958. Nature, 182:732.
Buyalo, O. D., Golovacheva, R. S., Loginova, L. G., Kharat'yan, E. F., and Ostrovskii, D. N., 1972. Mikrobiologiya (in press).
Calvin, M., Wang, H., Entine, G., Gill, D., Ferruti, P., Harpold, M., and Klein, M., 1969. Proc. Nat. Acad. Sci. (Washington), 63:1.
Campbell, J., and Bentley, R., 1968. Biochemistry, 7:3323.
Campbell, J., Coscia, C., Kelsey, M., and Bentley, R., 1967. Biochem. Biophys. Res. Commun., 28:25.
Card, G., Georgi, C., and Militzer, W., 1969. J. Bacteriol., 97:186.
Carr, N., 1964. Biochem. J., 91:28.
Carr, N., and Exell, G., 1965. Biochem. J., 96:688.
Castor, L., and Chance, B., 1959. J. Biol. Chem., 234:1587.
Cawthorne, M., Jeffries, L., Harris, M., Price, S., Diplock, A., and Green, J., 1967. Biochem. J., 104:35.
Cerbon, J., 1970. Biochim. Biophys. Acta, 211:389.

Cerletti, P., Strom, R., and Giordano, M., 1965. Biochem. Biophys. Res. Commun., 18:259.
Cerletti, P., Glovenco, M., Giordano, M., Glovenco, S., and Strom, R., 1967. Biochim. Biophys. Acta, 146:380.
Chaix, P., and Petit, J., 1965. Biochim. Biophys. Acta, 22:66.
Chaix, P., and Petit, J., 1957. Biochim. Biophys. Acta, 25:481.
Chance, B., 1951. Rev. Scient. Instrum., 22:619.
Chance, B., 1952. Nature, 169:215.
Chance, B., 1965. In: Morton, R. (Editor). Biochemistry of Quinones. New York, Academic Press, p. 460.
Chance, B., Estabrock, R., and Yonetani, T. (Editors), 1966. Hemes and Heme Proteins, New York, Academic Press, p. 624.
Chang, Y.-Y., and Kennedy, E., 1967. J. Lipid Res., 8:447.
Changeux, J.-P., Thiery, J., Tung, Y., and Kittel, C., 1967. Proc. Nat. Acad. Sci. (Washington), 57:335.
Chapman, D., 1966. Ann. New York Acad. Sci., 137:747.
Chapman, D., 1969. Lipids, 4:251.
Chapman, D., and Dodd, G., 1971. In: L. Rothfield (Editor). Structure and Function of Biological Membranes. New York, Academic Press, p. 13.
Chapman, D., and Kamat, V., 1968. In: J. Jarnefelt (Editor). Regulatory Functions of Biological Membranes. Amsterdam, Elsevier, p. 99.
Chapman, D., and Morrison, A., 1966. J. Biol. Chem., 241:5044.
Chapman, D., and Salsbury, N., 1970. In: Recent Progress in Surface Science, Vol. 3, p. 291.
Chapman, D., and Wallach, D., 1968. In: Chapman, D. (Editor). Biological Membranes. New York, Academic Press, p. 115.
Chapman, D., Kamat, V., De Gier, J., and Penkett, S., 1968. J. Mol. Biol., 31:101.
Cheah, K., 1969. Biochim. Biophys. Acta, 180:320.
Cheah, K., 1970a. Biochim. Biophys. Acta, 197:84.
Cheah, K., 1970b. Biochim. Biophys. Acta, 205:148.
Cheah, K., 1970c. Biochim. Biophys. Acta, 216:43.
Cherni, N. E., 1967. Izvest. Akad. Nauk SSSR, Seriya Biol., 2:302.
Cherni, N. I., Solov'eva, Zh. V., Fedorov, V. D., and Kondrat'eva, E. N., 1969. Mikrobiologiya, 38:479.
Chirgadze, Yu. N., 1965. Infrared Spectra and Structure of Polypeptides and Proteins (in Russian), Moscow, Nauka.
Cho, K., and Salton, M., 1964. Biochim. Biophys. Acta, 84:773.
Cho, K., and Salton, M., 1966. Biochim. Biophys. Acta, 116:73.
Cho, K., Doy, C., and Mercer, E., 1967. J. Bacteriol., 94:196.
Choules, G., and Bjorklund, R., 1970. Biochemistry, 9:4759.
Chung, A., 1970. J. Bacteriol., 102:438.
Clark-Walker, G., and Lascelles, J., 1970. Arch. Biochem. Biophys., 136:153.
Clark-Walker, G., Rittenberg, B., and Lascelles, J., 1967. J. Bacteriol., 94:1648.
Claus, G., and Roth, L., 1964. J. Cell. Biol., 20:217.
Cohen, M., and Wainio, W., 1963. J. Biol. Chem., 238:879.
Cohen-Bazire, G., and Kunizawa, R., 1963. J. Cell Biol., 16:401.
Cohen-Bazire, G., Pfennig, N., and Kunizawa, R., 1964. J. Cell Biol., 22:207.
Cohen-Bazire, G., Kunizawa, R., and Poindexter, J., 1966. J. Gen. Microbiol., 42:301.
Cohn, D., 1956. J. Biol. Chem., 221:413.
Cohn, D., 1958. J. Biol. Chem., 233:299.
Cole, J., 1968. Biochim. Biophys. Acta, 162:356.
Cole, J., and Aleem, M., 1970. Biochem. Biophys. Res. Commun., 38:736.

Cole, J., and Wimpenny, J., 1968. Biochim. Biophys. Acta, 162:39.
Cole, R., 1965. Bacteriol. Rev., 29:326.
Cole, R., Popkin, T., Prescott, B., Chanock, R., and Razin, S., 1971. Biochim. Biophys. Acta, 233:76.
Coleman, R., Finean, J., Knutton, S., and Zimbrick, A., 1970. Biochim. Biophys. Acta, 119:81.
Conti, S., Jacobs, N., and Gray, C., 1968. J. Bacteriol., 96:554.
Corner, T., and Marquis, R., 1969. Biochim. Biophys. Acta, 183:544.
Cota-Robles, R., 1966. J. Ultrastr. Res., 16:626.
Coval, M., Horio, T., and Kamen, M., 1961. Biochim. Biophys. Acta, 51:246.
Cox, G., Snowswell, A., and Gibson, F., 1968. Biochim. Biophys. Acta, 153:1.
Cox, G., Young, Y., McCann, L., and Gibson, F., 1969. J. Bacteriol., 99:450.
Cox, G., Newton, N., Gibson, F., Snowswell, A., and Hamilton, J. A., 1970. Biochem. J., 117:551.
Cramer, W., and Phillips, S., 1970. J. Bacteriol., 104:819.
Crane, F., 1965. In: Morton, R. (Editor). Biochemistry of Quinones. New York, Academic Press, p. 183.
Crane, F., 1968. In: Singer, T. (Editor). Biological Oxidation. New York, Interscience Publishers, p. 533.
Crane, F., Stiles, J., Prezbindowski, K., Ruzicka, F., and Sun, F., 1968. In: Järnefelt, J. (Editor). Regulatory Functions of Biological Membranes. Amsterdam, Elsevier, p. 21.
Crespi, H., and Katz, J., 1969. Nature, 224:560.
Criddle, R., 1966. In: Goodwin, T. (Editor). Biochemistry of Chloroplasts. New York, Academic Press, p. 203.
Criddle, R., Bock, R., Green, D., and Tisdale, H., 1962. Biochemistry, 1:827.
Cronan, J., 1968. J. Bacteriol., 95:2054.
Cronan, J., and Vagelos, R., 1972. Biochim. Biophys. Acta, 265:25.
Cronewett, C., and Wagner, R., 1965. Proc. Nat. Acad. Sci. (Washington), 54:1643.
Cruden, D., and Stamier, R., 1970. Arch. Mikrobiol., 72:115.
Cundliffe, E., 1970. J. Mol. Biol., 52:467.
Current Problems in the Chemistry of Peptides and Proteins (in Russian), 1969. Moscow, Nauka.
Curtis, P., 1969. Biochim. Biophys. Acta, 183:239.
Cusanovich, M., and Bartsch, R., 1969. Biochim. Biophys. Acta, 189:249.
Cusanovich, M., and Kamen, M., 1968. Biochim. Biophys. Acta, 153:376.
Cutinelli, C., Galdiero, F., and Tufano, M., 1969. J. Bacteriol., 100:123.
Daniel, R., 1970. Biochim. Biophys. Acta, 216:328.
Danielli, J., and Davson, H., 1935. J. Cell Comp. Physiol., 5:495.
Daniels, M., 1971. Biochem. J., 122:197.
Daron, H., 1970. J. Bacteriol., 101:145.
Das, M., Haak, F., and Crane, F., 1965. Biochemistry, 4:859.
Davidson, C., and Hartree, E., 1968. Nature, 220:502.
Davies, S., and Whittenbury, R., 1970. J. Gen. Microbiol., 61:227.
Davis, D., and Inesi, G., 1971. Biochim. Biophys. Acta, 241:1.
Davis, K., and Hatefi, Y., 1971a. Biochemistry, 10:2509.
Davis, K., and Hatefi, Y., 1971b. Biochemistry, 10:2517.
Davson, H., and Danielli, J., 1952. The Permeability of Natural Membranes. London, Cambridge Univ. Press.
Dawson, R., 1968. In: Chapman, D. (Editor). Biological Membranes. New York, Academic Press, p. 203.
Day, C., and Levy, R., 1969. J. Theoret. Biol., 23:387.

Dearborn, D., and Wetlaufer, D., 1969. Proc. Nat. Acad. Sci. (Washington), 62:179.
Deborin, G. A., 1967. Protein-Lipid Surfaces of the Cell as the Prototype of Biological Membranes. Author's Abstract of Doctoral Dissertation.
Deeb, S., and Hager, L, 1964. J. Biol. Chem., 239:1024.
De Groot, G., and Stouthamer, A., 1970. Arch. Microbiol., 74:340.
De Ley, J., 1960. Nature, 188:331.
De Ley, J., 1964. Annual Rev. Microbiol., 18:17.
De Ley, J., and Dochy, R., 1960a. Biochim. Biophys. Acta, 40:277.
De Ley, J., and Dochy, R., 1960b. Biochim. Biophys. Acta, 42:538.
De Ley, J., and Kersters, K., 1964. Bacteriol. Rev., 28:164.
De Ley, J., and Schel, J., 1959. Biochim. Biophys. Acta, 35:154.
De Ley, Y., and Schel, Y., 1965, cited by Bartsch, R., 1968. Annual Rev. Microbiol. 22:181.
Demus, H., and Mehl, E., 1970. Biochim. Biophys. Acta, 211:148.
Den'ko, E. I., 1970. Uspekhi Sovr. Biol., 70:41.
De Siervo, A., and Salton, M., 1971. Biochim. Biophys. Acta, 259:280.
De Voe, J., and Oginsky, E., 1969a. J. Bacteriol., 98:1355.
De Voe, J., and Oginsky, E., 1969b. J. Bacteriol., 98:1368.
Dickerson, R., and Geis, J., 1969. The Structure and Action of Proteins. New York, Harper and Row.
Din, G., Suzuki, J., and Lees, H., 1967. Canad. J. Biochem., 45:1523.
Doi, R., and Halvorson, H., 1961. J. Bacteriol., 81:51.
Dolin, M., 1963. In: Metabolism of Bacteria (Russian translation). Moscow, IL, p. 316.
Doman, N. G., and Tikhonova, G. V., 1965. Uspekhi Sovr. Biol., 60:238.
Dos Santos Mota, J., Op den Kamp, J., Verheij, H., and van Deenen, L., 1970. J. Bacteriol., 104:611.
Doss, M., and Philipp-Dormston, W., 1971. Hoppe-Seylers Z. Physiol. Chem., 352:43.
Downey, R., 1964. J. Bacteriol., 88:904.
Downey, R., Georgi, C., and Militzer, W., 1962. J. Bacteriol., 83:1140.
Drabikowska, A., 1970. Acta Biochem. Polonica, 17:89.
Dranovskaya, E. A., and Kushnarev, V. M., 1968. Zh. Mikrobiol., Epidemiol., Immunobiol., 12:3.
Drews, G., and Giesbrecht, P., 1965. Arch. Mikrobiol., 52:242.
Drott, H., Lee, C., and Yonetani, T., 1970. J. Biol. Chem., 245:5875.
Drucker, H., Campbell, L., and Woody, R., 1970a. Biochemistry, 9:1519.
Drucker, H., Trousil, E., Campbell, L., Barlow, G., and Margoliash, E., 1970b. Biochemistry, 9:1515.
Drucker, H., Trousil, E., and Campbell, L., 1970c. Biochemistry, 9:3395.
Dunphy, P., Gutnick, D. Phillips, P., and Brodie, A., 1967. Arch. Biochem. Biophys., 122:252.
Dunphy, P., Gutnick, D. Phillips, P., and Brodie, A., 1968. J. Biol. Chem., 243:398.
Durner, G., and Mach, F., 1970. Z. allg. Mikrobiol., 10:537.
Dus, K., and Kamen, M., 1963. Biochem. Z., 338:364.
Dus, K., Sletten, K., and Kamen, M., 1968. J. Biol. Chem., 243:5507.
Dus, K., Katagiri, M., Vu. C., Erbes, D., and Gunsalus, I., 1970. Biochem. Biophys. Res. Commun,, 40:1423.
Dworkin, M., and Niederpruem, D., 1964. J. Bacteriol., 87:316.
Eagon, R., and Carson, J., 1965. Canad. J. Microbiol., 11:193.
Edebo, L., 1961a. Acta Pathol. Microbiol. Scand., 52:372, 384.
Edebo, L., 1961b. Acta Pathol. Microbiol. Scand., 53:121.
Edwards, M., and Stevens, R., 1963. J. Bacteriol., 86:414.
Eilerman, L., Pandit-Hovenkamp, H., and Kolk, A., 1970. Biochim. Biophys. Acta, 197:25.

Eimjellen, K., Steensland, H., and Traetteberg, J., 1967. Arch. Mikrobiol., 59:82.
Eisenberg, R., Yu. L., and Wolin, M., 1970a. J. Bacteriol., 102:161.
Eisenberg, R., Yu. L., and Wolin, M., 1970b. J. Bacteriol., 102:172.
Elfvin, L., 1961. J. Ultrastr. Res., 5:374.
Elfvin, L., 1963. J. Ultrastr. Res., 8:283.
Ellar, D., 1970. In: Organization and Control in Procaryotic and Eucaryotic Cells. London, Cambridge Univ. Press, p. 167.
Ellar, D., and Lundgren, D., 1966. J. Bacteriol., 92:1748.
Ellar, D., Lundgren, D., and Slepecky, R., 1968. J. Bacteriol., 94:1189.
Ellar, D., Munoz, E., and Salton, M., 1971. Biochim. Biophys. Acta, 225:140.
Elworthy, P., Florence, A., and Macfarlane, G., 1968. Solubilization by Surface Active Agents. London, Chapman and Hall.
Engelman, D., 1970. J. Mol. Biol., 47:115.
Engelman, D., 1971. J. Mol. Biol., 58:153.
Engelman, D., and Morowitz, H., 1968. Biochim. Biophys. Acta, 150:385.
Engelman, D., Terry, T., and Morowitz, H., 1967. Biochim. Biophys. Acta, 135:381.
Erickson, S., 1971. Biochim. Biophys. Acta, 245:63.
Erickson, S., and Parker, G., 1969. Biochim. Biophys. Acta, 180:56.
Ermachenko, V. A., 1967. Growth and Development of *Nitrosomas europea*. Author's Abstract of Candidate's Dissertation, Moscow.
Erma ienko, V. A., Lisenkova, L. L., and Lozinov, A. B., 1966. Mikrobiologiya, 25:242.
Ernster, L., Lee, J., Norling, B., and Persson, B., 1969. Europ. J. Biochem., 9:299.
Erokhin, Yu. E., and Sinegub, O. A., 1970a. Molekul. Biol., 4:401.
Erokhin, Yu. E., and Sinegub, O. A., 1970b. Molekul. Biol., 4:541.
Erwin, J., and Bloch, K., 1964. Science, 143:1008.
Esfahami, M., and Wakil, S., 1972. Federat. Proc., 31:413.
Esfahami, M., Barnes, E., and Wakil, S., 1969. Proc. Nat. Acad. Sci. (Washington), 64:1057.
Esfahami, M., Limbrick, A., Knutton, S., Oka, T., and Wakil, S., 1971. Proc. Nat. Acad. Sci. (Washington), 68:3180.
Evans, D., 1970. J. Bacteriol.,,104:1203.
Exterkate, F., Vrensen, G., and Veerkamp, J., 1970. Biochim. Biophys. Acta, 219:141.
Falcone, A., Shug, A., and Nicholas, D., 1962. Biochem. Biophys. Res. Commun., 9:126.
Falcone, A., Shug, A., and Nicholas, D., 1963. Biochim. Biophys. Acta, 77:199.
Falk, J., 1964. Porphyrins and Metalloporphyrins, BBA Library, Vol. 2. Amsterdam, Elsevier.
Falk, Y., Leniberg, R., and Morton, R. (eds.),,1961. Hematin Enzymes, Vol. 2, Pergamon Press, Oxford, p. 432.
Fang, F., and Burris, R., 1968. Bacteriol., 96:298.
Farias, R., Londero, L., and Trucco, R., 1972. J. Bacteriol., 109:471.
Farrell, J., and Campbell, L., 1969. In: Rose, A., and Wilkinson, J. (Editors). Advances in Microbial Physiology, Vol. 3, p. 83.
Faust, P., and Vandemark, P., 1970. Arch. Biochem. and Biophys., 137:392.
Feigenblum, E., and Krasna, A., 1970. Biochim. Biophys. Acta, 198:157.
Feingold, D., 1970. J. Membr. Biol., 3:372.
Felter, R., Kennedy, S., Colwell, R., and Chapman, G., 1970. J. Bacteriol., 102:552.
Feo, F., Gabriel, L., and Terranova, T., 1962. Hoppe-Seylers Z. Physiol. Chem., 329:188.
Fernández-Morán, Oda, T., Balir, P., and Green, D., 1964. J. Cell Biol., 22:63.
Ferrandes, B., Frehel, C., and Chaix, P., 1970. Biochim. Biophys. Acta, 223:293.
Fewson, C., and Nicholas, D., 1961a. Biochim. Biophys. Acta, 48:210.
Fewson, C., and Nicholas, D., 1961b. Biochim. Biophys. Acta, 49:335.
Fields, K., and Luria, S., 1969. J. Bacteriol., 97:57.
Fiil, A., and Branton, D., 1969. J. Bacteriol., 98:1320.

Finean, J., 1969. Quart. Rev. Biophys., 2:1.
Finean, J., Coleman, R., Knutton, S., Limbrick, A., and Thompson, J., 1968. J. Gen. Physiol., 51:195.
Fisher, R., and Guillory, R., 1969. FEBS Letters, 3:27.
Fisher, R., and Sanadi, R., 1971. Biochim. Biophys. Acta, 245:34.
Fisher, R., Lam, K., and Sanadi, D., 1970. Biochem. Biophys. Res. Commun., 39:1026.
Fishman, D., and Weinbaum, G., 1967. J. Cell. Biol., 32:524.
Fitz-James, P., 1960. J. Biophys. Biochem. Cytol., 8:507.
Fitz-James, P., 1962. J. Bacteriol., 84:104.
Fitz-James, P., 1964. J. Bacteriol., 87:1483.
Fitz-James, P., 1968. In: Guze, L. (Editor). Microbial Protoplasts, Spheroplasts and L-Forms. Baltimore, p. 124.
Flatmark, T., and Robinson, A., 1968. In: Okunuki, K. (Editor). Structure and Function of Cytochromes. Tokyo, Univ. Tokyo Press, p. 318.
Fleischer, S., Brierley, G., Klouwen, H., and Slautterback, D., 1962. J. Biol. Chem., 237: 3264.
Fleischer, S., Fleischer, B., and Stoeckemus, W., 1967. J. Cell. Biol., 32:193.
Fleischer, S., Zahler, W., and Ozawa, H., 1968. Biochem. Biophys. Res. Commun., 32:1031.
Forget, P., 1971. Europ. J. Biochem., 18:442.
Fox, C., Law, J., Tsukagashi, N., and Wilson, G., 1970. Proc. Nat. Acad. Sci. (Washington), 67:598.
Francis, M., and Phizackerley, P., 1965. Biochem. J., 95:25.
Franics, M., Hughes, D., Kornberg, H., and Phizackerley, P., 1963. Biochem. J., 89:430.
Franklin, R., Dutta, A., Dahlberg, J., and Braunstein, S., 1971. Biochim. Biophys. Acta, 253:521.
Freer, J., and Levinson, H., 1967. J. Bacteriol., 94:441.
Freimer, E., 1963. J. Exper. Med., 117:377.
Frerman, F., and White, D., 1967. J. Bacteriol., 94:1868.
Frey-Wyssling, A., and Mühlethaler, K., 1968. The Ultrastructure of the Plant Cell (Russian translation). Moscow, Mir.
Frieden, J., 1964. In: Horizons in Biochemistry (Russian translation). Moscow, Mir, p. 354.
Friis, P., Daves, D., and Folkers, K., 1966. J. Amer. Chem. Soc., 88:4754.
Frydman, B., and Rapoport, H., 1963, J. Amer. Chem. Soc., 85:823.
Fuchs, G., 1966. Arch. Microbiol., 54:253.
Fujita, T., 1966a. J. Biochem., 60:204.
Fujita, T., 1966b., J. Biochem., 60:329.
Fujita, F., and Sato, R., 1963. Biochim. Biophys. Acta, 77:690.
Fujita, T., and Sato, R., 1966. J. Biochem., 60:568.
Fujita, M., Ishikawa, S., and Shimazono, N., 1966. J. Biochem., 59:104.
Fujita, T., Itagaki, E., and Sato, R., 1963. J. Biochem., 53:282.
Fukui, Y., Nachbar, M., and Salton, M., 1971. Biochim. Biophys. Acta, 241:30.
Fulco, A., 1969. Biochim. Biophys. Acta, 187:169.
Fulco, A., Levy, R., and Bloch, K., 1964. J. Biol. Chem., 239:998.
Fulco, A., 1970. J. Biol. Chem., 245:2985.
Gale, E., 1971. J. Gen. Microbiol., 68:1.
Gale, P., Arison, B., Trenner, N., Page, A., Folkers, K., Jr., Brodie, A., 1963. Biochemistry, 2:200.
Gallin, I., and Van Demark, P., 1964. Biochem. Biophys. Res. Commun., 17:630.
Garcia, A., Vernon, L., Ke, B., and Mollenhauer, H., 1966. Biochemistry, 5:2399.

Garcia, A., Vernon, L., Ke, B., and Mollenhauer, H., 1968a. Biochemistry, 7:319.
Garcia, A., Vernon, L., Ke, B., and Mollenhauer, H., 1968b. Biochemistry, 7:326.
Garyaev, P. P., 1970. Uspekhi Sovr. Biol., 70:166.
Gauthier, G., Clark-Walker, G., Garrard, W., and Lascelles, J., 1970. J. Bacteriol., 102:797.
Geller, D., 1962. J. Biol. Chem., 237:2947.
Gel'man, N. S., 1967. Uspekhi Sovr. Biol., 64:379.
Gel'man, N. S., 1969. Uspekhi Sovr. Biol., 68:3.
Gel'man, N. S., Zhukova, I. G., Lukoyanova, M. A., and Oparin, A. I., 1959. Biokhimiya, 24:481.
Gel'man, N. S., Zhukova, I. G., and Oparin, A. I., 1960a. Dokl. Akad. Nauk SSSR, 135:200.
Gel'man, N. S., Zhukova, I. G., and Oparin, A. I., 1960b. Dokl. Akad. Nauk SSSR, 133:1209.
Gel'man, N. S., Zhukova, I. G., and Zaitseva, N. I., 1962. Dokl. Akad. Nauk SSSR, 145:206.
Gel'man, N. S. Lukoyanova, M. A., Zhukova, I. G., and Oparin, A. I., 1963a. Biokhimiya, 28:801.
Gel'man, N. S., Zhukova, I. G., and Oparin, A. I., 1963b. Biokhimiya, 28:122.
Gel'man, N. S., Lukoyanova, M. A., and Ostrovskii, D. N., 1966. The Respiratory Apparatus of Bacteria (in Russian). Moscow, Nauka.
Gel'man, N. S., Biryuzova, V. I., Lukoyanova, M. A., and Simakova, I. M., 1968. Izvest. Akad. Nauk SSSR, Seriya Biol., 6:810.
Gel'man, N., Tikhonova, G., Simakova, I., Lukoyanova, M., Taptykova, S., and Mikelsaar, H., 1970. Biochim. Biophys. Acta, 223:321.
Gendel', L. Ya., Gol'fel'd, M. G., Kol'tover, V. K., Rozantsev, E. G., and Suskina, V. I., 1968. Biofizika, 13:1114.
Gershanovich, V. N., Andreeva, I. V., Burd, G. I., and Zuev, V. A., 1966. Mikrobiologiya, 35:132.
Gest, H., San Pietro, A., and Vernon, L. (Editors), 1963. Bacterial Photosynthesis. Yellow Springs, Antioch Press.
Ghosh, B., and Carroll, K., 1968. J. Bacteriol., 95:367.
Ghosh, B., and Murray, R., 1967. J. Bacteriol., 93:411.
Ghosh, B., and Murray, R., 1969. J. Bacteriol., 97:426.
Ghuysen, J., Tipper, D., and Strominger, J., 1969. In: Florkin, M., and Stotz, E. (Editors). Comprehensive Biochemistry, Vol. 26A. Amsterdam, Elsevier, p. 53.
Gibian, M., and Rynd, J., 1969. Biochem. Biophys. Res. Commun., 34:594.
Gibson, J., 1961. Biochem. J., 79:151.
Gibson, K., 1965. Biochemistry, 4:2042, 2052.
Giesbrecht, P., and Drews, G., 1962. Arch. Mikrobiol., 43:152.
Giesbrecht, P., and Drews, G., 1966. Arch. Mikrobiol., 54:297.
Giesbrecht, P., and Ruska, H., 1968. Klin. Wochenschr., 46:575.
Gilby, A., Few, A., and McQuillen, K., 1958. Biochim. Biophys. Acta, 29:21.
Glaser, M., Simpkins, H., Singer, S., Sheetz, M., and Chan, S., 1970. Proc. Nat. Acad. Sci. (Washington), 65:721.
Glauert, A., 1962. Brit. Med. Bull., 18:245.
Glauert, A., 1968. J. Roy. Microscop. Soc., 88:49.
Glauert, A., and Lucy, J., 1969. In: Dalton, A., and Hagenau, F. (Editors). The Membranes. New York, Academic Press, p. 1.
Glauert, A., and Thornley, M., 1969. Annual Rev. Microbiol., 23:159.
Glover, J., 1965. In: Morton, R. (Editor). Biochemistry of Quinones. New York, Academic Press, p. 207.
Glover, J., and Threlfall, D., 1962. Biochem. J., 85:14.

Godson, G., and Butler, J., 1964. Biochem. J., 93:573.

Godson, G., Hunter, G., and Butler, J., 1961. Biochem. J., 81:59.

Goldfine, H., 1968. Annual Rev. Biochem., 37:303.

Goldfine, H., and Ellis, M., 1964. J. Bacteriol., 87:8.

Goldfine, H., and Hagen, P.-O., 1968. J. Bacteriol., 95:367.

Goldfine, H., Wagner, M., Oda, T., and Shug, A., 1963. Biochim. Biophys. Acta, 73:367.

Goldman, D., Wagner, M., Oda T., and Shug, A., 1963. Biochim. Biophys. Acta 73:367.

Golovacheva, R. S., Biryuzova, V. I., Kalyuzhnaya, A. P., and Zaitseva, G. N., 1968. Dokl. Akad. Nauk SSSR, 181:1260.

Golovacheva, R. S., Kalyuzhnaya, A. P., Biryuzova, V. I., and Zaitseva, G. N., 1969. Mikrobiologiya, 38:679.

Gordon, A., Lombardi, F., and Kaback, H., 1972. Proc. Nat. Acad. Sci. (Washington), 69:358.

Gordon, R., and MacLeod, R., 1966. Biochem. Biophys. Res. Commun., 24:684.

Gordon, R., and MacLeod, R., 1967. J. Bacteriol., 93:1465.

Gorchein, A., 1968a. Proc. Roy. Soc. London, B, 170:265.

Gorchein, A., 1968b. Proc. Roy. Soc. London, B, 170:279.

Gorchein, A., Neuberger, A., and Tait, G., 1968a. Proc. Roy. Soc. London, B, 170:299.

Gorchein, A., Neuberger, A., and Tait, G., 1968b. Proc. Roy. Soc. London, B, 170:211.

Gould, R., and Lennarz, W., 1970. J. Bacteriol., 104:1135.

Graham, J., and Wallach, D., 1969. Biochim. Biophys. Acta, 193:225.

Graham, J., and Wallach, D., 1971. Biochim. Biophys. Acta, 241:180.

Granboulan, P., and Leduc, E., 1967. J. Ultrastr. Res., 20:111.

Granda, J., and Scanu, A., 1966. Biochemistry, 5:3301.

Grassowicz, N., and Ariel, M., 1963. J. Bacteriol., 85:293.

Gray, C., and O'Hara, J., 1968. In: Okunuki, K. (Editor). Structure and Function of Cytochromes. Tokyo, Univ. Tokyo Press, p. 441.

Gray, C., Winipenny, J., Hughes, D., and Mossman, M., 1966. Biochim. Biophys. Acta., 117:22.

Gray, E., 1967. Biochim. Biophys. Acta, 138:550.

Gray, G., and Thurman, P., 1967. Biochim. Biophys. Acta, 135:947.

Gray, G., and Wilkinson, S., 1965a. J. Gen. Microbiol., 39:385.

Gray, G., and Wilkinson, S., 1965b. J. Appl. Bacteriol., 28:153.

Green, A., and Mascarenhas, J., 1964. Science, 144:1455.

Green, D., 1962. In: Meisel', M. H. (Editor). Structural Components of the Cell (Russian translation). Moscow, IL, p. 76.

Green, D., and Baum, H., 1970. Energy and the Mitochondrion. New York, Academic Press.

Green, D., and Brierley, G., 1965. In: Morton, R. (Editor). Biochemistry of Quinones. New York and London, Academic Press, p. 405.

Green, D., and Fleischer, S., 1963. Biochim. Biophys. Acta, 70:554.

Green, D., and Fleischer, S., 1964a. In: Dawson and Rhodes (Editors). Metabolism and Physiological Significance of Lipids. New York, Wiley, p. 581.

Green, D., and Fleischer, S., 1964b. In: Horizons in Biochemistry (Russian translation). Moscow, Mir, p. 293.

Green, D., and Goldberger, R., 1968. Molecular Aspects of Life (Russian translation). Moscow, Mir.

Green, D., and Hechter, O., 1965. Proc. Nat. Acad. Sci. (Washington), 53:318.

Green, D., and Perdue, J., 1966. Proc. Nat. Acad. Sci. (Washington), 55:1295.

Green, D., and Salton, M., 1970. Biochim. Biophys. Acta, 211:139.

Green, D., Allman, D., Bachmann, E., Baum, H., Kopaczyk, K., Korman, E., Lipton, S., MacLennan, D., McConnell, D., Perdue, J., Riesue, J., and Tzagoloff, A., 1967. Arch. Biochem. Biophys, 119:312.

Green, G., and Wharton, D., 1963. Biochem. Z., 338:335.
Greenlees, J., and Wainio, W., 1959. J. Biol. Chem., 234:658.
Grigera, J., and Cereijido, M., 1971. J. Membr. Biol., 4:148.
Gross, R., and Coles, N., 1968. J. Bacteriol., 95:1322.
Grossbard, E., and Preston, R., 1957. Nature, 179:448.
Grula, E., and Savoy, C., 1971. Biochem. Biophys. Res. Commun., 43:325.
Grula, E., Butler, T., King, R., and Smith, G., 1967. Canad. J. Microbiol., 13:1499.
Guerin, M., Leduc, M., and Azerad, R., 1970. Europ. J. Biochem., 15:421.
Gulik-Krzywicki, T., Shechter, E., Luzzati, V., and Faure, M., 1969. Nature, 223:1116.
Gulik-Krzywicki, T., Shechter, E., Iwatsubo, M., Ranck, J., and Luzzati, V., 1970.
    Biochim. Biophys. Acta, 219:1.
Gumpert, J., and Nermut, M., 1967. Z. allg. Mikorbiol., 7:7.
Gutman, M., Schejter, M., and Avi-Dor, Y., 1968. Biochim. Biophys. Acta, 162:506.
Gvozdev, R. I., Sadkov, A. P., Yakovlev, V. A., and Alfimova, E. A., 1968. Izvest. Akad.
    Nauk SSSR, Seriya Biol., 6:838.
Gvozdez, R. I., Tat'yanenko L. V., and Starchenkov, E. P., 1971. Biokhimiya, 36:118.
Hachimori, A., Muramatsu, N., and Nosoh, Y., 1970. Biochim. Biophys. Acta, 206:426.
Hadjipetrou, L., and Stouthamer, A., 1965. J. Gen. Microbiol., 38:29.
Hadjipetrou, L., Gerrits, I., Teulings, F., and Stouthamer, A., 1964. J. Gen. Microbiol.,
    36:139.
Haest, C., De Gier, J., and Van Deenen, L., 1969. Chem. Phys. Lipids, 4:413.
Hagen, P.-O., Goldfine, H., and Williams, P., 1966. Science, 151:1543.
Hall, J., 1971. J. Theoret. Biol., 30:429.
Hallberg, P., and Hauge, J., 1965. Biochim. Biophys. Acta, 95:80.
Hammond, R., and White, D., 1969. J. Bacteriol., 100:573.
Hammond, R., and White, D., 1970. J. Bacteriol., 103:611.
Hancock, J., and Meadow, P., 1969. Biochim. Biophys. Acta, 187:366.
Harold, F., 1964. J. Bacteriol., 88:1416.
Harold, F., 1970. In: Rose, A., and Wilkinson, J. (Editors). Advances in Microbial
    Physiology, Vol. 4. London, Academic Press, p. 46.
Harold, F., and Baarda, J., 1969. J. Bacteriol., 96:2025.
Harold, F., Baarda, J., Baron, C., and Abrams, A., 1969a. J. Biol. Chem., 244:2261.
Harold, F., Baarda, J., Baron, C., and Abrams, A., 1969b. Biochim. Biophys. Acta,
    183:129.
Harold, F., Pavlasova, E., and Baarda, J., 1970. Biochim. Biophys. Acta, 196:235.
Hatch, F., and Bruce, A., 1968. Nature, 218:1166.
Hatchikian, E., and Le Gall, J., 1972. Biochim. Biophys. Acta, 267:479.
Hatchinson, F., and Pollard, E., 1961. In: Mechanisms in Radiobiology. New York,
    Academic Press, p. 71.
Hatefi, Y., 1963. Advan. Enzymol., 25:275.
Hatefi, Y., and Stempel, K., 1969. J. Biol. Chem., 244:2350.
Hatefi, Y., Haavik, A., and Griffiths, D., 1962. J. Biol. Chem., 237:1676.
Hauge, J., 1960. Biochim. Biophys. Acta, 45:250.
Hauge, J., 1961a. J. Bacteriol., 82:609.
Hauge, J., 1961b. Arch. Biochem. Biophys., 94:308.
Hauge, J., 1964. J. Biol. Chem., 239:3630.
Hauge, J., and Hallberg, P., 1964. Biochim. Biophys. Acta, 81:251.
Hauge, J., and Murer, E., 1964. Biochim. Biophys. Acta, 81:244.
Haywood, A., 1971. Proc. Nat. Acad. Sci. (Washington), 68:435.
Hechter, G., 1965. Federat. Proc., 24:91.
Heinen, W., 1967. Arch. Biochem. Biophys., 120:86.
Hempfling, W., 1970. Biochim. Biophys. Acta, 205:169.

Hempfling, W., and Vishniac, W., 1965. Biochem. Z., 342:272.
Henderson, R., and Nankiville, D., 1966. Biochem. J., 98:587.
Hendler, R., and Nanninga, N., 1970. J. Cell. Biol., 46:114.
Hengstenberg, W., 1970. FEBS Letters, 8:277.
Henneman, D., and Umbreit, W., 1964. J. Bacteriol., 87:1274.
Heptinstall, S., Archibald, A., and Baddiley, P., 1970. Nature, 225:519.
Heydeman, M., and Azouley, E., 1963. Biochim. Biophys. Acta, 77:545.
Higashi, Y., and Strominger, J., 1970. J. Biol. Chem., 245:3691.
Higashi, T., Bogin, E., and Brodie, A., 1969. J. Biol. Chem., 244:500.
Higgins, M., and Shockman, G., 1970. J. Bacteriol., 103:244.
Highton, P., 1969. J. Ultrastr. Res., 26:130.
Highton, P., 1970. J. Ultrastr. Res., 31:247, 260.
Hines, W., Freeman, B., and Pearson, G., 1964. J. Bacteriol., 87:1492.
Hirata, H., and Brodie, A., 1972. Biochem. Biophys. Res. Commun., 47:633.
Hirata, H., and Fukui, S., 1968. J. Biochem., 63:780.
Hirata, H., Fukui, S., and Ishikawa, S., 1969. J. Biochem., 65:843.
Hochstein, L., and Dalton, B., 1968. J. Bacteriol., 95:37.
Hoeniger, J., and Headley, C., 1969. J. Bacteriol., 96:1835
Hoeniger, J., Stuart, P., and Holt, S., 1968. J. Bacteriol., 96:1818.
Hoffmann, K., 1962. Fatty Acid Metabolism in Microorganisms. New York, Wiley.
Holland, J., 1969a. In: Mechanism of Action of Antibiotics (Russian translation). Moscow,
    Mir. p. 646.
Holland, J., 1969b. In: Mechanism of Action of Antibiotics (Russian translation). Moscow.
    Mir. p. 650.
Holland, J. B., Samson, A., Holland, E., and Senior, B., 1971. In: Wolstenholme, G., and
    Knight, J. (Editors). Growth Control in Cell Cultures. Ciba Foundation Symposium,
    Edinburgh, Churchill-Livingstone., p. 221.
Holmes, P., and Halvorson, H., 1963. Canad. J. Microbiol., 9:904.
Holmes, P., and Halvorson, H., 1965. J. Bacteriol., 90:316.
Holt, S., and Canale-Parola, E., 1967. J. Bacteriol.,' 93:399.
Holt, S., and Leadbetter, E., 1967. Arch. Microbiol., 57:199.
Holt, S., Conti, S., and Fuller, R., 1966. J. Bacteriol., 91:311.
Hooper, A., and Nason, A., 1965. J. Biol. Chem., 240:4044.
Hooper, A., Erickson, R., and Terry, K., 1972. J. Bacteriol., 110:430.
Hopper, D., Chapman, P., and Dagley, S., 1970. J. Bacteriol., 104:1197.
Hori, K., 1961. J. Biol. Chem., 50:440.
Hori, K., 1963. J. Biochem., 53:354.
Horio, T., 1958a. J. Biochem., 45:195.
Horio, T., 1958b. J. Biochem., 45:267.
Horio, T., and Kamen, M., 1961. Biochim. Biophys. Acta, 48:266.
Horio, T., and Kamen, M., 1970. Annual Rev. Microbiol., 24:399.
Horio, T., and Taylor, C., 1965. J. Biol. Chem., 240:1772.
Horio, T., Higashi, T., Matsubara, H., Kusai, K., Nakai, M., and Okunuki, K., 1958a.
    Biochim. Biophys. Acta, 29:297.
Horio, T., Higashi, T., Nakai, M., Kusai, K., and Okunuki, K., 1958b. Nature, 182:1307.
Horio, T., Higashi, T., Sasagawa, M., Kusai, K., Nakai, M., and Okunuki, K., 1960.
    Biochem. J., 77:194.
Horio, T., Higashi, T., Yamanaka, T., Matsubara, H., and Okunuki, K., 1961a. J. Biol.
    Chem., 236:944.
Horio, T., Sekuzu, Y., Higashi, T., and Okunuki, K., 1961b. In: Falk, V., Lemberg, M.,
    and Morton, R. (Editors). Haematin Enzymes, Vol. 1. London, Pergamon Press,
    p. 302.

Horth, C., McHale, D., Jeffries, J., Price, S., Diplock, A., and Green, J., 1966. Biochem.,
    J., 100:424.
Houtsmuller, U., and Van Deenen, L., 1963. Biochem. J., 88:43.
Houtsmuller, U., and Van Deenen, L., 1964. Biochim. Biophys. Acta, 84:96.
Houtsmuller, U., Van Deenen, L., 1965. Biochim. Biophys. Acta, 106:564.
Hsia, J.-C., Schneider, H., and Smith, I., 1970. Biochim. Biophys. Acta, 202:399.
Hubbel, W., and McConnell, H., 1969. Proc. Nat. Acad. Sci. (Washington), 63:16.
Hughes, A., Stow, M., Hancock, J., and Baddiley, J., 1971. Nature New Biology, 229:53.
Hughes, D., and Cunningham, V., 1963. Biochem. Soc. Sympos., 23:8.
Hunt, A., Rodgers, A., and Hughes, D., 1959. Biochim. Biophys. Acta, 34:354.
Hurst, A., and Stubbs, J., 1969. J. Bacteriol., 97:1466.
Husson, F., and Luzzati, V., 1963. Nature, 197:822.
Ibbott, F., and Abrams, A., 1964a. Federat. Proc., 23:222.
Ibbott, F., and Abrams, A., 1964b. Biochemistry, 3:2008.
Ikawa, M., 1963. J. Bacteriol., 85:773.
Ikawa, M., 1967. Bacteriol. Rev., 31:54.
Imai, K., Asano, A., and Sato, R., 1967. Biochim. Biophys. Acta, 143:462.
Imai, K., Asano, A., and Sato, R., 1968a. J. Biochem., 63:207.
Imai, K., Asano, A., and Sato, R., 1968b. J. Biochem., 63:219.
Imai, Y., Imai, K., Ikeda, K., Hamaguchi, K., and Horio, T., 1969. J. Biochem., 65:629.
Inouye, M., and Guthrie, J., 1969. Proc. Nat. Acad. Sci. (Washington), 64:957.
Inouye, M., and Pardee, A., 1970a. J. Bacteriol., 101:770.
Inouye, M., and Pardee, A., 1970b. J. Biol. Chem., 245:5813.
Ishida, M., and Mizushima, S., 1969. J. Biochem., 66:33.
Ishikawa, S., 1970 J. Biochem., 67:297.
Ishikawa, S., and Lehninger, A., 1962. J. Biol. Chem., 237:2401.
Ishimoto, M., Koyama, J., Ohmura, T., and Nagai, Y., 1954. J. Biochem., 41:537.
Itagaki, E., 1964. J. Biochem., 55:432.
Itagaki, E., and Hager, L., 1966. J. Biol. Chem., 241:3687.
Itagaki, E., and Hager, L., 1968. Biochem. Biophys. Res. Commun., 32:1013.
Itagaki, E., and Sato, R., 1962. J. Japan Biochem. Soc., 34:480.
Itagaki, E., Fujita, T., and Sato, R., 1961. Biochem. Biophys. Res. Commun., 5:30.
Itagaki, E., Fujita, T., and Sato, R., 1962. J. Biochem., 52:131.
Itagaki, E., Palmer, G., and Hager, L., 1967. J. Biol. Chem., 242:2272.
Ito, A., and Sato, R., 1968. J. Biol. Chem., 243:4922.
Ivarie, R., and Pene, J., 1970. J. Bacteriol., 104:839.
Ivleva, I. N., Sadkov, A. N., and Yakovlev, V. A., 1969. Izvest. Akad. Nauk SSSR,
    Seriya Biol., 5:688.
Iwasaki, H., and Shidara, S., 1969a. J. Bacteriol., 66:775.
Iwasaki, H., and Shidara, S., 1969b. Plant Cell Physiol., 10:291.
Iwasaki, Y., 1960. Plant Cell Physiol., 1:195, 207.
Iwasaki, Y., 1966. Plant Cell Physiol., 7:199.
Jackson, J., and Crofts, A., 1969. Abstrs. Communs. 6 FEBS Meeting, No. 969.
Jackson, F., and Lawton, V., 1958. Nature, 181:1539.
Jackson, F., and Lawton, V., 1959. Biochim. Biophys. Acta, 35:76.
Jacob, F., Brenner, S., and Cuzin, F., 1963. Cold Spring Harbor Sympos. Quant. Biol.,
    28:329.
Jacob, F., Ryter, A., and Cuzin, F., 1966. Proc. Roy. Soc. London, B, 164:267.
Jacobs, N., and Conti, S., 1965. J. Bacteriol., 89:675.
Jacobs, N., and Wolin, M., 1963. Biochim. Biophys. Acta, 69:18.
Jacobs, N., Maclosky, E., and Conti, S., 1967a. J. Bacteriol., 93:278.
Jacobs, N., Maclosky, E., and Jacobs, J., 1967b. Biochim. Biophys. Acta, 148:645.

Jeffries, L., Cawthorne, M., Harris, M., Diplock, A., Grenn, J., and Price, S., 1967.
    Nature, 215:257.
Jeffries, L., Cawthorne, M., Harris, M., Cook, B., and Diplock, A., 1968. J. Gen. Microbiol.,
    54:365.
Jenkinson, T., Kamat, V., and Chapman, D., 1969. Biochim. Biophys. Acta, 183:427.
Ji, T., Hess, J., and Benson, A., 1968. Biochim. Biophys. Acta, 150:676.
Jirgensons, B., 1970. Biochim. Biophys. Acta, 200:9.
John, P., and Hamilton, W., 1970. FEBS Letters, 10:246.
John, P., and Whatley, F., 1970. Biochim. Biophys. Acta, 216:342.
Jones, C., and King, H., 1964. Biochem. J., 91:10.
Jones, C., and Redfearn, E., 1966. Biochim. Biophys. Acta, 113:467.
Jones, C., and Redfearn, E., 1967a. Biochim. Biophys. Acta, 143:340.
Jones, C., and Redfearn, E., 1967b. Biochim. Biophys. Acta, 143:354.
Jones, C., Ackrell, B., and Erickson, S., 1971a. Biochim. Biophys. Acta, 245:54.
Jones, C., Erickson, S., and Ackrell, B., 1971b. FEBS Letters, 13:33.
Jones, R., 1967. Biochem. J., 103:114.
Jones, T., and Kennedy, E., 1969. J. Biol. Chem., 244:5981.
Jose, A., and Wilson, A., 1959. Proc. Nat. Acad. Sci. (Washington), 45:692.
Joyce, G., Hammond, R., and White, D., 1970. J. Bacteriol., 104:323.
Judah, I., and Ahmed, K., 1964. Biol. Rev., 39:160.
Jurtshuk, P., and Harper, L., 1968. J. Bacteriol., 96:678.
Jurtshuk, P., and Old, L., 1968. J. Bacteriol., 95:1790.
Jurtshuk, P., and Schlech, B., 1969. J. Bacteriol., 97:1507.
Jurtshuk, P., Sekuzu, I., and Green, D., 1961. Biochem. Biophys. Res. Commun., 6:76.
Jurtshuk, P., Sekuzu, I., and Green, D., 1963. J. Biol. Chem., 238:3595.
Jurtshuk, P., May, A., Pope, L., and Aston, P., 1969. Canad. J. Microbiol., 15:597.
Kaback, H., 1969. Proc. Nat. Acad. Sci. (Washington), 63:724.
Kaback, H., 1970a. Annual Rev. Biochem., 39:561.
Kaback, H., 1970b. In: Kleinzeller, A., and Bronner, F. (Editors). Current Topics in
    Membranes and Transport. New York, Academic Press.
Kaback, H., 1972. Biochim. Biophys. Acta, 265:367.
Kaback, H., and Barnes, E., 1971. J. Biol. Chem., 246:5523.
Kaback, H., and Deuel, T., 1969. Arch. Biochem. Biophys., 132:118.
Kaback, H., and Stadtman, E., 1966. Proc. Nat. Acad. Sci. (Washington), 55:921.
Kaback, H., and Stadtman, E., 1968. J. Biol. Chem., 243:1390.
Kagawa, Y., 1969. J. Biochem., 65:925.
Kahane, J., and Razin, S., 1969a. Biochim. Biophys. Acta, 183:79.
Kahane, J., and Razin, S., 1969b. J. Bacteriol., 100:187.
Kahane, J., and Razin, S., 1970. FEBS Letters, 10:261.
Kakefuda, T., Holden, J., and Utech, N., 1967. J. Bacteriol., 93:472.
Kalmanson, A. E., 1963. Uspekhi Sovr. Biokhim., 5:289.
Kalra, V., Murti, C., and Brodie, A., 1971. Arch. Biochem. Biophys., 147:734.
Kalra, V., Aithal, H., and Brodie, A., 1972. Biochim. Biophys. Res. Commun., 46:979.
Kamen, M., 1962. In: Mechanism of Photosynthesis. Proceedings of the 5th International
    Biochemical Congress, Symposium VI (in Russian). Moscow, Izd. AN SSSR, p. 253.
Kamen, M., and Horio, T., 1970. Annual Rev. Biochem., 39:673.
Kamen, M., Bartsch, R., Horio, T., and de Klerk, H., 1963. In: Colowick, S., and Kaplan,
    N. (Editors). Methods in Enzymology, Vol. 6. New York, Academic Press, p. 391.
Kamen, M., Dus, K., Flatmark, K., and de Klerk, H., 1970. In: King, T., Klingenberg, M.,
    and Dekker (Editors). Treatise on Electron and Coupled Energy Transfer in Biological
    Systems, Chapter 5. New York.

Kaneda, F., 1971. Biochem. Biophys. Res. Commun., 43:298.

Kaneda, T., 1967. J. Bacteriol., 93:894.

Kaneda, T., 1968. J. Bacteriol., 95:2210.

Kanemasa, Y., Akamatsu, Y., and Nojima, S., 1967. Biochim. Biophys. Acta, 144:382.

Kaneshiro, T., and Marr, A., 1962. J. Lipid Res., 3:184.

Kashket, E., and Brodie, A., 1960. Biochim. Biophys. Acta, 40:550.

Kashket, E., and Brodie, A., 1963. Biochim. Biophys. Acta, 78:52.

Kashket, E., and Wong, P., 1969. Biochim. Biophys. Acta, 193:219.

Katagiri, M., Ganguli, B., and Gunsalus, I., 1968. J. Biol. Chem., 243:3543.

Kates, M., 1964. Adv. Lipid Res., 2:17.

Kates, M., 1966. Annual Rev. Microbiol., 20:13.

Kates, M., and Wessef, M., 1970. Annual Rev. Biochem., 39:323.

Kates, M., Palameta, B., Joo, C., Kushner, D., and Gibbons, N., 1966. Biochemistry, 5:4092.

Kates, M., Joo, C., Palameta, B., and Shier, T., 1967. Biochemistry, 6:3329.

Kats, L. N., 1966. Zh. Mikrobiol., Épidemiol., Immunobiol., 7:84.

Kats, L. N., 1971. Submicroscopic Organization of Surface and Membrane Structures of the Bacterial Cell in Connection with their Function. Author's Abstract of Doctoral Dissertation, Moscow.

Kats, L. N., and Kharat'yan, E. F., 1969. Mikrobiologiya, 38:278.

Kats, L. N., and Konstantinova, N. D., 1966. Dokl. Akad. Nauk SSSR, 169:950.

Kats, L. N., and Tordzhyan, I. Kh., 1968. Mikrobiologiya, 37:890.

Kauzmann, W., 1959. Adv. Protein Chem., 14:1.

Kavanau, J., 1965. Structure and Function in Biological Membranes. San Francisco, Holden-Day.

Kawata, T., 1967. J. Gen. Appl. Microbiol., 13:405.

Kawata, T., Inoue, T., and Takagi, A., 1963. Japan J. Microbiol., 7:23.

Kearney, E., and Goldman, D., 1970. Biochim. Biophys. Acta, 197:197.

Keilin, D., 1966. The History of Cell Respiration and Cytochrome. London, Cambridge Univ. Press.

Keilin, D., and Harpley, C., 1941. Biochem. J., 35:688.

Keith, A., Waggoner, A., and Griffith, O., 1968. Proc. Nat. Acad. Sci. (Washington), 61:819.

Kellenberger, E., and Ryter, A., 1964, In: Siegel, B. M. (Editor). Modern Developments in Electron Microscopy. New York, Academic Press, p. 335.

Kennedy, E., 1969. J. Gen. Physiol., 54 (Part 2):91.

Kennel, S., and Kamen, M., 1971a. Biochim. Biophys. Acta, 234:458.

Kennel, S., and Kamen, M., 1971b. Biochim. Biophys. Acta, 253:153.

Kennell, D., and Kotoulas, A., 1967. J. Bacteriol., 93:367.

Kenny, G., and Grayton, J., 1965. J. Immunol., 95:19.

Kepes, A., 1971. J. Membr. Biol., 4:87.

Kerr, C., and Miller, R., 1968. J. Biol. Chem., 243:2963.

Kersters, K., Wood, W., and De Ley, J., 1965. J. Biol. Chem., 240:965.

Kerwar, K., Gordon, A., and Kaback, H., 1972. J. Biol. Chem., 247:298.

Ketchum, P., and Holt, S., 1970. Biochim. Biophys. Acta, 196:141.

Ketchum, P., Sanders, H., Gryder, J., and Nason, A., 1969. Biochim. Biophys. Acta, 189:360.

Kharat'yan, E. F., Kats, L. N., and Oparin, A. I., 1967. Dokl. Akad. Nauk SSSR, 175:1154.

Kidwai, A., and Murti, K., 1965. Indian J. Biochem., 2:217.

Kiehn, D., and Holland, J., 1968. Proc. Nat. Acad. Sci. (Washington), 61:1370.

Kiesow, L., 1964. Proc. Nat. Acad. Sci. (Washington), 52:980.

Kiesow, L., 1967. In: Sanadi, D. (Editor). Current Topics in Bioenergetics, Vol. 1. New York, Academic Press, p. 196.

Kijimoto, Sh., 1968a. Annual Rept. Biol. Works, 16:1.
Kijimoto, Sh., 1968b. Annual Rept. Biol. Works, 16:19.
Kikuchi, G., and Motokawa, Y., 1968. In: Okunuki, K., Kamen, M., and Sekuzu, I. (Editors). Structure and Function of Cytochromes. Tokyo, Tokyo Univ. Press, p. 174.
Kikuchi, G., Saito, Y., Motokawa, Y., 1965. Biochim. Biophys. Acta, 94:1.
Kimura, T., and Tobari, J., 1963. Biochim. Biophys. Acta, 73:399.
King, T., and Cheldelin, V., 1957. J. Biol. Chem., 224:579.
King, T., and Cheldelin, V., 1958. Biochem. J., 68:31.
Kinsky, S., 1969. In: Mechanism of Action of Antibiotics (Russian translation). Moscow, Mir, p. 125.
Kleczkowska, H., Witerowa, A., and Bagdasarian, G., 1965. Acta Microbiol. Polon., 14:117.
Kline, E., and Mahler, H., 1965. Ann. New York Acad. Sci., 119:905.
Klotz, J., 1964. In: Horizons in Biochemistry (Russian translation). Moscow, Mir, p. 399.
Klungsoyr, L., King, T., and Cheldelin, V., 1957. J. Biol. Chem., 227:135.
Knobloch, K., Ushague, M., and Aleem, M., 1971. Arch. Mikrobiol., 76:114.
Knoche, H., and Shively, J., 1969. J. Biol. Chem., 244:4773.
Knook, D., and Planta, R., 1971. J. Bacteriol., 105:483.
Knowles, C., and Redfearn, E., 1966. Biochem. J., 99:33.
Knowles, C., and Redfearn, E., 1968. Biochim. Biophys. Acta, 162:348.
Knowles, C., and Redfearn, E., 1969. J. Bacteriol., 97:756.
Knowles, C., and Smith, L., 1970. Biochim. Biophys. Acta, 197:152.
Kobozev, G. V., and Troitskii, G. V., 1967. Molekul. Biol., 1:202.
Kobyakov, V. V., 1969. Study of the Amide Absorption Bands of Polypeptides and Proteins in the Region of Frequencies of the N–H Valency Oscillation and the Use of These Bands for the Analysis of Peptide Chain Conformations. Author's Abstract of Candidate's Dissertation, Pushchino-na-Oke.
Koch, A., and Boniface, J., 1971. Biochim. Biophys. Acta, 225:239.
Kocur, F., 1970. Biochim. Biophys. Acta, 202:277.
Kocur, M., Martinec, T., and Mazanec, K., 1968a. Antonie van Leeuwenhoek J. Microbiol. Serol., 34:19.
Kocur, M., Martinec, T., and Mazanec, K., 1968b. J. Gen. Microbiol., 52:343.
Kodama, T., and Morí, T., 1969. J. Biochem., 65:621.
Kodama, T., and Shidara, S., 1969. J. Biochem., 65:351.
Kodicek, E., 1963. In: Gibbons, N. (Editor). Recent Progress in Microbiology. Toronto, Univ. Toronto Press, p. 23.
Koga, Y., and Kusaka, I., 1970. Europ. J. Biochem., 16:407.
Kolber, A., and Stein, W., 1967. Curr. Modern. Biol., 1:249.
Kol'tover, V., Goldfield, M., Hendel, L., and Rozantsev, E., 1968. Biochem. Biophys. Res. Commun., 32:421.
Kol'tover, V. K., Raikman, L. M., Yasaitis, A. A., and Blyumenfel'd, L. A., 1971. Dokl. Akad. Nauk SSSR, 197:219.
Kondrat'eva, E. N., 1963. Photosynthesizing Bacteria (in Russian), Moscow, Izd. AN SSSR.
Konev, S. V., Aksentsev, S. L., and Chernitskii, E. A., 1970. Cooperative Transitions of Proteins in the Cell (in Russian). Minsk, Nauka i Tekhnika.
Koostra, W., and Smith, P., 1969. Biochemistry, 8:4794.
Korn, E., 1969. Annual Rev. Biochem., 38:263.
Kornberg, H., and Phizackerley, P., 1961. Biochem. J., 79:10.
Korngold, R., and Kushner, D., 1968. Canad. J. Microbiol., 14:253.
Korte, T., and Hengstenberg, W., 1971. Europ. J. Biochem., 25:295.
Kotel'nikova, A. V., 1969. Uspekhi Sovr. Biol., 68:155.
Kotel'nikova, A. V., and Ivanova, E. V., 1964. Dokl. Akad. Nauk SSSR, 157:710.

Kraut, J., Singh, S., and Alden, R., 1968. In: Okunuki, K., Kamen, M., and Sakuzu, I.
    (Editors). Structure and Function of Cytochromes. Tokyo, Univ. Tokyo Press, p. 252.
Kretovich, V. L., Melik-Sarkisyan, S. S., and Matus, V. K., 1972. Biokhimiya, 37:711.
Kröger, A., and Dadak, V., 1969. Europ. J. Biochem., 11:328.
Kröger, A., and Klingenberg, M., 1967. In: Sanadi, D. (Editor). Current Topics in
    Bioenergetics, Vol. 2. New York, Academic Press, p. 152.
Kröger, A., Dadak, V., Klingenberg, M., and Diemer, F., 1971. Europ. J. Biochem., 21:322.
Krogstad, D., and Howland, J., 1966. Biochim. Biophys. Acta, 118:189.
Kroon, A., 1964. Biochim. Biophys. Acta, 91:145.
Kudlai, D. G., and Likhoded, V. G., 1966. Bacteriocin Production (in Russian). Leningrad,
    Meditsina.
Kuehn, G., McFadden, B., Johanson, R., Hill, J., and Shumway, L., 1969. Proc. Nat. Acad.
    Sci. (Washington), 62:407.
Kufe, D., and Howland, J., 1968. Biochim. Biophys. Acta, 153:291.
Kuligowska, E., and Erecinska, M., 1967. Bull. Acad. Polon. Sci., 15:323.
Kundig, W., and Roseman, S., 1969. Federat. Proc., 28:463.
Kundig, W., and Roseman, S., 1971. J. Biol. Chem., 246:1393 and 1407.
Kunugita, K., and Matsuhashi, M., 1970. J. Bacteriol., 104:1017.
Kurup, C., and Brodie, A., 1967a. J. Biol. Chem., 242:197, 2909.
Kurup, C., and Brodie, A., 1967b. Biochem. Biophys. Res. Commun., 28:862.
Kurup, C., Vaidyanathan, C., and Ramasarma, T., 1966. Arch. Biochem. Biophys., 113:548.
Kushnarev, V. M., 1966a. Zh. Mikrobiol., Epidemiol., Immunobiol., 1:98.
Kushnarev, V. M., 1966b. Dokl. Akad. Nauk SSSR, 170:714.
Kushnarev, V. M., 1966c. Dokl. Akad. Nauk SSSR, 171:207.
Kusunose, M., and Kusunose, E., 1959. Annual Rept. Japan. Soc. Tubercul., 4:17.
Ladbrooke, B., Jenkinson, T., Kamat, V., and Chapman, D., 1968. Biochim. Biophys.
    Acta, 164:101.
Lahav, M., Chin, T., and Lennarz, W., 1969. J. Biol. Chem., 244:5890.
Lam, Y., and Nicholas, D., 1969a. Biochim. Biophys. Acta, 180:459.
Lam, Y., and Nicholas, D., 1969b. Biochim. Biophys. Acta, 172:450.
Lampen, J., 1967. J. Gen. Microbiol., 48:249.
Lang, D., Felix, J., and Lundgren, D., 1972. J. Bacteriol., 110:968.
Langenberg, K., Bryant, M., and Wolfe, R., 1968. J. Bacteriol., 95:1124.
Langridge, R., Shinagawa, H., and Pardee, A., 1970. Science, 169:59.
Lanyi, J., 1968. Arch. Biochem. Biophys., 128:716.
Lanyi, J., 1969. J. Biol. Chem., 244:2864.
Lanyi, J., 1971. J. Biol. Chem., 246:4552.
Lanyi, J., 1972. J. Biol. Chem., 247:3001.
Lanyi, K., and Stevenson, J., 1970. J. Biol. Chem., 245:4074.
Lark, K., 1966. Bacteriol. Rev., 30:3.
Lascelles, J., 1965. In: Structure and Function in Microorganisms. Soc. Gen. Microbiol.,
    Vol. 15, p. 32.
Lascelles, J., 1968. Adv. Microbiol. Physiol., 2:1.
Lascelles, J., and Szilagyi, J., 1965. J. Gen. Microbiol., 38:55.
Leadbetter, E., and Holt, S., 1968. J. Gen. Microbiol., 52:299.
Lederer, E., 1967. Chem. Phys. Lipids, 1:294.
Leene, W., and Van Iterson, W., 1965. J. Cell Biol., 27:241.
Lees, H., 1958. Biochemistry of Autotrophic Bacteria (Russian translation). Moscow, IL.
Lees, H. 1960. Ann. Rev. Microbiol. 14:83.
Lees, H., and Simpson, J., 1957. Biochim. J., 65:297.
Le Gall, Mazza, G., and Dragoni, N., 1965. Biochim. Biophys. Acta, 99:385.

Lenard, J., 1969. Trans. New York Acad. Sci., Ser. 11, 31:872.

Lenard, J., and Singer, S., 1966. Proc. Nat. Acad. Sci. (Washington), 56:1828.

Lenard, J., and Singer, S., 1968. J. Cell. Biol., 37:117.

Lenaz, G., 1968. Ital. J. Biochem., 17:131.

Lenaz, G., Sechi, A., Masotti, L., and Castelli, G., 1969a. Biochem. Biophys. Res. Commun., 34:392.

Lenaz, G., Sechi, A., Parenti-Castelli, G., and Masotti, L., 1969b. Abstrs. 6 FEBS Meeting, No. 201.

Lennarz, W., 1964. J. Biol. Chem., 239:PC 3110.

Lennarz, W., 1966. Adv. Lipid Res., 4:175.

Lennarz, W., and Scher, M., 1972. Biochim. Biophys. Acta, 265:417.

Lennarz, W., and Talamo, B., 1966. J. Biol. Chem., 241:2707.

Lester, R., and Crane, F., 1959. J. Biol. Chem., 234:2169.

Lester, R., White, D., and Smith, S., 1964. Biochemistry, 3:949.

Levchuk, T. P., Ermachenko, V. A., and Lozinov, A. V., 1967. Mikrobiologiya, 36:24.

Lewis, V., Weaver, R., and Hollins, D., 1968. J. Bacteriol., 96:1.

Liberman, E A., 1970. Biofizika, 15:2.

Lickfeld, K., 1967. Zbl. Bakteril. Abt. II, 121:479.

Lieberman, E., and Skulachev, V., 1970. Biochim. Biophys. Acta, 216:30.

Lieberman, M., and Baker, J., 1965. Annual Rev. Plant Physiol., 16:343.

Likhtenshtein, G. I., Pivovarov, A. P., Bobodzhanov, P. Kh., Rozantsev, E. G., and Smolina, N. B., 1968. Biofizika, 13:396.

Likhtenshtein, G. I., Pobodzhanov, P. Kh., Rozantsev, E. G., and Suskina, V. I., 1969. Molekul. Biol., 2:344.

Likhtenshtein, G. I., Grebenshchikov, Yu. B., Bobodzhanov, P. Kh., and Kokhanov, Yu. V., 1970. Molekul. Biol., 4:682.

Ling, G. N., 1965. Federat. Proc., 24:S103.

Linnane, A., and Wrigley, C., 1963. Biochim. Biophys. Acta, 77:408.

Lisenkova, L. L., 1967. Comparative Quantitative Study of Cytochromes of Chemoautotrophic and Heterotrophic Microorganisms. Author's Abstract of Candidate's Dissertation. Moscow.

Lisenkova, L. L., and Khmel', I. A., 1967. Mikrobiologiya, 36:905.

Lisenkova, L. I., and Lozinov, A. B., 1966. Priklad. Biokhim. Mikrobiol., 2:175.

Lisenkova, L. L., and Mokhova, E. N., 1964. Mikrobiologiya, 33:918.

Loginova, L. G., Golovacheva, R. S., and Egorova, L. A., 1966. Life of Microorganisms at High Temperatures (in Russian). Moscow, Nauka.

Löw, H., and Afzelius, B., 1964. Exper. Cell Res., 35:431.

Lozinov, A. B., and Ermachenko, V. A., 1960. Mikrobiologiya, 29:523.

Lozinov, A. B., and Ermachenko, V. A., 1962. Mikrobiologiya, 31:972.

Lozinov, A. B., and Lisenkova, L. L., 1966. In: Abstracts of Proceedings of the 9th International Congress of Microbiology (in Russian). Moscow, p. 71.

Lozinov, A. B., Levchuk, T. P., and Ermachenko, V. A., 1966. In: Abstracts of Proceedings of the 9th International Congress of Microbiology (in Russian). Moscow, p. 97.

Lucy, J., 1969. FEBS Letters, 3:297.

Lukoyanova, M. A., 1964. Organization of Respiratory Enzymes in the Membranes of Micrococcus lysodeikticus. Author's Abstract of Candidate's Dissertation. Moscow.

Lukoyanova, M. A., and Biryuzova, V. I., 1965. Biokhimiya, 3:529.

Lukoyanova, M. A., and Taptykova, S. D., 1968. Biokhimiya, 33:888.

Lukoyanova, M. A., Gel'man, N. S., and Biryuzova, V. I., 1961. Biokhimiya, 26:916.

Lukoyanova, M. A., Biryuzova, V. I, Simakova, I. M., and Gel'man, N. S., 1967. Biokhimiya, 32:816.

Lukoyanova, M. A., Simakova, I. M., Tikhonova, G. V., Biryuzova, V. I., Mikel'saar, Kh. N., and Gel'man, N. S., 1972. Izvest. Akad. Nauk SSSR, Seriya Biol., 2:220.

Luria, S., 1969. In: Wolstenholme, G., and O'Connor, M. (Editors). Bacterial Episomes and Plasmids, Ciba Foundation Symposium. London, p. 288.

Luria, S., 1970. Science, 168:1166.

Luzzati, V., 1962. In: Harris, R. (Editor). The Interpretation of Ultrastructure. New York, Academic Press, p. 366.

Luzzati, V., and Husson, F., 1962. J. Biophys. Biochem. Cytol., 12:207.

L'vov, N. P., Lyubimov, V. I., Kirshteine, B. E., Veinova, M. K., Ganelin, V. L., Sergeev, N. S., and Kretovich, V. L., 1970. Dokl. Akad. Nauk SSSR, 194:719.

Lyalikova, I. I., 1959. Physiology and Ecology of *Thiobacillus ferrooxidans* in Connection with its Role in the Oxidation of Sulfide Ores. Author's Abstract of Candidate's Dissertation. Moscow.

Lyubimov, V. I.,1969. The Biochemistry of Fixation of Molecular Nitrogen (in Russian). Moscow, Nauka.

Macfarlane, M., 1961a. Biochem. J., 79:4.

Macfarlane, M., 1961b. Biochem. J., 80:45.

Macfarlane, M., 1962a. Biochem. J., 82:40.

Macfarlane, M., 1962b. Nature, 196:136.

Macfarlane, M., 1964a. Adv. Lipid Res., 2:91.

Macfarlane, M., 1964b. In: Metabolism and Physiological Significance of Lipids. Wiley, p. 399.

Machinist, J., and Singer, T., 1965. J. Biol. Chem., 240:3182.

Maddy, A., 1967. In: Warren, K. B. (Editor). Formation and Fate of Cell Organelles. New York, Academic Press, p. 255.

Maddy, A., and Malcolm, B., 1965. Science, 150:1616.

Madsen, N., 1960. Canad. J. Biochem. Physiol., 38:481.

Maeda, A., and Nomura, M., 1966. J. Bacteriol., 91:685.

Mahoney, R., and Edwards, M., 1966. J. Bacteriol., 92:487.

Malavolta, E., Delwiche, C., and Burge, M., 1960. Biochem. Biophys. Res. Commun., 2:445.

Mal'tseva, N. N., 1963. Physiological Features of *Bacillus megaterium*. De Bary. Author's Abstract of Candidate's Dissertation. Kiev.

Margoliash, E., 1961. Annual Rev. Biochem., 30:549.

Margoliash, E., 1966. In: Chance, B., Estabrook, T., and Yonetani, T. (Editors). Hemes and Hemoproteins. New York, Academic Press, p. 371.

Margoliash, E., and Schejter, A., 1966. Adv. Protein Chem., 21:113.

Margoliash, E., Frohwirt, N., and Wiener, E., 1959. Biochem. J., 71:559.

Maroc, J., de Klerk, H., and Kamen, M., 1968. Biochim. Biophys. Acta, 162:621.

Marquez, E., and Brodie, A., 1970. J. Bacteriol., 103:260.

Marquis, R., and Gerhardt, P., 1964. J. Biol. Chem., 239:3361.

Marr, A., and Ingraham, J., 1962. J. Bacteriol., 84:1260.

Marshall, C., and Brown, A., 1968. Biochem. J., 110:441.

Martin, E., and MacLeod, R., 1970. J. Bacteriol., 105:1160.

Mathews, M., and Sistrom, W., 1959. J. Bacteriol., 78:778.

Mavis, R., and Vagelos, P. R., 1972. J. Biol. Chem., 247:652.

McClare, C., 1967. Nature, 216:766.

McConnell, D., Tzagoloff, A., MacLennan, D., and Green, D., 1966. J. Biol. Chem., 241:2373.

McElhaney, R., and Tourtellotte, M., 1969. Science, 164:433.

McElhaney, R., and Tourtellotte, M., 1970. Biochim. Biophys. Acta, 202:120.

McFeters, C., and Ulrich, J., 1972. J. Bacteriol., 110:777.

McFeters, C., Wilson, D., and Stobel, G., 1970. Canad. J. Microbiol., 16:1221.

McLennan, D., Tzagoloff, A., and Rieske, J., 1965. Arch, Biochem. Biophys., 109:383.

Medveczky, N., and Rosenberg, H., 1970. Biochim. Biophys. Acta, 211:158.

Melchior, D., Morowitz, H., Sturtevant, J., and Tsong, T., 1970. Biochim. Biophys. Acta, 219:114.

Metcalfe, J., Metcalfe, S., and Engelman, D., 1971a. Biochim. Biophys. Acta, 241:412.

Metcalfe, S., Metcalfe, J., and Engelman, D., 1971b. Biochim. Biophys. Acta, 241:422.

Meyer, T., Bartsch, R., Cusanovich, M., and Mathewson, J., 1968. Biochim. Biophys. Acta, 153:854.

Mikelsaar, N.,1970. Eesto Loodus, p. 546.

Miki, K., and Okunuki, K., 1969a. J. Biochem., 66:831.

Miki, K., and Okunuki, K., 1969b. J. Biochem., 66:845.

Miller, R., and Kerr, C., 1967. Canad. J. Biochem., 45:1283.

Mi Mari, S., and Rapoport, H., 1968. Biochemistry, 7:2650.

Mindich, L., 1970. J. Mol. Biol., 49:415, 433.

Mirsky, R., 1969. Biochemistry, 8:1164.

Mirsky, R., 1970. Arch. Biochem. Biophys., 139:97.

Mitchell, P., 1961. Nature, 191:144.

Mitchell, P., 1962. J. Gen. Microbiol., 29:25.

Mitchell, P., 1963. In: Brown, H. (Editor). Cell Interface Reactions. New York, Scholar's Library.

Mitchell, P., 1966. Chemiosmotic Coupling in Oxidative and Photosynthetic Phosphorylation. Glynn Research Ltd. Bodmin.

Mitchell, P., 1970. In: Organization and Control in Procaryotic and Eucaryotic Cells. London, Cambridge Univ. Press, p. 121.

Mitchell, P., and Moyle, J., 1956. J. Gen. Microbiol., 15:512.

Mitchell, P., and Moyle, J., 1959. Biochem., J., 72:21.

Mitchell, P., and Moyle, J., 1967. Nature, 213:137.

Mitsuhashi, S., Kojima, Y., and Yagi, K., 1956. J. Biochem., 43:337.

Miura, T., and Mizushima, S., 1968. Biochim. Biophys. Acta, 150:159.

Miura, T., and Mizushima, S., 1969. Biochim. Biophys. Acta, 193:268.

Miyata, I., Kawasaki, T., and Nose, Y., 1967. Biochem. Biophys. Res. Commun., 27:601.

Miyata, M., and Mori, T., 1969. J. Biochem., 66:463.

Mizuno, S., Yoshida, E., Takahashi, H., and Mazuo, B., 1961. Biochim. Biophys. Acta, 49:369.

Mizushima, S., 1968. J. Biochem., 63:317.

Mizushima, S., and Ito, M., 1968. J. Biochem., 63:681.

Mizushima, S., Ishida, M., and Kitahara, K., 1966a. J. Biochem., 59:374.

Mizushima, S., Ishida, M., and Miura, T., 1966b. J. Biochem., 60:256.

Mizushima, S., Miura, T., and Ishida, M., 1967. J. Biochem., 61:146.

Moffitt, W., and Young, J., 1956. Proc. Nat. Acad. Sci. (Washington), 42:596.

Moor, H., 1969. Internat. Rev. Cytol., 25:391.

Moor, H., Mühlethaler, K., Waldner, H., and Frey-Wyssling, A., 1961. J. Biophys. Biochem. Cytol., 10:1.

Mori, T., and Hirai, K., 1968. In: Structure and Function of Cytochromes. Tokyo, Tokyo Univ. Press, p. 681.

Moriarty, D., and Nicholas, D., 1970. Biochim. Biophys. Acta, 216:130.

Morowitz, H., and Terry, T., 1969. Biochim. Biophys. Acta, 183:276.

Morowitz, H., Tourtelotte, M., Guild, W., Castro, E., and Woese, C., 1962. J. Mol. Biol., 4:93.

Morrison, M., 1968. In: Okunuki, K., Kamen, M., and Sekuzu, I. (Editors). Structure and Function of Cytochromes. Tokyo, Tokyo Univ. Press.
Morton, R. (Editor), 1965. Biochemistry of Quinones. New York and London, Academic Press.
Morton, R., 1971. Biol. Rev., 46:47.
Moses, V., and Prevost, C., 1966. Biochem. J., 169:336.
Mosolov, V. V., 1971. Proteolytic Enzymes (in Russian). Moscow, Nauka.
Moss, F., 1956. Austral. J. Exper. Biol. Med. Sci., 34:395.
Motokawa, Y., and Kikuchi, G., 1966. Biochim. Biophys. Acta, 120:274.
Moyed, H., and O'Kane, D., 1956. J. Biol. Chem., 218:831.
Munoz, E., Nachbar, M., Schor, M., and Salton, M., 1968. Biochem. Biophys. Res. Commun., 32:539.
Munoz, E., Salton, M., Ng, M., and Schor, M., 1969. Europ. J. Biochem., 7:490.
Murray, R., 1963. In: Mazia, D., and Tyler, A. (Editors). The General Physiology of Cell Specialization. New York, McGraw-Hill, p. 28.
Murray, R., 1968. In: Guze, L. (Editor). Microbial Protoplasts, Spheroplasts and L-Forms. Baltimore, Williams and Wilkins, p. 1.
Murray, R., and Watson, S., 1965. J. Bacteriol.. 89:1594.
Murthy, P. Bogin, E., Higashi, T., and Brodie A., 1969. J. Biol. Chem., 244:3117.
Nachbar, M., and Salton, M., 1970a. In: Blank, M. (Editor). Surface Chemistry of Biological Systems. New York, Plenum Press, p. 175.
Nachbar, M., and Salton, M., 1970b. Biochim. Biophys. Acta, 223:309.
Nagata, Y., Mizuno, S., and Maruo, B., 1966. J. Biochem., 59:404.
Nagata, Y., Shibuya, I., and Maruo, B., 1967. J. Biochem., 61:622.
Nagata, Y., Yamanaka, T., and Okunuki, K., 1970. Biochim. Biophys. Acta, 221:668.
Nagel de Zwaig, R., and Luria, S., 1967. J. Bacteriol., 94:1112.
Naik, M., and Nicholas, D., 1966a. Biochim. Biophys. Acta, 45:751.
Naik, M., and Nicholas, D., 1966b. Biochim. Biophys. Acta, 118:195.
Nakayama, T., 1961. J. Biochem., 49:240.
Nakayama, T., and De Ley, J., 1965. Antonie van Leeuwenhoek J. Microbiol. Serol., 31:205.
Nanninga, N., 1968. J. Cell Biol., 39:251.
Nanninga, N., 1970a. Dissecting a Bacterium. Amsterdam, Academisch Proefschrift.
Nanninga, N., 1970b. J. Bacteriol., 101:297, 1971.
Nanninga, N., 1971. J. Cell.Biol., 48:219.
Nasir, M., and Murti, C., 1965. Indian.J. Biochem., 2:217.
Nass, S., 1969. Internat. Rev. Cytol., 25:55.
Neale, E., and Chapman, C., 1970. J. Bacteriol., 104:518.
Needleman, S., and Blair, T., 1969. Proc. Nat. Acad. Sci. (Washington), 63:1227.
Nemethy, G., 1967. Angew. Chem., 79:260.
Nermut, M., 1965. Folio Microbiol., 10:104.
Nesbitt, J., and Lennarz, W., 1965. J. Bacteriol., 89:1020.
Neujahr, H., 1970. Biochim. Biophys. Acta, 203:261.
Newton, B., 1960. In: The Strategy of Chemotherapy (Russian translation). Moscow, IL, p. 78.
Newton, J., and Kamen, M., 1963. In: Metabolism of Bacteria (Russian translation). Moscow, IL, p. 389.
Newton, N., 1967. Biochem. J., 105:21.
Newton, N., 1969. Biochim. Biophys. Acta, 185:316.
Nicholas, D., 1964. 6 Internat. Congr. Biochem. New York. Abstr. X:784.
Niederpruem, D., and Doudoroff, M., 1965. J. Bacteriol., 89:697.

Nikaido, H., 1968. Adv. Enzymol., 31:77.
Nishizuka, Y., and Hayashi, O., 1962. J. Biol. Chem., 237:2721.
Nomenclature of Enzymes, 1966. Itogi Nauki, Seriya Biol. Khimiya.
Nomenclature of Quinones with Isoprenoid Side Chain, 1966. J. Biol. Chem., 241:2989.
Nomura, M., 1969. In: Mechanism of Action of Antibiotics (Russian translation).
   Moscow, Mir, p. 658.
Nomura, M., and Witten, C., 1967. J. Bacteriol., 94:1093.
Norton, J., Bulmer, G., and Sokatch, J., 1963. Biochim. Biophys. Acta, 78:136.
Nose, K., Ono, M., and Mizuno, D., 1970. J. Bacteriol., 101:102.
Nossal, P., Kleech, D., and Morton, D., 1956. Biochim. Biophys. Acta, 22:412.
O'Brien, J., 1967. J. Theoret. Biol., 15:307.
Oelze, J., and Drews, G., 1969. Arch. Mikrobiol., 69:12.
Oelze, J., and Drews, G., 1970a. Biochim. Biophys. Acta, 203:189.
Oelse, J., and Drews, G., 1970b. Biochim. Biophys. Acta, 219:131.
Oelze, J., Schroeder, J., and Drews, G., 1970. J. Bacteriol., 101:669.
Ohnishi, K., 1966. J. Biochem., 59:1.
Ohnishi, T., 1963. J. Biochem., 53:71.
Ohnishi, T., and Mori, T., 1960. J. Biochem., 48:406.
Ohnishi, T., and Mori, T., 1962. Nature, 193:488.
Ohsumi, Y., and Maeda, A., 1972. J. Biochem. (Tokyo), 71:911.
Oishi, K., Kim, R., Aida, K., and Uemura, T., 1970. J. Gen. Appl. Microbiol., 16:301.
Okada, Y., and Okunuki, T., 1969. J. Biochem., 65:581.
Okayama, S., Yamamoto, M., Nishikawa, K., and Horio, T., 1968. J. Biol. Chem., 243:2995.
Okunuki, K., 1963. In: Orekhovich, V. (Editor). Analytical Methods in Protein Chemistry
   (Russian translation). Moscow, IL, p. 37.
Okunuki, K., 1966. In: Florkin, M., and Stotz, E. (Editors). Comprehensive Biochemistry,
   Vol. 14. Amsterdam, Elsevier, p. 232.
Okunuki, K., Sekuzu, I., Orii, Y., Tsudzuki, T., and Matsumara, Y., 1968. In: Okunuki, K.
   et al. (Editors). Structure and Function of Cytochromes. Tokyo, Tokyo Univ. Press,
   p. 35.
Okuyama, H., 1969. Biochim. Biophys. Acta, 176:125.
Okuyama, H., Kankura, T., and Nojima, S., 1967. J. Biochem., 61:732.
O'Leary, W., 1962. Bacteriol. Rev., 26:421.
Omura, T., Siekevitz, P., and Palade, G., 1967. J. Biol. Chem., 242:2389.
Onishi, H., and Kushner, D., 1966. J. Bacteriol., 91:646.
Ono, Y., and Nojima, S., 1969. J. Biochem., 65:979.
Ono, Y., and White, D., 1970. J. Bacteriol., 104:712.
Op den Kamp, J., Houtsmuller, P., and Van Deenen, L., 1965. Biochim. Biophys. Acta,
   106:438.
Op den Kamp, J., Van Iterson, W., and Van Deenen, L., 1967. Biochim. Biophys. Acta,
   135:862.
Op den Kamp, J., Redai, J., and Van Deenen, L., 1969. J. Bacteriol., 99:298.
Oparin, A. I., 1957. The Origin of Life on Earth (in Russian). Moscow, Izd. AN SSSR.
Oparin, A. I., 1960. Life, Its Nature, Origin, and Development (in Russian). Moscow,
   Izd. AN SSSR.
Oparin, A. I., 1966. In: The Origin of Prebiological Systems (Russian translation). Moscow,
   Mir.
Oparin, A. I., Gel'man, N. S., and Zhukova, I. G., 1965a. Dokl. Akad. Nauk SSSR, 161:237.
Oparin, A. I., Lukoyanova, M. A., Shvets, V. I., Gel'man, N. S., and Torkhovskaya, T. I.,
   1965b. Zh. Evolyuts. Biokhim. i Fiziol., 1:7.
Oppenheimer, J., and Marcus, L., 1970. J. Bacteriol., 101:286.

Orlando, J., 1967. Biochim. Biophys. Acta, 143:634.

Orlando, J., and Horio, T., 1963. Biochim. Biophys. Acta, 50:367.

Osnitskaya, L., Threlfall, D., and Goodwin, T., 1964. Nature, 204:80.

Ostrovskii, D. N., 1964. the Study of Oxidative Phosphorylation in the Membranes of *Micrococcus lysodeikticus*. Author's Abstract of Candidate's Dissertation. Moscow.

Ostrovskii, D. N., and Gel'man, N. S., 1963. Dokl. Akad. Nauk SSSR, 148:945.

Ostrovskii, D. N., and Gel'man, N. S., 1965. Biokhimiya, 30:772.

Ostrovskii, D. N., and Sofronova, M. Yu., 1972. Proceedings of the 4th International Biophysical Congress (in Russian). Moscow (in press).

Ostrovskii, D. N., Kharat'yan, E. F., and Gel'man, N. S., 1964. Biokhimiya, 29:154.

Ostrovskii, D. N., Pereverzev, N. A., Zhukova, I. G., Trutko, S. M., and Gel'man, N. S., 1968a. Biokhimiya, 33:319.

Ostrovskii, D. N., Zhukova, I. G., and Gel'man, N. S., 1968b. Biokhimiya, 33:612.

Ostrovskii, D. N., Tsfasman, I. M., and Gel'man, N. S., 1969. Biokhimiya, 34:993.

Ostrovskii, D. N., Bystrov, V. F., Zhukova, I. G., Chervin, I. I., and Yakovlev, G. I., 1972. Biokhimiya (in press).

Ota, A., 1965. J. Biochem., 58:137.

Otsuka, S., and Shio, J., 1968. J. Gen. and Appl. Microbiol., 14:135.

Ovchinnikov, Yu. A., Ivanov, V. T., Evstratov, A. B., Laine, I. A., Malenkov, G. G., Shkrob, A. M., and Shemyakin, M. M., 1969. In: Abstracts of Proceedings of Symposia of the Second All-Union Biochemical Congress (in Russian). Tashkent, FAN, p. 284.

Overath, P., Schairer, H., and Stoffel, N., 1970. Proc. Nat. Acad. Sci. (Washington), 67:606.

Owen, P., and Freer, J., 1970. Biochem. J., 120:237.

Packer, L., 1958. Arch. Biochem. and Biophys., 78:54.

Palmer, G., Horgan, D., Tisdall, H., Singer, T., and Beinert, H., 1968. J. Biol. Chem., 243:844.

Pandit-Hovenkamp, H., and Eilermann, L., 1968. Abstract of the Fifth FEBS Meeting, p. 50.

Pandya, K., and King, H., 1962. Biochem. Y. 85:15p.

Pandya, K., and King, H., 1966. Arch. Biochem. and Biophys., 114:154.

Pandya, K., Bishop, D., and King, H., 1961. Biochem. J., 78:35.

Pangborn, J., Marr, A., and Robrish, S., 1962. J. Bacteriol., 84:669.

Panos, C., 1968. In: Guse, L. (Editor). Microbial Protoplasts, Spheroplasts and L-Forms. p. 154.

Panos, C., Cohen, M., and Fagan, G., 1966. Biochemistry, 5:1461.

Pappenheimer, A., 1955. In: Colowick, S., and Kaplan, N. (Editors). Methods in Enzymology, Vol. 2. New York, Academic Press, p. 744.

Pappenheimer, A., and Hendee, E., 1949. J. Biol. Chem., 180:597.

Pappenheimer, A., Howland, J., and Miller, P., 1962. Biochim. Biophys. Acta, 64:229.

Pardee, A., 1966. J. Biol. Chem., 241:5886.

Pardee, A., 1967. Science, 156:1627.

Pardee, A., 1969. J. Gen. Physiol., 54:2, 79.

Park, C., and Berger, L., 1967. J. Bacteriol., 93:221.

Park, R., 1966. Internat. Rev. Cytol., 20:67.

Parsons, D., 1966. Information Exchange Group No. 1, Sci. Memo. 672.

Patch, C., and Landman, O., 1971. J. Bacteriol., 107:345.

Pate, J., and Ordal, E., 1967. J. Cell Biol., 35:1.

Patterson, D., Weinstein, M., Nixon, R., and Gillespie, D., 1970. J. Bacteriol., 101:584.

Patterson, P., and Lennarz, W., 1969. Federat. Proc. 28:403.

Patterson, P., and Lennarz, W., 1970. Biochem. Biophys. Res. Commun., 40:408.

Patterson, P., and Lennarz, W., 1971. J. Biol. Chem., 246:1062.

Pavlasova, E., and Harold, F., 1969. J. Bacteriol., 98:198.
Pavlova, I. B., and Larina, I. A., 1969. Zh. Mikrobiol., Epidemiol., Immunobiol., 8:9.
Pavlova, I. B., and Pershina, Z. G., 1966. Zh. Mikrobiol., Epidemiol., Immunobiol.,4:8.
Pavlova, I. B., and Sergeeva, T. N., 1969. Zh. Mikrobiol., Epidemiol., Immunobiol., 4:12.
Peck, H., 1968. Annual Rev. Microbiol., 22:489.
Peeters, T., and Aleem, M., 1970. Arch. Mikrobiol., 71:319.
Pendleton, J. R., Kim, K. S., and Bernheimer, A., 1972. J. Bacteriol., 110:722.
Pennock, J., 1967. In: Pridham, J. (Editor). Terpenoids in Plants. New York, Academic
    Press, p. 129.
Penrose, W., Nichoalds, G., Piperno, J., and Oxender, D., 1968. J. Biol. Chem., 243:5921.
Penrose, W., Zand, R., and Oxender, D., 1970. J. Biol. Chem., 245:1432.
Peterson, J., 1970. J. Bacteriol., 103:714.
Pfister, R., and Lundgren, D., 1964. J. Bacteriol., 88:1119.
Phillips, M. C., Ladbrooke, B., and Chapman, D., 1970. Biochim. Biophys. Acta, 196:35.
Phillips, P., Revsin, B., Drell, E., and Brodie, A., 1970. Arch. Biochem. Biophys. 139:59.
Phizackerley, P., and Francis, M., 1966. Biochem. J., 101:524.
Phizackerley, P., MacDougall, J., and Francis, M., 1966. Biochem. J., 99:21.
Pinchot, G., 1957a. J. Biol. Chem., 229:1.
Pinchot, G., 1957b. J. Biol. Chem., 229:11.
Pinchot, G., 1957c. J. Biol. Chem., 229:25.
Pinchot, G., 1957d. Biochim. Biophys. Acta, 23:660.
Plackett, P., Smith, P., and Mayberry, W., 1970. J. Bacteriol., 104:798.
Poglazov, B. F., 1970. The Assembly of Biological Structures (in Russian). Moscow,
    Nauka.
Polglase, W., Pun, W., and Withaar, J., 1966. Biochim. Biophys. Acta, 118:425.
Pollack, J., Razin, S., and Cleverdon, R., 1965. J. Bacteriol., 90:617.
Pollack, J., Linder, R., and Salton, M., 1971. J. Bacteriol., 107:230.
Poltorak, O. M., and Chukhrai, E. S., 1969. In: Abstracts of Proceedings of Symposia of
    the Second All-Union Biochemical Congress (in Russian). Tashkent, FAN, p. 286.
Pontefract, R., and Thatcher, T., 1970. J. Ultrastr. Res., 30:78.
Pontefract, R., Bergeron, G., and Thatcher, F., 1969. J. Bacteriol., 97:367.
Popkin, T., Theodore, T., and Cole, R., 1971. J. Bacteriol., 107:907.
Postgate, J., 1956. J. Gen. Microbiol., 14:545.
Powls, R., and Redfearn, E., 1969. Biochim. Biophys. Acta, 172:429.
Prakash, O., Rao, R., and Sadana, J., 1966. Biochim. Biophys. Acta, 118:426.
Prezbindowski, K., Ruzicka, F., Sun, F., and Crane, F., 1969. Exper. Cell Res., 57:385.
Privalov, P. L., and Monaselidze, D. R., 1963. Biofizika, 8:420.
Proctor, H., Norris, J., and Ribbons, D., 1969. J. Appl. Bacteriol., 32:118.
Ptisyn, O. B., 1967. Uspekhi Sovr. Biol., 63:3.
Pullman, M., and Schatz, G., 1967. Annual Rev. Biochem., 36:652.
Pumphrey, A., and Redfearn, E., 1960. Biochem. J., 76:61.
Quigley, J., and Cohen, S., 1969. J. Biol. Chem., 244:2450.
Racker, E., 1967. Bioenergetic Mechanisms (Russian translation). Moscow, Mir.
Racker, E., 1969a. In: Molecules and Cells (Russian translation), No. 4, Moscow, Mir,
    p. 150.
Racker, E., 1969b. J. Gen. Physiol., 54:38.
Radcliffe, B., and Nicholas, D., 1970. Biochim. Biophys. Acta, 205:273.
Rajagopalan, K., and Handler, P., 1968. In: Singer, T. (Editor). Biological Oxidation.
    New York, Interscience Publishers, p. 302.
Ramaiah, A., and Nicholas, D., 1964. Biochim. Biophys. Acta, 86:459.
Randle, C., Albro, P., and Dittmer, J., 1969. Biochim. Biophys. Acta, 187:214.

Ray, P., and Brock, T., 1971. J. Gen. Microbiol., 66:133.
Razin, S., 1964. J. Gen. Microbiol., 36:451.
Razin, S., 1969. Annual Rev. Microbiol., 23:317.
Razin, S., 1972. Biochim. Biophys. Acta, 265:241.
Razin, S., and Barash, V., 1969. FEBS Letters, 3:217.
Razin, S., and Cleverdon, R., 1965. J. Gen. Microbiol., 41:409.
Razin, S., and Kahane, J., 1969. Nature, 223:863.
Razin, S., and Rottem, S., 1968. J. Bacteriol., 94:1807.
Razin, S., Argaman, M., and Avigan, J., 1963. J. Gen. Microbiol., 33:477.
Razin, S., Morowitz, H., and Terry, T., 1965. Proc. Nat. Acad. Sci. (Washington), 54:219.
Razin, S., Tourtellotte, M., McElhaney, R., and Pollack, J., 1966. J. Bacteriol., 91:609.
Razin, S., Neeman, Z., and Ohad, I., 1969. Biochim. Biophys. Acta, 193:277.
Razin, S., Prescott, B., and Chanock, R., 1970. Proc. Nat. Acad. Sci. (Washington), 67:590.
Reaveley, D., 1968. Biochem. Biophys. Res. Commun., 30:649.
Reaveley, D., and Rogers, H., 1969. Biochem. J., 113:67.
Rebel, G., and Mandel, P., 1965. Biochim. Biophys. Acta, 98:380.
Rebel, G., Sensenbrenner, B., and Mandel, P., 1964. Bull. Soc. Chim. Biol., 46:1113.
Redwood, W., Muldner, H., and Thompson, T., 1969. Proc. Nat. Acad. Sci. (Washington), 64:989.
Reed, D., 1969. J. Biol. Chem., 244:4936.
Rees, M., 1968. Biochemistry, 7:353, 366.
Rees, M., and Nason, A., 1965. Biochem. Biophys. Res. Commun., 21:248.
Reeves, K. P., 1968. J. Bacteriol., 96:1700.
Reiss-Husson, F., and Luzzati, V., 1967. Adv. Biol. Med. Phys., 11:87.
Remsen, C., 1966. Arch. Mikrobiol., 54:266.
Remsen, C., 1968. Arch. Mikrobiol., 61:40.
Remsen, C., and Lundgren, D., 1966a. J. Bacteriol., 92:1748.
Remsen, C., and Lundgren, D., 1966b. J. Bacteriol., 92:1765.
Remsen, C., Valois, F., and Watson, S., 1967. J. Bacteriol., 94:422.
Remsen, C., Watson, S., Waterbury, J., and Truper, H., 1968. J. Bacteriol., 95:2374.
Repaske, R., 1954. J. Bacteriol., 68:555.
Repaske, R., 1958. Biochim. Biophys. Acta, 30:225.
Repaske, R., 1966. Biotechnol. Bioengng, 8:217.
Repaske, R., and Josten, J., 1958. J. Biol. Chem., 233:466.
Repaske, R., and Lizotte, C., 1965. J. Biol. Chem., 240:4774.
Revsin, B., and Brodie, A., 1969. J. Biol. Chem., 244:23101.
Revsin, B., Marquez, E., and Brodie, A., 1970a. Arch. Biochem. Biophys., 139:114.
Revsin, B., Marquez, E., and Brodie, A., 1970b. Arch. Biochem. Biophys., 136:563.
Reyn, A., Birch-Andersen, A., and Lapage, S., 1966. Canad. J. Microbiol., 12:1125.
Reynolds, B., and Reeves, P., 1969. J. Bacteriol., 100:301.
Ribbons, D., Harrison, J., and Wadcinski, A., 1970. Annual Rev. Microbiol., 24:135.
Rieske, J., 1969. In: The Mechanism of Action of Antibiotics (Russian translation). Moscow, Mir, p. 509.
Rizza, V., Sinclair, P., White, D., and Courant, P., 1968. J. Bacteriol., 96:665.
Rizza, V., Tucker, A., and White, D., 1970. J. Bacteriol., 101:84.
Robertson, J., 1959. Biochem. Soc. Symp., 16:3.
Robertson, J., 1960. Progr. Biophysics Biophys. Chem., 10:34.
Robertson, J., 1967. In: Allen, J. (Editor). Molecular Organization and Biological Functions. p. 65.
Robrish, S., and Marr, A., 1962. J. Bacteriol., 83:158.
Rodwell, A., 1965. J. Gen. Microbiol., 40:227.

Rodwell, A., Razin, S., Rottem, S., and Argaman, M., 1967. Arch. Biochem. Biophys., 122:621.

Rolfe, B., and Onodera, K., 1971. Biochem. Biophys. Res. Commun., 44:767.

Romanov, V. I., Taptykova, S. D., and Kretovich, V. L., 1970. Dokl. Akad. Nauk SSSR, 193:1189.

Romeo, D., Cramer, R., and Rossi, F., 1970. Biochem. Biophys. Res. Commun., 41:582.

Rosenberg, R., and Guidotti, G., 1968. J. Biol. Chem., 243:1985.

Rosenberg, R., and Guidotti, G., 1969. J. Biol. Chem., 244:5118.

Rossi, C., and Lehninger, A., 1964. J. Biol. chem., 239:3971.

Rothblat, G., Ellis, D., and Kritchevsky, D., 1964. Biochim. Biophys. Acta, 84:340.

Rothfield, L., and Finkelstein, A., 1968. Annual Rev. Biochem., 37:463.

Rothfield, L., and Horecker, B., 1964. Proc. Nat. Acad. Sci. (Washington), 52:939.

Rothfield, L., Weiser, M., and Endo, A., 1969. J. Gen. Physiol., 54:27.

Rothfield, L., Romeo, D., and Hinckley, A., 1972. Federat. Proc., 31:12.

Rottem, S., and Panos, C., 1969. J. Gen. Microbiol., 59:317.

Rottem, S., and Razin, S., 1967. J. Bacteriol., 94:359.

Rottem, S., Stein, O., and Razin, S., 1968. Biochim. Biophys. Acta, 125:46.

Rottem, S., Hubbell, W., Hayflick, L., and McConnell, H., 1970. Biochim. Biophys. Acta, 219:104.

Rottem, S., Kalkstein, A., and Citri, N., 1971. Biochim. Biophys. Acta, 241:593.

Rouser, G., Nelson, G., Fleischer, S., and Simon, G., 1968. In: Chapman, D. (Editor). Biological Membranes. New York, Academic Press.

Rozantsev, E. G., 1970. Free Iminoxyl Radicals (in Russian). Moscow, Khimiya.

Rubenstein, K., Nass, M., and Cohen, S., 1970. J. Bacteriol., 104:443.

Rubin, L. B., Dubrovin, V. A., Adamova, I. P., and Shvinka, Yu. E., 1970. Mikrobiologiya, 29:264.

Rudney, H., 1969. Biochem. J., 113:21.

Ruiz-Herrera, J., and De Moss, J., 1969. J. Bacteriol., 99:720.

Ryter, A., 1965. Ann. Inst. Pasteur, 108:40.

Ryter, A., 1968. Bacteriol. Rev., 32:39.

Ryter, A., 1969. Current Topics Microbiol. Immunol., 49:151.

Ryter, A., and Jacob, F., 1963. C. R. Acad. Sci., 257:3060.

Ryter, A., and Jacob, F., 1964. Ann. Inst. Pasteur, 107:389.

Ryter, A., and Jacob, F., 1966. Ann. Inst. Pasteur, 110:801.

Ryter, A., and Landman, O., 1964. J. Bacteriol., 88:457.

Ryter, A., and Landman, O., 1968. In: Guze, L. (Editor). Microbial Protoplasts, Spheroplasts and L-Forms. Baltimore, Williams and Wilkins, p. 110.

Ryter, A., Frehel, C., and Ferrandes, B., 1967. C. R. Acad. Sci., 265:1259.

Saito, K., and Mukoyama, K., 1971. J. Biochem. (Tokyo), 69:83.

Sale, A., and Hamilton, W., 1968. Biochim. Biophys. Acta, 163:37.

Salem, L., 1964. In: Pullman, B. (Editor). Electronic Aspects of Biochemistry, p. 293.

Salton, M., 1964. The Bacterial Cell Wall, Elsevier.

Salton, M., 1967a. Trans. New York Acad. Sci., 29:764.

Salton, M., 1967b. In: Davis, B., and Warren, L. (Editors). The Specificity of Cell Surfaces. Englewood Cliffs, Prentice-Hall, p. 71.

Salton, M., 1968. In: Guze, L. (Editor). Microbial Protoplasts and L-Forms, p. 144.

Salton, M., and Chapman, J., 1962. J. Ultrastr. Res., 6:489.

Salton, M., and Ehtisham-ud-Din, A., 1964. Austral. J. Exper. Biol. Med. Sci., 43:255.

Salton, M., and Freer, J., 1965. Biochim. Biophys. Acta, 107:531.

Salton, M., and Freer, J., 1968. Biochem. Biophys. Res. Commun., 33:909.

Salton, M., and Netschey, A., 1965. Biochim. Biophys. Acta, 107:539.

Salton, M., and Schmitt, M., 1967. Biochim. Biophys. Acta, 135:196.
Salton, M., Schmitt, M., and Trefts, P., 1967. Biochem. Biophys. Res. Commun., 29:728.
Samuilov, V. D., 1969. Uspekhi Sovr. Biol., 68:232.
Samuilov, V. D., Nikiforov, E. A., and Nikiforova, V. I., 1971. Dokl. Akad. Nauk SSSR, 196:723.
Sands, D., Gleason, F., and Hildebrand, D., 1967. J. Bacteriol., 94:1785.
San Pietro, A. (Editor), 1965. Non-Heme Iron Proteins: Role in Energy Conversion. Yellow Springs, Antioch Press.
Sapshead, L., and Wimpenny, J., 1972. Biochim. Biophys. Acta, 267:388.
Sargent, M., and Lampen, J., 1970. Arch. Biochem. Biophys., 136:167.
Sasaki, T., 1964. J. Biochem., 55:225.
Sasaki, T., Motokawa, V., and Kikuchi, G., 1970. Biochim. Biophys. Acta, 197:284.
Sato, R., 1956. In: McElroy, W., and Glass, B. (Editors). Inorganic Nitrogen Metabolism. Baltimore, Johns Hopkins Press, p. 163.
Sazykin, Yu. C., 1968. Antibiotics as Inhibitors of Biochemical Processes (in Russian). Moscow, Nauka.
Scanu, A., Pollard, H., Hirz, R., and Kothary, K., 1969. Proc. Nat. Acad. Sci. (Washington), 62:171.
Schachman, H., Pardee, A., and Stanier, R., 1952. Arch. Biochem. Biophys., 38:245.
Schaeffer, P., 1952. Biochim. Biophys. Acta, 9:261.
Scharff, K., Hendler, R., Nanninga, N., and Burgess, A., 1972. J. Cell. Biol., 53:1.
Schatz, G., and Saltzgaber, J., 1969. Biochim. Biophys. Acta, 180:186.
Schick, J., and Drews, G., 1969. Biochim. Biophys. Acta, 183:215.
Schlessinger, D., 1963. J. Mol. Biol., 7:569.
Schlessinger, D., Marchesi, V., and Kwan, B., 1965. J. Bacteriol., 90:456.
Schmitz, R., 1967. Arch. Mikrobiol., 56:238.
Schnaitman, C., 1969a. Biochem. Biophys. Res. Commun., 37:1.
Schnaitman, C., 1969b. Proc. Nat. Acad. Sci. (Washington), 63:412.
Schnaitman, C., 1970a. J. Bacteriol., 104:882.
Schnaitman, C., 1970b. J. Bacteriol., 104:890.
Schnaitman, C., and Greenawalt, J., 1966. J. Bacteriol., 92:780.
Schnebli, H., Vatter, A., and Abrams, A., 1970. J. Biol. Chem., 245:1122.
Scholes, P., and King, H., 1963. Biochem. J., 87:10.
Scholes, P., and King, H., 1965a. Biochem. J., 97:755.
Scholes, P., and King, H., 1965b. Biochem. J., 97:766.
Scholes, P., and Mitchell, P., 1970. J. Bioenerg., 1:309.
Scholes, P., and Smith, L., 1968a. Biochim. Biophys. Acta, 153:350.
Scholes, P., and Smith, L., 1968b. Biochim. Biophys. Acta, 153:363.
Scholes, P., McLain, G., and Smith, L., 1971. Biochemistry, 10:2072.
Schonfeldt, N., 1965. Nonionic Detergents (Russian translation). Moscow, Khimiya.
Schötz, F., Abo-Elnaga, I., and Kandler, O., 1965. Z. Naturforsch., 20b:790.
Schröder, J., and Drews, G., 1968. Arch. Mikrobiol., 64:59.
Schulz, R., and Asunman, S., 1970. Recent Progress Surface Sci., 3:291.
Schwartz, A. M., and Perry, J. W., 1949. Surface-Active Agents, Their Chemistry and Technology (Russian translation, 1953). New York. Moscow, IL.
Scocca, J., and Pinchot, G., 1968. Arch. Biochem. Biophys., 124:206.
Seddon, B., and Fynn, G., 1971. Arch. Mikrobiol., 77:252.
Segel, W., and Goldman, D., 1963. Biochim. Biophys. Acta, 73:380.
Shah, D., 1970b. Adv. Lipid Res., 8:347.
Shah, S., and King, H., 1965. Biochem. J., 94:13.
Shah, S., and King, H., 1966. J. Gen. Microbiol., 44:1.

Shapiro, A., Vinuela, E., and Maizel, J., 1967. Biochim. Biophys. Acta, 28:815.
Shapiro, B., Siccardi, A., Hirata, Y., and Jacob, F., 1970. J. Mol. Biol., 52:75.
Shaw, M., 1968. J. Bacteriol., 95:221.
Shaw, N., 1970. Bacteriol. Rev., 34:365.
Shaw, N., and Baddiley, J., 1968. Nature, 217:142.
Shaw, N., Smith, P., and Verheij, H. M., 1970. Biochem. J., 120:439.
Sheard, B., 1969. Nature, 223:1057.
Shechter, E., Gulik-Krzywicki, T., and Kaback, H., 1972. Biochim. Biophys. Acta, 274:446.
Shemyakin, M., Antonov, V., Bergelson, L., Ivanov, V., Malenkov, G., Ovchinnikov, Yu.,
    and Shkrob, A., 1969. In: Tosteson, D. (Editor). The Molecular Basis of Membrane
    Function. Englewood Cliffs, Prentice-Hall.
Shen, P., Coles, E., Foote, J., and Stenesi, H., 1970. J. Bacteriol., 103:479.
Sheraga, H., 1963. In: Neurath, H. (Editor). The Proteins. New York, Academic Press,
    p. 478.
Shetna, Y., 1970. Biochim. Biophys. Acta, 205:58.
Shetna, Y., Wilson, P., and Beinert, H., 1966. Biochim. Biophys. Acta, 113:225.
Shetna, Y., Der Vartanian, D., and Beinert, H., 1968. Biochem. Biophys. Res. Commun.,
    31:862.
Shibata, K., Takamiya, A., Jagendorf, T., and Fuller, R. (Editors), 1968. Comparative
    Biochemistry and Biophysics of Photosynthesis. Tokyo, Tokyo Univ. Press.
Shibko, S., and Pinchot, G., 1961. Arch. Biochem. Biophys., 94:257.
Shibuya, J., Honda, H., and Maruo, B., 1967. Agric. Biol. Chem., 31:633.
Shibuya, J., Honda, H., and Maruo, B., 1968. J. Biochem., 64:571.
Shinoda, K., Tamamushi, B. I., Nakagawa, T., and Isemura, T., 1963. Colloidal Sur-
    factants: Some Physicochemical Properties (Russian translation, 1966). New York.
Shively, J., and Benson, A., 1967. J. Bacteriol., 94:1679.
Shively, J., Decker, G., and Greenawalt, J., 1970. J. Bacteriol., 101:618.
Shkrob, A. M., Malenkov, G. G., Mel'nik, E. I., Oreshnikova, N. A., Gulanyan, S. A.,
    Ryabova, I. D., Vinogradova, E. I., and Shemyakim, M. M., 1969. In: Abstracts of
    Proceedings of Symposia of the Second All-Union Biochemical Congress (in Rus-
    sian). Tashkent, FAN, p. 585.
Shockman, G., Kolb, J., Bakay, B., Conover, M., and Toennis, G., 1963. J. Bacteriol.,
    85:168.
Short, S., and White, D., 1970. J. Bacteriol., 104:126.
Short, S., and White, D. 1972 J. Bacteriol., 109:820.
Short, S., White, D., and Aleem, M., 1969. J. Bacteriol., 99:142.
Short, S., White, D., and Kaback, H., 1972. J. Biol. Chem., 247:298.
Shul'ga, A. V., and Tongur, V. S., 1966. Biokhimiya, 31:696.
Shull, F., Fralick, J., Stratton, L., and Fisher, W., 1971. J. Bacteriol., 106:626.
Sih, C., and Bennett, R., 1962. Biochim. Biophys. Acta, 56:584.
Silva, B., 1967. Exper. Cell Res., 46:245.
Simakova, I. M., 1970. The Role of Noncovalent Interactions in Stabilization of the
    Membrane of *Micrococcus lysodeikticus*. Author's Abstract of Candidate's Dis-
    sertation. Moscow.
Simakova, I. M., Lukoyanova, M. A., Biryuzova, V. I., and Gel'man, N. S., 1968.
    Biokhimiya, 33:1047.
Simakova, I. M., Lukoyanova, M. A., Biryuzova, V. I., and Gel'man, N. S., 1969.
    Biokhimiya, 34:1271.
Sinclair, P., and White, D., 1970. J. Bacteriol., 101:365.
Sinesky, M., 1971. J. Bacteriol., 106:449.
Singer, S. J., 1971. In: Rothfield, L. (Editor). Structure and Function of Biological
    Membranes. New York, Academic Press, p. 145.

Singer, T., 1966. In: Florkin, M., and Stotz, E. (Editors). Comprehensive Biochemistry, Vol. 14. Amsterdam, Elsevier, p. 127.
Singer, T., 1968a. In: Biological Oxidations. New York, Academic Press.
Singer, T. (Editor), 1968b. Biological Oxidation, New York, Academic Press.
Singer, T., and Lara, F., 1958. Proc. Intern. Symp. Enzyme. Chem. (Tokyo–Kyoto, 1957). Tokyo, Maruzen, p. 203.
Singha, D., and Gaby, W., 1964. J. Biol. Chem., 239:3668.
Sjöstrand, F., 1963. J. Ultrastr. Res., 9:561.
Sjöstrand, F., and Barajas, L., 1968. J. Ultrastr. Res., 25:121.
Sjöstrand, F., and Barajas, L., 1970. J. Ultrastr. Res., 32:293.
Skopinskaya, S. N., Biryuzova, V. I., Taptykova, S. D., Proisser, E., Gel'man, N. S., and Lukoyanova, M A., 1972. Mikrobiologiya 41:1068.
Skov, K., and Williams, G., 1968. Okunuki, K., et al. (Editors). Structure and Function of Cytochromes. Tokyo, Tokyo Univ. Press, p. 346.
Skulachev, V. P., 1962. Relationship between Oxidation and Phosphorylation in the Respiratory Chain (in Russian). Moscow, Izd. AN SSSR.
Skulachev, V. P., 1969. The Accumulation of Energy in the Cell (in Russian). Moscow, Nauka.
Skulachev, v. P., 1972. The Transformation of Energy in Biological Membranes (in Russian). Moscow, Nauka.
Slater, E., 1953. Nature, 172:975.
Sletten, K., Dus, K., de Klerk, H., and Kamen, M., 1968. J. Biol. Chem., 243:5492, 5507.
Sleytr, U., 1970. Arch. Mikrobiol., 72:238.
Small, D., Penkett, S., and Chapman, D., 1969. Biochim. Biophys. Acta, 176:178.
Smalley, A., Johrling, P., and Van Demark, P., 1968. J. Bacteriol., 96:1595.
Smarda, J., and Taubeneck, U., 1968. J. Gen. Microbiol., 52:161.
Smith, L., 1954a. Bacteriol. Rev., 18:106.
Smith, L., 1954b. Arch. Biochem. Biophys., 50:255.
Smith, L., 1954c. Arch. Biochem. Biophys., 50:299.
Smith, L., 1955. J. Biol. Chem., 215:847.
Smith, L., 1962. Biochim. Biophys. Acta, 62:145.
Smith, L., 1963. In: Metabolism of Bacteria (Russian translation). Moscow, IL, p. 360.
Smith, L., 1968. In: Singer, T. (Editor). Biological Oxidation. New York, Academic Press, p. 56.
Smith, L., and Minnaert, K., 1965. Biochim. Biophys. Acta, 105:1.
Smith, L., and Newton, N., 1968. In; Okunuki, K., et al. (Editors). Structure and Function of Cytochromes. Tokyo, Tokyo Univ. Press, p. 153.
Smith, L., Newton, N., and Scholes, P., 1966. In: Chance, B. et al. (Editors). Hemes and Hemeproteins, p. 395.
Smith, L., and White, D., 1962. J. Biol. Chem. 237:1332.
Smith, L., White, D., Sinclair, P., and Chance, B., 1970. J. Biol. Chem., 245:5096.
Smith, P., 1963. J. Gen. Microbiol., 32:307.
Smith, P., 1964. Bacteriol. Rev., 28:97.
Smith, P., 1969. J. Bacteriol., 99:480.
Smith, P., and Henrikson, C., 1965. J. Lipid Res., 6:106.
Smith, P., and Koostra, W., 1967. J. Bacteriol., 93:1853.
Smith, P., and Mayberry, W., 1968. Biochemistry, 7:2706.
Smith, P., and Rothblat, G., 1960. J. Bacteriol., 80:842.
Smith, P., Koostra, W., and Henrikson, C., 1965. J. Bacteriol., 90:282.
Smith, P., Koostra, W., and Mayberry, W., 1970. J. Bacteriol., 100:1166.
Snyder, C., and Rapoport, H., 1970. Biochemistry, 9:2033.
Sofronova, M Yu., Ostrovskii, D. N., and Gel'man, N. S., 1971. Biokhimiya, 36:977.

Soininen, R., Sojonen, H., and Ellfolk, N., 1970. Acta Chem. Scand., 24:2314.

Sokawa, Y., 1965. J. Biochem., 57:706.

Spiegelman, S., Aronson, A., and Fitz-James, P., 1958. J. Bacteriol., 75:102.

Stanier, R., 1964. In: Gunsalus, I., and Stanier, R. (Editors). The Bacteria, Vol. 5. New York, Academic Press.

Steed, P., and Murray, R., 1966. Canad. J. Microbiol., 12:263.

Steim, J., and Fleischer, S., 1967. Proc. Nat. Acad. Sci. (Washington), 58:1292.

Steim, J., Tourtellotte, M., Reinert, J., McElhaney, R., and Rader, R., 1969. Proc. Nat. Acad. Sci. (Washington), 63:104.

Stein, W., 1969. J. Gen. Physiol., 54:81.

Stephens, W., and Starr, M., 1963. J. Bacteriol., 86:1070.

Stevenson, J., and Brown, R., 1969. Anst. Commun. 6 FEBS Meeting.

Stoeckenius, W., and Engelman, D., 1969. J. Cell.Biol., 42:613.

Stoeckenius, W., and Kunau, W., 1968. J. Cell Biol., 38:337.

Stoeckenius, W., and Rowen, R., 1967. J. Cell Biol., 34:365.

Stolp, H., and Starr, M., 1965. Annual Rev. Microbiol., 19:79.

Stopkie, R., and Weber, M., 1967. Biochem. Biophys. Res. Commun., 28:1034.

Stouthamer, A., 1961. Biochim. Biophys. Acta, 48:484.

Stouthamer, A., 1962. Biochim. Biophys. Acta, 56:19.

Straat, P., and Nason, A., 1965. J. Biol. Chem., 240:1412.

Strominger, J., and Ghuysen, J., 1967. Science, 156:213.

Sugimura, T., and Okabe, K., 1962. J. Biochem., 52:235.

Superstein, A., 1970. Antonie van Leeuwenhoek J. Microbiol. Serol., 36:335.

Sutherland, I., 1963. Biochim. Biophys. Acta, 73:162.

Sutherland, I., and Wilkinson, J., 1963. J. Gen. Microbiol., 30:105.

Sutton, D., and Starr, M., 1960. J. Bacteriol., 80:104.

Swank, R., and Burris, R., 1969. Biochim. Biophys. Acta, 180:473.

Sweetman, A., and Griffiths, D., 1971a. Biochem. J., 121:117.

Sweetman, A., and Griffiths, D., 1971b. Biochem. J., 121:125.

Sykes, J., Gibbon, J., and Hoare, D., 1965. Biochim. Biophys. Acta, 109:409.

Taber, H., and Morrison, M., 1964. Arch. Biochem. Biophys., 105:367.

Tabor, C., 1962. J. Bacteriol., 83:1101.

Tabor, C., and Tabor, H., 1966. J. Biol. Chem., 241:3714.

Tabor, H., Tabor, C., and Rosenthal, S., 1961. Annual Rev. Biochem., 30:579.

Tait, M., and Franks, F., 1961. Nature, 230:91.

Takacs, B., and Holt, S., 1971. Biochim. Biophys. Acta, 233:278.

Takagi, A., Nakamura, K., and Ueda, M., 1965. Japan. J. Microbiol., 9:131.

Takahashi, K., Titani, K., and Minakami, S., 1959. J. Biochem., 46:1323.

Takayama, K., MacLennan, D., Tzagoloff, A., and Stoner, C., 1966. Arch. Biochem. Biophys., 114:223.

Takamiya, K., Nishimura, M., and Takamiya, A., 1967. Plant Cell Physiol., 8:79.

Takeyoshi, N., 1961. J. Biochem., 49:240.

Tanford, C., 1970. Adv. Protein Chem., 24:2.

Tang, S., Coleman, J., and Myer, Y., 1968. J. Biol. Chem., 243:4286.

Tani, J., and Hendler, R., 1964. Biochim. Biophys. Acta, 80:279, 307.

Taniguchi, S., and Kamen, M., 1965. Biochim. Biophys. Acta, 96:395.

Taniguchi, S., and Ohmachi, K., 1960. J. Biochem., 48:50.

Tano, T., Kagawa, H., and Imai, K., 1968. Agric. Biol. Chem., 32:279.

Taptykova, S. D., 1968. Comparative Study of Cytochromes from Members of the Class Actinomycetes. Author's Abstract of Candidate's Dissertation. Moscow.

Tauschel, H. D., and Drews, G., 1967. Arch. Mikrobiol., 59:381.

Tchan, Y., and Webber, A., 1966. Arch. Mikrobiol., 54:215.
Terry, T., 1966. Doctoral Thesis, Yale Univ. (cited by S. Razin, 1969).
Terry, T., Engelman, D., and Morowitz, H., 1967. Biochim. Biophys. Acta, 135:391.
Thiele, O., and Busse, D., 1968. Experientia, 24:112.
Thore, A., Keister, D., and San Pietro, A., 1969. Arch. Mikrobiol., 67:378.
Thorne, K., and Kodicek, E., 1962. Biochim. Biophys. Acta, 59:306.
Threlfall, D., 1967. In: Pridham, J. (Editor). Terpenoids in Plants. New York, Academic Press, p. 191.
Throm, E., Oelze, J., and Drews, G., 1970. Arch. Mikrobiol., 72:361.
Tien, H., and Diana, L., 1968. Chem. Phys. Lipids, 2:55.
Tikhonova, G. V., 1967. Investigation of Electron Transport Systems of Iron-Oxidizing Bacteria *Thiobacillus ferrooxidans.* Author's Abstract of Candidate's Dissertation. Moscow.
Tikhonova, G. V., Lisenkova, L. L., Doman, N. G., and Skulachev, V. P., 1967. Biokhimiya, 32:725.
Tikhonova, G. V., Simakova, I. M., Lukoyanova, M. A., Taptykova, S. D., Mikel'saar, Kh. N., and Gel'man, N. S., 1970. Biokhimiya, 35:1123.
Tillack, T., Carter, R., and Razin, S., 1970. Biochim. Biophys. Acta, 219:123.
Tissières, A., 1951. Biochem. J., 50:279.
Tissières, A., 1956. Biochem. J., 64:582.
Tissières, A., and Burris, R., 1956. Biochim. Biophys. Acta, 20:436.
Tobari, J., 1964. Biochem. Biophys. Res. Commun., 15:50.
Tobari, J., and Kimura, T., 1966. J. Biochem., 60:470.
Torchinskii, Yu. M., 1967. Uspekhi Biol. Khimii, 8:61.
Tordzhyan, I. Kh., and Kats, L. N., 1969. Dokl. Akad. Nauk SSSR, 186:1192.
Tordzhyan, I. Kh., and Kats, L. N., 1970. Dokl. Akad. Nauk SSSR, 195:969.
Tourtellotte, M., Jensen, R., Gander, G., and Morowitz, H., 1963. J. Bacteriol., 86:370.
Tourtellotte, M., Branton, D., and Keith, A., 1970. Proc. Nat. Acad. Sci. (Washington), 66:909.
Tremblay, G., Daniels, M., and Schaechter, M., 1969. J. Mol. Biol., 40:65.
Trudinger, P., 1958. Biochim. Biophys. Acta, 30:211.
Trudinger, P., 1961. Biochem. J., 78:680.
Trudinger, P., 1964. Biochem. J., 90:640.
Tsfasman, I. M., Ostrovskii, D. N., and Gel'man, N. S., 1972a. Biokhimiya, 37:92.
Tsfasman, I. M., Ostrovskii, D. N., and Gel'man, N. S., 1972b. Biokhimiya, 37:389.
Tsien, H., and Laudelout, H., 1968. Arch. Mikrobiol., 61:280.
Tsien, H., and Laudelout, T., 1971. Arch. Mikrobiol., 75:266.
Tsien, H., Lambert, R., and Laudelout, H., 1968. Antonie Leeuwenhoek, J. Microbiol. Serol., 34:483.
Tsukagoshi, N., and Fox, C. F., 1971. Biochemistry, 10:3309.
Tsukagoshi, N., Tamura, G., and Arima, K., 1970. Biochim. Biophys. Acta, 197:204, 211.
Tsukagoshi, N., Fielding, P., and Fox, C., 1971. Biochem. Biophys. Res. Commun., 44:497.
Tucker, A., and White, D., 1970a. J. Bacteriol., 102:498.
Tucker, A., and White, D., 1970b. J. Bacteriol., 102:508.
Tucker, A., and White, D., 1970c. J. Bacteriol., 103:329.
Tucker, R., 1960. J. Gen. Microbiol., 23:267.
Tzagoloff, A., McConnell, D., and MacLennan, D., 1968. J. Biol. Chem., 243:4117.
Uchida, T., and Yoneda, M., 1967. Biochim. Biophys. Acta, 145:210.
Unemoto, T., Hayashi, M., and Miyaki, K., 1965. Biochim. Biophys. Acta, 110:319.
Urakami, C., and Umetani, K., 1968. Biochim. Biophys. Acta, 164:64.
Urban, P., and Klingenberg, M., 1969. Europ. J. Biochem., 9:519.

Urry, D., Masotti, L., and Krivacic, J., 1970. Biochem. Biophys. Res. Commun., 41:521.

Ushijima, T., 1967. Japan. J. Microbiol., 11:275.

Vaituzis, Z., and Doetsch, R., 1969. J. Bacteriol., 100:512.

Vambutas, V., and Salton, M., 1970a. Biochim. Biophys. Acta, 203:83.

Vambutas, V., and Salton, M., 1970b. Biochim. Biophys. Acta, 203:94.

Van Deenen, L., 1965. Phospholipids and Biomembranes. London, Pergamon Press.

Van Deenen, L., 1966. Ann. New York Acad. Sci., 137:717.

Van Deenen, L., 1968. In: Jarnefelt, J. (Editor). Regulatory Functions of Biological Membranes. Amsterdam, Elsevier, p. 72.

Van Demark, P., and Smith, P., 1964. J. Bacteriol., 88:122.

Van den Bosch, H., Williamson, J., and Vagelos, P., 1970. Nature, 228:338.

Van Gelder, B., Urry, D., and Beinert, H., 1968. In: Okuriuki (Editor), Structure and Function of Cytochromes. Tokyo, Univ. Tokyo Press, p. 335.

Van Golde, L., and Van Deenen, L., 1967. Chem. Phys. Lipids, 1:157.

Van Golde, L., McElhaney, R., and Van Deenen, L., 1971. Biochim. Biophys. Acta, 231:245.

Van Gool, A., and Laudelout, H., 1966. Biochim. Biophys. Acta, 113:41.

Van Iterson, W., 1961. Biophys. Biochem. Cytol., 9:183.

Van Iterson, W., 1965. Bacteriol. Rev., 29:277.

Van Iterson, W., 1969a. In: Lima-de-Faria, A. (Editor). Handbook of Molecular Cytology. Amsterdam, North Holland Publishing Co., p. 150.

Van Iterson, W., 1969b. In: Lima-de-Faria, A. (Editor). Handbook of Molecular Cytology. Amsterdam. North Holland Publishing Co., p. 175.

Van Iterson, W., 1969c. In: Lima-de-Faria, A. (Editor). Handbook of Molecular Cytology. Amsterdam, North Holland Publishing Co., p. 198.

Van Iterson, W., and Op den Kamp, J., 1969. J. Bacteriol., 99:304.

Vanderkooi, G., and Green, D., 1970. Proc. Nat. Acad. Sci. (Washington), 66:615.

Vanderkooi, G., and Sundaralingam, M., 1970. Proc. Nat. Acad. Sci. (Washington), 67:233.

Van't Riet, J., and Planta, R., 1969. FEBS Letters, 5:249.

Vasil'eva, T. P., Groshev, V. V., and Shestakov, S. V., 1970. Biokhimiya, 35:989.

Verhoeven, W., and Takeda, Y., 1956. In: McElroy, W., and Glass, B. (Editors), Inorganic Nitrogen Metabolism. Baltimore, Johns Hopkins Press, p. 159.

Vernon, L., 1956. J. Biol. Chem., 222:1035.

Vernon, L., and Garcia, A., 1967. Biochim. Biophys. Acta, 143:144.

Vernon, L., and Mangum, J., 1960. Arch. Biochem. Biophys., 90:103.

Vernon, L., Mangum, J., Beck, J., and Shafia, F., 1960. Arch. Biochem. Biophys., 88:227.

Verzhbinskaya, N. A., and Pershina, L. I., 1971. Dokl. Akad. Nauk SSSR, 196:455.

Vladimirov, Yu. A., 1965. Photochemistry and Luminescence of Proteins (in Russian). Moscow, Nauka.

Voelz, H., 1965. Arch. Mikrobiol., 51:60.

Vorbeck, M., and Marinetti, G., 1965. Biochemistry, 4:296.

Voss, J., 1967. J. Gen. Microbiol., 48:391.

Vu, C., and Gunsalus, I., 1970. Biochem. Biophys. Res. Commun., 40:1431.

Waggoner, A., and Stryer, L., 1970. Proc. Nat. Acad. Sci. (Washington), 67:579.

Wainio, W., and Aronoff, M., 1955. Arch. Biochem. Biophys., 57:115.

Wainio, W., Kirschbaum, J., and Shore, J., 1968. In: Okunuki, K., Kamen, M., and Sekuzu, J. (Editors). Structure and Function of Cytochromes. Tokyo, Tokyo Univ. Press, p. 713.

Walker, P., and Baillie, A., 1968. J. Appl. Bacteriol., 31:108.

Wallace, P., Huang, M., and Linnane, A., 1968. J. Cell Biol., 37:221.

Wallace, W., and Nicholas, D., 1969. Biol. Rev., 44:359.

Wallach, D., 1968. Proc. Nat. Acad. Sci. (Washington), 61:868.

Wallach, D., 1969. J. Gen. Physiol., 54:3.
Wallach, D., and Kamat, V., 1964. Proc. Nat. Acad. Sci. (Washington), 52:721.
Wallach, D., and Zahler, P., 1966. Proc. Nat. Acad. Sci. (Washington), 56:1552.
Wallach, D., and Zahler, P., 1968. Biochim. Biophys. Acta, 150:186.
Wallach, D., and Gordon, A., 1968. In: Jarnefelt, J. (Editor). Regulatory Functions of Biological Membranes. Amsterdam, Elsevier, p. 87.
Wallach, D., Graham, J., and Ferbach, B., 1969. Arch. Biochem. Biophys., 131:322.
Wallach, D., Ferber, E., Selin, D., Weidelkamm, F., and Fischer, H., 1970. Biochim. Biophys. Acta, 203:67.
Walter, H., and Eagon, R., 1964. J. Bacteriol., 88:25.
Ward, J., and Perkins, H., 1968. Biochem. J., 106:391.
Waring, W., and Werkman, C., 1944. Arch. Biochem., 4:75.
Watanabe, T., Murthy, P., and Brodie, A., 1965. Federat. Proc., 24:362.
Watkinson, R., Hussay, H., and Baddiley, J., 1971. Nature New Biology, 229:57.
Weber, M., Matschiner, J., and Peck, H., 1970. Biochem. Biophys. Res. Commun., 38:197.
Webster, D., and Hackett, D., 1966. J. Biol. Chem., 241:3308.
Weibull, C., 1953. J. Bacteriol., 66:688.
Weibull, C., 1956. Sympos. Soc. Gen. Microbiol., 6:111.
Weibull, C., 1957. Acta Chem. Scand., 11:881.
Weibull, C., 1965. J. Bacteriol., 89:1151.
Weibull, C., and Bergstrom, L., 1958. Biochim. Biophys. Acta, 30:340.
Weibull, C., Beckman, H., and Bergstrom, L., 1959. J. Gen. Microbiol., 20:519.
Weibull, C., Greenawalt, I., and Low, H., 1962. J. Biol. Chem., 237:847.
Weier, E., and Benson, A., 1966. In: Goodwin, I. T. (Editor). Biochemistry of Chloroplasts, Vol. 1. New York, Academic Press, p. 92.
Weigand, R., Shively, J., and Greenawalt, J., 1970. J. Bacteriol., 102:240.
Weimberg, R., 1963. Biochim. Biophys. Acta, 67:349.
Weinbaum, G., Fisehman, D., and Okuda, S., 1970. J. Cell Biol., 45:493.
Whale, T., and Jones, O., 1970. Biochim. Biophys. Acta, 223:146.
Whatley, F., 1962. Plant Physiol., 37:8.
Whistance, G., and Threlfall, D., 1968. Biochem. J., 108:505.
Whistance, G., Dillon, J., and Threlfall, D., 1969. Biochem. J., 11:461.
Whistance, G., Brown, R., and Threlfall, D., 1970. Biochem. J., 117:119.
White, D., 1962. J. Bacteriol., 83:851.
White, D., 1963. J. Bacteriol., 85:84.
White, D., 1964. J. Biol. Chem., 239:2055.
White, D., 1965a. J. Biol. Chem., 240:1387.
White, D., 1965b. J. Bacteriol., 89:299.
White, D., 1966. Antonie Leeuwenhoek J. Microbiol. Serol., 32:139.
White, D., 1967. J. Bacteriol., 93:567.
White, D., 1968. J. Bacteriol., 96:1159.
White, D., and Cox, R., 1967. J. Bacteriol., 93:1079.
White, D., and Frerman, F., 1967. J. Bacteriol., 94:1854.
White, D., and Frerman, F., 1968. J. Bacteriol., 95:2198.
White, D., and Sinclair, P., 1971. Adv. Microb. Physiol., 5:173.
White, D., and Smith, L., 1962. J. Biol. Chem., 237:1332.
White, D., and Smith, L., 1964. J. Biol. Chem., 239:3956.
White, D., and Tucker, A., 1969a. J. Bacteriol., 97:199.
White, D., and Tucker, A., 1969b. J. Lipid Res., 10:220.
White, D., Albright, F., Lennarz, W., and Schnaitman, C., 1971. Biochim. Biophys. Acta, 249:636.
White, D., Lennarz, W., and Schnaitman, K., 1972. J. Bacteriol., 109:686.

Whittaker, P., 1971. Microbios, 4:65.
Wiebe, W., and Chapman, G., 1968a. J. Bacteriol., 95:1862.
Wiebe, W., and Chapman, G., 1968b. J. Bacteriol., 95:1874.
Wilkins, M., Blaurock, A., and Engelman, D., 1971. Nature New Biology, 230:72.
Wilkinson, S., 1969. Biochim. Biophys. Acta, 187:492.
Wilson, G., and Fox, C., 1971a. Biochem. Biophys. Res. Commun., 44:503.
Wilson, G., and Fox, C., 1971b. J. Mol. Biol., 55:49.
Wilson, G., Rose, S., and Fox, C., 1970. Biochem. Biophys. Res. Commun., 28:617.
Wimpenny, J., 1969. In: Microbiological Growth, Cambridge, p. 161.
Wimpenny, I., Ranlett, M., and Gray, C., 1963. Biochim. Biophys. Acta, 73:170.
Wirtz, K., and Zilversmit, D., 1969. Biochim. Biophys. Acta, 193:105.
Wohtczak L. Zaluska, H., and Drahota, Z. 1965. Biochim. Biophys. Acta, 98:8.
Wolin, M., 1966. J. Bacteriol., 91:1781.
Wood, D., and Tristram, H., 1970. J. Bacteriol., 104:1045.
Worcel, A., and Goldman, D., 1967. Arch. Biochem. Biophys., 118:420.
Work, E., 1967. Proc. Roy. Soc. London, B, 167:446.
Wrigglesworth, J., and Packer, L., 1969a. Arch. Biochem. Biophys., 128:790.
Wrigglesworth, J., and Packer, L., 1969b. Arch. Biochem. Biophys., 133:194.
Yagi, T., 1969. J. Biochem., 66:473.
Yagi, T., and Maruyama, K., 1971. Biochim. Biophys. Acta, 243:214.
Yakovlev, V. A., and Mitsova, I. Z., 1970. Izvest. Akad. Nauk SSSR, Seriya Biol., 2:283.
Yakovlev, V. A., Vorob'ev, L. V., Levina, L. A., Linde, V. R., Slepko, G. I., and Syrtsova,
    L. A., 1965. Biokhimiya, 30:1167.
Yamada, Y., Aida, K., and Uemura, T., 1966. Agric. Biol. Chem., 30:95.
Yamada, Y., Aida, K., and Uemura, T., 1967. J. Biochem., 61:636.
Yamada, Y., Uemura, T., and Aida, K., 1968. Agric. Biol. Chem., 32:786.
Yamada, Y., Aida, K., and Uemura, T., 1969. J. Gen. Appl. Microbiol., 15:181.
Yamaguchi, T., Tamura, G., and Arima, K., 1966. Biochim. Biophys. Acta, 124:413.
Yamaguchi, T., Tamura, G., and Arima, K., 1967. J. Bacteriol., 93:483.
Yamamoto, N., Hatakeyama, H., Nishikawa, K., and Horio, T., 1970. J. Biochem., 67:587.
Yamanaka, T., 1964. Nature, 204:253.
Yamanaka, T., 1967. Nature, 213:1183.
Yamanaka, T., and Okunuki, K., 1963a. Biochem. Z., 338:62.
Yamanaka, T., and Okunuki, K., 1963b. Biochim. Biophys. Acta, 67:379.
Yamanaka, T., and Okunuki, K., 1964. J. Biol. Chem., 239:1813.
Yamanaka, T., and Okunuki, K., 1968. J. Biochem., 63:341.
Yamanaka, T., Kijimoto, S., Okunuki, K., and Kusai, K., 1962a. Nature, 194:759.
Yamanaka, T., Ota, K., and Okunuki, K., 1962b. J. Biochem., 51:253.
Yamanaka, T., Kijimoto, S., and Okunuki, K., 1963. J. Biochem., 53:416.
Yamashita, S., and Ishikawa, S., 1965. J. Biochem., 57:232.
Yano, I., Furukawa, Y., and Kusunose, M., 1970. Biochim. Biophys. Acta, 210:105.
Yao, M., Walker, H., and Lillard, D., 1970. J. Bacteriol., 102:877.
Yates, M., 1970. J. Gen. Microbiol., 60:393.
Yates, M., and Nason, A., 1966. J. Biol. Chem., 241:4872.
Yonetani, T., 1959. J. Biochem., 46:917.
Yonetani, T., 1963. In: Boyer, B., Lardy, H., and Myrback, K. (Editors). The Enzymes,
    Vol. 8. New York, Academic Press, p. 42.
Yoneya, T., and Adams, E., 1961. J. Biol. Chem., 236:3272.
Yong, C., and King, T., 1970. J. Biol. Chem., 245:1331.
Young, J., Leppik, R., Hamilton, J., and Gibson, F., 1972. J. Bacteriol., 110:18.

Yu, L., and Wolin, M., 1970. J. Bacteriol., 103:467.
Yu, L., and Wolin, M., 1972. J. Bacteriol., 109:51 and 59.
Yudkin, M., 1962. Biochem. J., 82:40.
Yudkin, M., 1966. Biochem. J., 98:923.
Zahler, P., and Wallach, D., 1967. Biochim. Biophys. Acta, 135:371.
Zahler, P., Saito, A., and Fleischer, S., 1968. Biochem. Biophys. Res. Commun., 32:512.
Zaitseva, G. N., 1963. Nitrogen and Phosphorus Metabolism of *Azotobacter* during
    Its Development. Author's Abstract of Doctoral Dissertation. Moscow.
Zaitseva, G. N., 1965. The Biochemistry of *Azotobacter* (in Russian). Moscow, Nauka.
Zaitseva, G. N., Agatova, A. I., and Belozerskii, A. N., 1961. Biokhimiya, 26:338.
Zaitseva, G. N., Ngo Ke Syong, and Belozerskii, A. N., 1963. Biokhimiya, 28:172.
Zaitseva, G. N., Kalyuzhnaya, L. P., and Golovacheva, R. S., 1970. Dokl. Akad. Nauk
    SSSR, 195:1226.
Zavarzin, G. A., 1958. Mikrobiologiya, 27:401.
Zavarzin, G. A., 1964. Uspekhi Mikrobiol., 1:30.
Zav'yalov, V. P., Kiryukhin, I. F., and Troitskii, G. V., 1970. Molekul. Biol., 4:655.
Zhukova, I. G., 1969. A Study of a Complex of Malate Dehydrogenase and $NADH_2$
    Dehydrogenase from the Membranes of *Micrococcus lysodeikticus*. Author's Abstract
    of Candidate's Dissertation. Moscow.
Zhukova, I. G., Ostrovskii, D. N., Gel'man, N. S., and Oparin, A. I., 1966. Biokhimiya,
    31:1209.

Index

255